Dismantling Discontent: Buddha's Way Through Darwin's World

Charles Fisher
www.DismantlingDiscontent.com

Elite Books
Santa Rosa, CA 95403
www.EliteBooksOnline.com

Cataloging-in-Publication Data

Fisher, Charles, 1956-

Dismantling discontent: Buddha's way through Darwin's world / Charles Fisher.

p. cm.

Includes bibliographical references.

ISBN-13: 978-1-60070-031-6 (hardcover)

ISBN-13: 978-1-60070-032-3 (softcover)

1. Buddhism and science. 2. Evolution—Religious aspects—Buddhism. 3. Buddhism—Doctrines. I. Title.

BQ4570.S3F56 2007

294.3'365—dc22

2007015137

Typeset in ITC Galliard

Printed in USA

First Edition

Typeset by Karin Kinsey

Cover design by Victoria Valentine

10 9 8 7 6 5 4 3 2 1

Contents

Illustrations

Dedicated to

my mother, Elizabeth Fisher,

whose life so influenced me

and to

Maury Stein,

friend, benefactor, and comrad.

Preface

On the weekend of September 6, 2003, the Dalai Lama met with a number of scientists at the Massachusetts Institute of Technology (MIT) to investigate the human mind from the perspectives of both Tibetan Buddhism and Western science. The meetings were open to the public and attracted so many people that fifteen-hundred extra seats had to be added. Science and meditation found a popular meeting ground. *Time* magazine dedicated a special issue to this new reconciliation. Such harmony has not always been the case. In the past, most brain scientists either ignored meditation, expressed reservations about its relevance or were hostile to it as a source of scientific knowledge. And traditional Buddhists were not known for their interest in science. Public image of the discord has changed due to the charisma of the Dalai Lama, the popularity of stress reduction techniques, and the growing interest in meditation. Currently scientists are studying people while they meditate to see if meditation affects the brain in positive ways. The results are suggestive but still tentative.

This book not only addresses some of the issues discussed at the MIT conference but also attempts to place them within the larger framework of evolutionary biology. It explores how the vast changes in the way humans have lived since we began using language have affected the main focus of Buddhist meditation practice: human discontent and how to alleviate it.

This book also addresses another current concern. In the last few decades it has become apparent that humans are displacing the wilds and poisoning the earth. How this has come about is inextricably bound to the repertoire of behaviors with which evolution has endowed us. Our human ancestors grubbed around the forests much like other animals. They didn't influence things too much. Then changes in our hands and brains enabled us to sustain ourselves more productively and increasingly impact our surroundings. Science and meditation give us tools to examine this process. Using both, we will see how evolution left humans, as it did other animals, with an imperfect structure.

Looking at one flaw, our discontent, I will examine these changes and how the newly built environments they created again affected human consciousness. After examining the character of his own mind, the Buddha came up with a strategy to counter the destructive effects of human minds and hands. One aspect of this strategy is to disarm destructive mind states as they appeared in consciousness and another is to commit oneself to behaving without causing harm. We will look at both of these from the viewpoint of evolution and meditation.

The hominids who came before humans had full speech and lived in an animal rapport with nature without discontent as we now experience it. The universe as understood by scientific cosmology and evolutionary biology works both physically and organismically without any narrative thread like the thoughts which accompany most of our personal journey through life. In scientific descriptions of the universe there is no place for that comforting voice which narrates a big screen movie trip through space. Besides its big bangs, the history of the universe is silent. The same is true for the evolution of life.

The crowded, ancient Cambrian seas five-hundred million years ago with their trilobites and fantastic looking *Hallucigenia* and *Anomalocaris* were silent. Changes in nature happen. The universe may be expanding and contracting but there is no destination. Darwinian evolution may proceed lawfully but it is not going anywhere. The Buddha's universe had no beginning or end. It was silent. The Buddha looked at the noise made by the human mind and felt that it was the cause of our discontent and destructiveness. Behind that noise he found a profound silence. If we could only learn to abide there, he claimed, we might be able to stem our tendency to hurt ourselves and others.

In the twenty-five hundred years of Buddhist history this silence has been characterized in many different ways (although more often discussed than achieved). It has been called emptiness, the luminous mind, serenity and wisdom. It is my goal to show that the Buddha's silence has something in common with the silence of nature in which our hominids ancestors lived, and the silence of evolutionary processes. These collective silences do not negate the noisy mind. As the Buddha emphasized, the mind is also part of them. His aim was to teach people how to subsume the noise of the mind into the larger silence. My project is to show how all these pieces fit together, and especially to describe how our noisy mind is an accident of nature. My presentation is partially intellectual and partially pragmatic. If we can understand the evolutionary roots of discontent and understand how the world we have built around us contributes to it then we may both be better able to address our discontent and understand the limits of our efforts. Although we may romanticize a past where we feel humans lived more in harmony with nature and each other, the world we now occupy and the minds we now possess are what we have to work with. Looking at our minds from the perspectives of both meditation and modern science we can learn about destructive noise and its relationship to silence and better aim towards easing our discontent. Looking at how our world contributes to our discontent we can try to modify exacerbating trends or choose not to participate in activities which encourage more discontent.

The Buddha's silence may seem negative and opposed to Western notions of fulfillment through self-improvement and success but he claimed it to be liberating, allowing humans to engage life free of the discontent that often reduces our apparent achievements to another round of the suffering they promise to banish. The longevity and ease of living which modern technology and medicine have delivered also have their downsides. They can cause harmful biological processes, which would rarely have had the opportunity, to emerge. We see this painfully in the physical and mental difficulties of our extended old age. One of the Buddha's starting points was disease and old age. Without romanticizing the past we can learn how to address these by comparing how other animals live and how we lived prior to civilization's great advances. The new ways we have found to engage our minds have brought with them new levels of discontent. Looking at how these came about over time can give us a handle to address the difficulties they create. Our bodies, our minds, and our environment are the products of evolution and human actions. Understanding how that is and seeing the role that meditation can play helps us to choose how to act. In this lies the release from discontent which the Buddha found.

I began working formally on this book in the early 1990s when very few people were writing about the relationship between meditation and nature. The roots of the book actually began many years earlier when I went canoeing in northern Canada. I was so shaken by the wilderness that on returning to my apartment in Cambridge, Massachusetts where I taught university, I could not bear to go out for several weeks. I sat reading Colin Turnbull's book on the pygmies of the Congolese jungles for a graduate course I was to teach. Hiding in my apartment, I tried to figure out how I ever again could relate to the city; it seemed so out of contact with natural life and death.

I carried this contradiction around with me for another seven years as I balanced my duties as an academic with attempts to escape by homesteading in the north, wandering the jungles of Central America, working as a migrant laborer picking apples and gardening vegetables in semi-rural New England. I led a schizophrenic existence. Although I had never meditated before, in 1977, overwhelmed by the death of my father, I found myself sitting cross-legged at a month-long Buddhist retreat, meditating fourteen hours a day. In the years that followed, I continued to venture out into nature but meditation drew me in as a way to mellow my life's imbalances. Over time, I began to integrate both my interest in nature and my meditation practice into my teaching. As I learned more about natural history, both in the field and by reading, I grew to appreciate how the inner world of meditation related, often directly, to the outer world of nature. I realized that my two passions were interrelated sides of the same coin.

In the early 1990s, I was having tea with a friend, Rodney Smith, in Seattle. Rodney had been a Buddhist monk in South East Asia and ran a hospice. He also taught meditation. We were talking about possible topics for a Buddhist workshop I had been asked to give. At one point Rodney looked at me and said, "but you know so much about the woods around the workshop site!" The proverbial light bulb went off in my head—"nature and meditation." Although my time on the cushion and my wanderings in nature seemed different, I had really been seeking the same thing: the meaning of human life as it is embedded in nature. Although I had taught environmental history and had experienced the wilds, although I had meditated for years both in nature and at meditation centers, I had read little textual or scholarly Buddhism. So with gusto I plunged into the double topic of meditation and nature. As I read evolution, neurophysiology, Buddhist sacred texts, and related them to my meditation practice and the ethology, anthropology, history of science and technology that I had taught, an integrated subject matter began to emerge. I conceived a grand design: I would bring together natural history, the processes by which the brain works, meditation's observations on mind and wisdom, and the material history of civilizations. I also wanted to tie and contrast these to the history of meditation in forests since the time of the Buddha. It was an ambitious project. I wanted to use an understanding of natural history and the lessons learned from the Buddha's retreat into the forests to illuminate the character of modern minds and their discontent. But when I finally put everything together, I found that it was simply too much.

So I reworked the themes. I came to realize that the audiences for books on the workings of natural history and books on Buddhist history were different. After returning to the drawing board, I ended up with the book you hold in your hands and another book entitled *Meditation in the Wilds.* They are companion pieces but each can stand alone. This book, *Dismantling Discontent,* looks at Buddha's great topic of suffering from the perspectives of nature and meditation, metaphorically from the viewpoints of Darwin and Buddha, respectively. I believe that it takes both perspectives to understand the human condition, much of which we share with other species. We humans do animal things in our own particular ways, making some of the contradictions of our existence more extreme than other beasts. Darwin helps us see the continuity of our species with the others around us. It takes a heavy dose of Darwin to get meditators to see that meditation is a process of nature. Much of what meditation does may be explicable scientifically. It takes an even heavier dose of Buddha to get scientists to understand that meditation is a valuable and valid way of describing the mind. This book lays out the ideas from each area which are required to understand my attempts to synthesize the two approaches into a fruitful union. Although only a first step, the combined approach is, I think, both revealing and necessary.

In conclusion, let me say that my ultimate reason for writing this book is to help us humans learn to be at greater peace with ourselves, each other and the world in which we live. While there may be no relief from death or extinction, I believe we can blunt the edges of harm towards which we would otherwise be inclined. Meditation's inward view may contribute toward calming the powerful mind and hands with which nature has endowed us; meditation thus makes life, with all its contradictions, more bearable. This, at least, has been my experience.

Charlie Fisher
Woodacre, California

Foreword

by Lynn Margulis

The perennial arguments that pit religion against science are churlish and childish. If no scientifically minded person need view Religion as the enemy of Science, neither ought religious activists dismiss science as an unfeeling tool of the military-industrial establishment. Science is a way of knowing about the world around us. As investigation, mainly of nature, science remains the best mode of inquiry shared by the community of scholars worldwide. Most scientists tend to be tinkerers rather than thinkers or intellectuals. This is possible because honesty is built into the practice. A scientist may be a rude wife-beater or a callous fortune seeker but, qua scientist, he must open-mindedly investigate and report his investigations as carefully as possible. Among the long strings of names that author many scientific papers only one person usually is the scientist. The results garnered by scientists may be flawed but they still are the best sources of knowledge we have....yes, even about religion.

The Darwinian evolution revolution of 19th century England replaced a Victorian, Christian-accepted wisdom of the world with a dynamic view permanently open to responsible criticism. The wondrous universe made 6,000 years ago for Man by an alternately vengeful and benevolent God, a bearded white man with a strong will and miraculous powers, was permanently replaced in the minds of thoughtful people. In the modern view a vast cosmos began with a bang 12,000 million years ago in a burst of energy still detectable today everywhere by electromagnetic sensors, radio telescopes that pick up background radiation from all directions in the sky. Points of light are stars more than 100 of which are surrounded by planetary systems. The gossipy sex- and death-obsessed misnamed primate, *Homo sapiens* ("Man, the smart"), has played a tiny and only a very recent role in the business of the universe. City-dwelling, clothed, and upright "chimpanzees," our immediately family, on this scale, have only appeared on the evolutionary scene 0.011 million years ago. The God of our Judeo-Christian-Muslim forefathers had little to do with the development of our so-called dominant civilization. However, as Charlie Fisher shows so nicely in this book, that God and the attitudes toward nature that invented Him, have everything to do with our discontent.

I've known Charlie Fisher since we were fourth graders at the Laboratory School of the University of Chicago. I suspect he would agree with me that our school instilled respect and love for all genuine learning. More real science was done in that school by eager students than in many scientific institutions I later attended. Smart and open-minded, Charlie came from a cohesive, intellectual Jewish family with European roots. His older sister Berenice (who alas we called Bunny even when she asked us not to) I remember as a fine conversationalist with deep intellectual interests. Although I knew he had become a Professor of Sociology at Brandeis University in Waltham, Massachusetts I only found out about his love of the natural world and knowledge of Buddhism when, out of the blue, he asked by snail mail if I could help him with both his book-length manuscripts.

For Charlie had discovered what I had independently confronted for years. Scientists tend not to be intellectual, they tend not to read, and certainly they don't read or care about religion or philosophy.

Religare means re-ligate or "bind together again." "Religion" thus refers to the set of beliefs and practices held in common by groups of people. Everyone of course has a religion and a philosophy, often intertwined, but scientists by avoiding discussion of religion think they can avoid the cultural phenomenon itself. Charlie was in a quandary. After many years of work on these manuscripts, he had been many times rejected. Essentially in taking on meditation and Buddhism in a context of Darwinism he had written about a scientifically taboo topic. But I read with great interest his unpolished gem and saw that his viewpoint was from the scientific intellectual tradition of honesty, open-mindedness and reconstruction of the past from clues taken as representative. Furthermore, to me, an open-minded skeptic and ignoramus in Eastern religious traditions and philosophies, I actually found Charlie's descriptions of meditation, death, disease, and old age in the context of our biological natures informative and compelling.

You, too, will no doubt enjoy Charlie's nature-loving revelation of the very non-Victorian Buddhist view of our lives and their discontents. Charlie has written of meditation in the clear light of evolutionary history. Evolution, as Ernst Mayr has pointed out, is a fact, not a theory. In his unique and independent way Charlie sees this universe in an evolutionary context. The cosmos without design and without any help from a superior father figure first brought forth the phenomenon of life without people. Then later, the appearance of our human species in its families, tribes and bands, developed belief systems that were labeled "religions." Religions, inevitable as sexual intercourse and childbirth, are part of how our ancestors, and hopefully ourselves, cope with the inevitability of disease, old age and death. They too can be viewed scientifically—and practically—as my old friend Charlie Fisher does here with simple grace.

Lynn Margulis
Distinguished University Professor
Department of Geosciences
University of Massachusetts Amherst

Introduction

by Dorion Sagan

Alan Watts, who initiated the current interest in eastern thought, called Buddhism the "no-religion religion." Charlie Fisher takes this notion a step further: he interprets Buddhism more as a practice than a religion, one centering on meditation and capable of alleviating a host of psychological maladies related to the deleterious effects on us humans of the agricultural revolution which allowed our ancestors to occupy this planet as densely as the seeds they sowed. Combining history and scholarship with his personal experience as a meditator and teacher, he brings Buddhist practice down from its sometimes rarified conceptual heights and places it firmly within an evolutionary setting. To understand ourselves and Buddhism, he emphasizes, we must understand time—evolutionary time as elucidated by Charles Darwin.

We come from a long line of animal ancestors, including relatively sparse populations of hunter-gatherers more at home in their unaltered landscape than are we among our skyscrapers. Thus, just as we tend to eat too much chocolate than is good for us, responding to a natural provision of sweet receptors that attracted our primate ancestors to the rare sources of fructose, Vitamin C, and other nutrients they needed in their arboreal environment—so modern civilization leads to psychological deprivations and depressions, "junk food" for the mind. Much of our discontent, Fisher argues, can be traced to a disconnect between life as it was lived by our ancestors, steeped in concern for the needs of day-to-day survival, kept busy and free of mental clutter, and modern life, with its myriad distractions. Instead of sitting around the fire, burning trees that formed our primeval environment, we sit around TVs and computers linked to other minds in nebulous networks of dubious commercialism and human attention-getting. Our own success has begotten failure as the human-made has replaced the environment which originally formed us and landscapes have become crowded cities.

Fisher successfully argues that the Buddha was one of the first to understand, if intuitively, that human malaise, ennui, and anomie—those diseases of the soul—are exacerbated by civilization born of agriculture and technology. And if civilization with all its hustle and bustle and chatter is part of the problem, so recognizing what parts of us come from the wild, and meditating upon nature free of humanity's overweening presence can be part of the solution. We thus begin to understand the importance in Buddhism of non-conceptual modes of thought--of no-mind and beginner's mind--akin to wiping the dust from a dirty mirror or finding a place beyond overconceptualization. The general chatter (exacerbated in this age of media) and of oneself to oneself, the immersion in too exclusively human modes of being, has cast a pall over our happiness even as it has helped us to multiply at alarming rates. To recover, we need to move in the directions pointed to by the Buddha, beginning with a recognition of our insatiable desires for distraction and ever greater creature comforts.

As a pampered prince of an agricultural society, the young Buddha had the earthly pleasures akin to those available to people now living in the industrialized world. Uncertain life in the wilds was a real problem whose amelioration for the privileged of the Buddha's era led to monotony, excess, and alienation from nature. Having been born to material comfort, the Buddha saw how little peace of mind it bestowed. Neither his youth, his pleasure, nor his loved ones would last forever. Time and death are great levelers. I confess that I am not completely persuaded by the version of the Buddha's ideas that finds a kind of secular salvation in the pristine serenity of pre-agricultural nature. To me there is an overlap between Darwin's evolution and Buddha's world, but it is incomplete. Darwin seems to speak of the same cycling of generations from which release is requested and granted in the Buddhist attainment of nirvana. In short, it seems to me that Buddha's view of time was less linear than that of either Darwin or Christianity. To be released from the cycle of generations suggests a place beyond linear time, and evolution itself. Buddha's time, I would argue, is more akin to that of a mystic, or a philosopher. But then Buddha, also known as the practitioner of "the middle way" after retreating to meditate in the woods (and leaving his children behind to do so!) is said to have returned to a more quotidian existence, if not quite of the world then still in it. The "place" of a mystical consciousness of time within time is thus to me reminiscent of the story of the Zen Buddhist monk who, when asked the ultimate meaning of existence, replied, "The wind is high today." Perhaps the timeless realm I imagine to have impacted Buddha's thought is imaginary, a bit of literal preposterousness generated by the imperfectly abstracting human mind. But then it is also real, existing if only in human consciousnesses along the timeline of earthly evolution that begins almost four billion years ago with our, and Buddha's, bacterial ancestors. As poet Paul Valéry wrote, "I live in a world in my head in the world."

According to philosophical scholar Gananath Obeseyekere, the story is told that when the baby Krishna opened his mouth there could be seen inside a whole (or was it this very?) solar system. In the same way, a nonlinear interpretation of Buddhism, by allowing of strange geometries and "impossible" temporalities, does not obviate, but is made more understandable by Fisher's Buddhism embedded in a Darwinian world. There is a dialectic between the social world of humans and the more-than-human world of nature. By showing how meditation is part of the natural world, Fisher highlights what may be an initial overcompensation by Buddha to redress the luxurious circumstances of his birth. Yet the phrase "the middle way" can help us to attend to a gap, to compensate for an overcompensation in a process that is both over and never ends. For the desire to be free of all craving is itself a craving, and there is no reason to think that craving does not mark the awareness of any being dependent upon future food and energy sources. I thus recommend this book to any such beings, and especially to those who have consciously come to that precipice beyond which they can see that their desire to overcome craving is itself a craving to be overcome.

Dorion Sagan
Amherst, Massachusetts

Buddha and Darwin Biographies

Buddha

The composing of a biography or "The Life of the Buddha" did not happen until many centuries after the Buddha was supposed to have lived. Information about his life survives in orally transmitted story collections called *Sutras* or "Sayings of the Buddha" to which were later added tales of the Buddha's incarnations before he was born Gautama Siddhartha, a young prince who looked on life's suffering and discovered how to transcend it through meditation. His realizations via meditation earned him the titles Gautama Buddha or Sakyamuni, meaning, respectively, the Gautama who is awakened or the sage of the Sakyas, his people.

The Buddha lived for eighty-three years about five centuries before the time of Jesus. His contemporaries, in very different parts of the world, included Lao Tzu and the Rabbis of the Second Temple. The Buddha lived during a time of change in the Indian subcontinent. Agricultural societies were beginning to produce enough goods to support states, standing armies, and priesthood. The free time allowed "Brahmins" and other seekers to turn inward to explore the nature of the human mind. Into this ferment the Buddha injected a new way of understanding life based on meditation.

Meditation is a careful examination of how the mind works and the relationship of that to the body. Siddhartha's meditation led him to understand that discontent arises because humans ignore the extent to which they refuse to acknowledge the impermanence of existence. For Buddha, a thorough acceptance of impermanence goes hand in hand with an ethical universe. Understanding the nature of suffering leads one to live a life of non-harming, a life in which one minimizes the harmful impact one has on one's self and others. People mired in their own discontent only produce more suffering, both for themselves and the world around them.

The Buddha's path was thus one of meditation and non-harming. During the first third of his life he is said to have lived an ordinary existence; but he was a seeker. For the last two-thirds of his life he taught what he learned to an ever growing audience. His students formed the nucleus of a movement which spread his ideas from India to central Asia and from there eastward into Mongolia, China, Korea and Japan as well as south-eastward into Sri Lanka, Southeast Asia and Indonesia. In the thousand years after the Buddha, Buddhism became the first great world religion.

Darwin

If one thinks of life in terms of colorful novels such as *War and Peace,* or *Catcher in the Rye,* then Darwin's biography can hardly be compared to that of the Buddha, whose life was, in many ways, much more romantic and mythical. The Buddha is a distant figure. Yet what the Buddha touched upon can be seen to apply as much to

Darwin as to us. And Darwin's research into nature can be applied to the problems and solutions the Buddha posed.

Born in 1809 into a thoroughly bourgeois family, Charles Darwin led the life of a 19th century English gentleman. His passions and tragedies include those with which many of us can easily identify. At University he became intrigued with the raging arguments about the history of nature and humans' place in it. Then, in December of 1831 he set sail aboard the HMS Beagle as ship naturalist under a Captain Fitzroy; his mission: to study as much nature as he could while the ship was employed in mapping the coast of South America for the British Empire. Upon his return five years later he wrote a history of the voyage which was a publishing success and earned him worldwide fame as a naturalist. Living off his royalties and inherited wealth, he married, had a large family, and settled down to a life of further, if more sedentary, observations and cogitations about the workings of the natural world .

Darwin's life was one of science and family. What he saw during his travels laid the groundwork for the theory of evolution he eventually proposed. Both the incredible differences in distant species he observed and the subtle differences in closely related species led him to formulate ideas on how these differences came to be. For years he pursued these questions, writing scholarly monographs on subjects which advanced his understanding of their answers without yet revealing his ideas on evolution. While he pondered his hypotheses and tried to figure out how to present his ideas, a fellow naturalist collecting specimens way off in the Indonesian archipelago sent him a paper which he asked Darwin to present to a scholarly meeting. The paper had in it the kernel of Darwin's ideas. This induced Darwin to go public. The presentation of his ideas of evolution caused tremendous controversy and, for the rest of his life, he and his growing legions of supporters countered those who did not accept his ideas.

When one looks behind this swirl of research and public debate, one finds a very human Darwin, a man who suffered from chronic illnesses and whose growing family experiences, including the death of some of his children, severely tested and ultimately destroyed his orthodox Christian faith. Darwin was also, immensely anxious about how people would respond to his ideas—ideas that were not only scientifically novel but threatening to classical Christian biblical dating and notions of creation, as well as to the Christian conceit that humans are superior to the rest of nature. Drawing both on his scientific research and his personal life, Darwin presented a new paradigm of understanding human's place in nature. He saw human beings as continuous with other animals. The evidence he offers for this is based on careful observation of animal species and their environments and how the former change over long periods of time. Later in life, Darwin studied emotional expression, linking human facial muscles and behavior to those of animals. Darwin tried to understand the illnesses of his family in terms of his theories about nature.

Darwin died in 1882 at age seventy-three. He even tried to understand his own dying as part of natural history. The human beings whose suffering Buddha explored and sought to alleviate were animals in an impermanent world—a world of continuous biological change whose chief characteristics, or modus operandi, were revealed by Darwin and his careful studies. Although they might seem an unlikely pair, the thesis of this book is that the investigations of Darwin and the Buddha overlap; they are synergistic when it comes to understanding our place in nature and how we can adjust to a changing, impermanent existence. Together these two men give us the tools necessary for taking their inquiries further.

Unlike the legendary life of the Buddha, the details of Charles Darwin's life are well documented.

A Darwin Chronology

1809 Charles Robert Darwin is born in Shrewsbury, Shropshire, England.

1825-27 Darwin goes to Edinburgh University where he studies medicine.

1827-31 Darwin attends Christ's College, Cambridge University.

1829 Darwin makes an entomological tour of North Wales.

1831-36 The H.M.S. Beagle circumnavigates the world.
On the trip Darwin makes important fossil finds, meets the indigenous natives of Tierra del Fuego, sees a cataclysmic earthquake first-hand, and studies the geology, flora and fauna of the Galapagos Islands. Darwin makes detailed observations of finches and Galapagos tortoises, and writes a draft of a paper on the formation of coral reefs.

1837 Darwin starts first notebook on *Transmutation of Species.* He begins work on *The Voyage of H.M.S. Beagle.*

1838 Darwin formulates his theory of evolution by natural selection.

1839 Darwin marries Emma Wedgwood.

1839-45 Five children are born. The second dies at a month old.

1839 First volume of *Voyage of the Beagle* is published.

1846 Third volume of *Voyage of the Beagle* is published and Darwin began an eight-year study of barnacles.

1850 Another son is born.

1851 His favorite daughter Annie dies.

1856 Darwin begins to write a work putatively devoted to the origins of new species.

1856 A final child is born.

1858 Darwin receives the letter from Alfred Wallace, a fellow naturalist, in which Wallace presents a theory of evolution through natural selection similar to Darwin's. Writings of both men are read at the Linnaean Society.

1858 A son dies; Darwin begins writing *On the Origin of Species.*

1859 Darwin's *Origin of Species* is published.

1860 Darwin studies climbing and insectivorous plants but is unable to figure out their underlying mechanisms.

1862 Darwin publishes a monograph on insect fertilization of orchids explaining their exotic forms in terms of evolution and natural selection.

1867 Darwin begins investigating the expression of emotions.

1868 Darwin buttresses *Origin of Species* in *The Variation of Animals and Plants under Domestication.*

1872 *Origin of Species* 6th edition is published.

1872 *The Expression of the Emotions in Man and Animals* is published.

1882 Darwin finishes his final study, *The Formation of Vegetable Mold through the Action of Worms.*

1882 Charles Darwin dies.

Chapter One
The Burden of Discontent

Bhikkus, both formerly and now what I teach is suffering and the cessation of suffering.
—Gautama Buddha

Nothing is easier than to admit in words the truth of the universal struggle for life, or more difficult...than constantly to bear in mind.
—Charles Darwin[1]

Silently the sun moves to the west over a frozen marsh on a warm February day in New England. I emerge from the meditation hall during a retreat and hike over the icy, crusted snow down the sloping landscape, across frozen cornfields and through the woods to a marsh about a half a mile wide and a mile long. My footsteps encounter many different textures. Sometimes my insulated boot crashes through the crust, sometimes just my heel. Occasionally I fall when my foot slides out from under me.

Moving slowly and quietly as I near the marsh, I see a tetragnathid spider climbing an invisible thread toward a dead twig. Its body measures half an inch with long front legs. It seems to sense my presence as I push my myopic eyes close to it. When it reaches the twig, it hunkers down. Its dark body is invisible against the gray bark. As I near the marsh, I am greeted by fallen trees and standing, dead trees ringed with tooth marks. On the marsh, the smooth ice is punctuated by crusts of snow, tufts of grass and deer prints frozen from warmer days.

The marsh has a long history. Born as a stream at the end of the last ice age, it supported generations of beavers and their works until the coming of the Europeans. Then the beavers were hunted to extinction. The stream was dammed for power to mill grain and lumber. The marsh was drained to make fertile fields and abandoned in the nineteenth century after only several generations' use. Years later, the marsh was preserved by the Audubon Society. Around 1986, beavers returned and rebuilt their dam. The beavers multiplied and felled the surrounding trees until there were no more close by they could safely harvest. So the beavers moved on. Without their continual maintenance, the dam was breached and the pond began to silt in. After two hundred years of disrupted succession, the marsh is gradually becoming a meadow; someday it will be a wooded stream for the beavers to re-inhabit as if human interruption had never occurred.

The human mind evolved to its present form during the ice ages. Before human ancestors had speech and complex thought, they may have lived in a mental quiet closer to that of the beavers than we now experience in the hustle and bustle of modern daily life. As the mind evolved, humans were able to create technologies eventually used by both Native Americans and Europeans to drive the resident beavers and Natives almost to extinction. Then active minds and the work of hands farmed the soil to exhaustion, mined the water and moved on. Years later, other busy minds began to manage the marsh as habitat for beavers.

Calmed by the tranquility of the retreat and buttressed by hours of meditating next to the marsh, I can feel the stillness in contrast to the incessant chatter in my head. I sleep for a while on a rock in the warm winter sun. On waking, the sun has advanced towards the trees on the opposite shore. For a moment I am awed by the silence and inevitability of nature. Then I begin to think again. Words to write come pouring out. I feel an urgency to be on time for the evening sitting in the meditation hall, and the problems of my life again possess me.

How did this mind so filled with thoughts come into being? What is its relationship to the body? With any self-examination it is not hard to see that these two are interwoven, much as we might like it otherwise. Personally, if my body behaved itself without bothering me, if my mind would shut up and my emotions calm down, I might be a lot happier. I want to exercise with abandon, eat to excess, be lazy, enjoy the senses, wake rested and be nurtured and deeply touched by connection to others. Unfortunately, except for a period of apparent indestructibility in youth, the human body experiences problems which intrude into happiness. And even during youth, our feelings and thoughts create one problem after another.

In addition to our personal problems, if we look at the world around us, we see many troubling things. There is disturbing population growth, environmental degradation, a world of haves and have-nots and among those with material security, a concern with consumption and entertainment that seems out of balance with the problems in their own lives and the rest of humanity.

Twenty-five hundred years ago, Gautama Buddha investigated human nature using the tools he had at hand. These involved an examination of his own mind and body through meditation. Buddha discovered what he claimed was universal human discontent. For discontent is the norm. The human mind does not often rest peacefully in the bosom of life. It is plagued with restlessness, greed, dissatisfaction, laziness, fear and anger. Occasionally humans rise to peaks of love, generosity, equanimity and compassion, but mostly they struggle in the valley of discontent. This is true even in the best of times.

The Buddha discovered the causes of discontent by becoming still and carefully observing how human experience unfolds. He saw how deeply entwined are the mind and the body, how much the human mind reacts to stimuli impinging upon it, and how much we are affected by our web of relationships with other people. The Buddha's method of observation was meditation, and for thousands of years meditators have emulated the Buddha's example. In so doing, they have found a greater understanding of who they are, and how the responses of their minds affect themselves and others around them.

The objects of Buddha's contemplation are all parts of the natural world. Our bodies and our minds come from nature. Our pleasures, our fears, and our thoughts are natural. The constraints which pit my desires against the limitations of my body and what the world around offers me are natural. The processes of life such as birth, growing up, disease, old age and death are organic. We are indeed part of nature. And the best tools we have for understanding nature are those which developed in the spirit of evolutionary theory as developed by Charles Darwin.

In *The Origin of Species*, Charles Darwin set forth a way of looking at how organisms changed over time and fit into the natural world around them. Sometimes Darwin and his followers viewed nature as wonderfully complex with each trait that a species possessed fitting neatly into the functioning apparatus that was nature as a whole. At other times, natural historians found the evolution of certain traits quite puzzling. Some traits seemed downright counterproductive. Calling on the perspectives of both Buddha and Darwin, this book offers a window through which one can revisit Buddhist insights by way of natural history. Buddha calls our attention to human discontent and gives us a way of looking at it first hand, introspectively. This human trait makes endless trouble for its owner and the world we inhabit. It is a characteristic of mind which leaves us dissatisfied with existence as it unfolds at a particular moment and drives us to react in ways that compound the dissatisfaction. In the spirit of Darwin we will look at the origin of human discontent as part of the evolution of the human mind and explore how it resides in our being.

In this book I seek to uncover the origin of discontent and then see how our present world, created by incredible human cleverness, affects the character of that discontent. The Buddha provided a way of observing our inner landscape while Darwin and natural history explore the outer landscape; that is, the mechanisms through which the inner comes into existence and is maintained. Life, as we so often experience it, provides us with the problem of discontent, while both meditation and natural history offer ways to explore it.

The natural historian, observing humans in the context of nature, finds an animal driven by the basic needs of other animals—needs for food, protection from the elements, procreation and survival—connected to other members of the species. A number of scientists have noted that contemporary humans possess bodies and minds that evolved to succeed under stone-aged conditions. Our bodies are designed for strength and endurance while running, carrying weight, digging, climbing and dexterously handling objects. Our minds are those of strategic hunters and intensely social animals. We collectively plan actions and our reactions are deeply affected by the responses of others.

I can hardly guess what upset our hominid ancestors more than two-hundred thousand years ago. Romantic novels to the contrary, the environment of stone-age life seems utterly distant from the world we now inhabit. Citizens of the developed world have at their fingertips computers, television, cell phones, automobiles, airplanes and a myriad of household appliances. Our lives are so cushioned by conveniences that we may only glimpse the raw nature in which our ancestors developed as part of a tourist landscape. A beautiful sunset still inspires us, but few see the sun rise and set each day. We do not know the rain or snow anymore. Civilization attempts to banish discomfort. We are more interested in having light at night and looking at computers, on which our current survival may, in fact, depend, than sleeping when it becomes dark. Somewhere between two hundred and one hundred thousand years ago human hunter-gatherers evolved speech and complex thought. With that came the ability to build a new world and the discontent with which those humans met their lives as we meet ours.

Discontent is like other characteristics which natural historians find in some species that seem almost counterproductive to survival.

Those who sit down to meditate for even a short period of time or try to meditate for weeks or months become quickly aware how many different thoughts and feelings arise in rapid succession for almost no apparent reason. Here is an example of six days' worth of daily observations of what occurs while an ordinary college student sits still and pays attention:

> Thursday: I started off focusing on my breathing. Then I noticed the resistance that always comes up. I felt an underlying sense of fear, anger and loneliness. The resistance overpowered all feelings within...
>
> Friday: ...I had a lot of initial physical discomfort...I had a dull ache in my lower back, a tenseness in my jaws...I noticed my mind was racing with thoughts. I had a planning mind... I felt my body and mind begin to unwind. As I watched my mind, I noticed a feeling of apathy. The apathy then changed to restlessness.
>
> Saturday: ...I felt stuck—stuck in resistance. I felt resistance to resistance. I wanted to see what would happen if I told myself to lighten up and laugh at the resistance. It helped. I felt some compassion for myself. A little spark of sadness came up after the compassion. Then my mind began to wander.
>
> Sunday: ...when I felt anxiety I felt jumpy and restless. When I felt fear, I felt a pit in my stomach.
>
> Monday: ...Struggled with sleepiness and laziness...My mind wandered continuously...
>
> Tuesday: ...whenever I feel anger, it feels like a toxin in my body...

In trying to do the simple task of observing one's breath, it is possible to see the large variety of thoughts and feeling which sweep through us mostly unnoticed. From a rational standpoint these thoughts and feelings do not appear helpful in carrying out the tasks of survival (which in our world may mean work and relationships).

We will look at this discontent as one of the downsides of the mind's ability to think.

Darwin felt that spontaneous heritable changes, geology, weather and competition among organisms help shape a given species. He encouraged natural historians to look closely at the context in which a species lived and how its capabilities fit into that context. For the evolution of horses, that might entail details of jaws, teeth and feet and transitions from forest to savanna, each habitat populated by different kinds of trees and grasses. Analogously, discontent has had expression in societies from hunter-gatherers through agriculturists to post-industrial civilizations. We will look at how the mind and its discontent may have evolved and examine its expressions in its changing contexts. In this investigation, we will look at the differences in the way discontent gets generated and see how the improvement in our material conditions and progress of our civilization may have added to our suffering rather than diminishing it.

Before proceeding, let's look at the Buddha's first encounter with discontent. The story is found in anecdotes in the *Sutras* or "Sayings of the Buddha," a part of the Buddhist sacred texts (not written down, however, until hundreds of years after the Buddha's death). The *Sutras* often start with the phrase, "Thus I have heard." In a north central area of what is now India, twenty-five hundred years ago, the Buddha, named Siddhartha Gautama at birth, was a king's son. Frightened by a prediction that his son would abandon the crown, the king kept the young prince surrounded by comfort and pleasure, unaware of the hardships that existed outside the palace walls. Thus protected and satisfied, thought the king, nothing would disturb the prince's peace of mind. Siddhartha grew up, married, and had a child. Then, without his father's knowledge, Siddhartha went on four successive chariot rides outside the palace. What he saw each day are called the "four signs." They were his undoing.

"Now the young lord saw...an aged man as bent as a roof gable, decrepit, leaning on a staff, tottering as he walked, afflicted and long past his prime." In his innocence, Siddhartha asked, "That man... what has he done, that his hair is not like that of other men, nor his body?" His charioteer answered, "He is what is called an aged man, my lord." And so, Siddhartha went home brooding on this, "Shame then, verily be upon this thing called birth, since to one born, old age shows itself like this." On the next day, he saw a sick man "weltering in his own water." Siddhartha asked, "but am I too...subject to fall ill...?" He went home even more upset. "Shame then, verily be upon this thing called birth, since to one born, decay shows itself like that....." Then he saw a dead person who, worst of all, could no longer relate to his loved ones. Finally, he saw a recluse with a shaven head, wearing yellow robes. Siddhartha was told that the recluse had "gone forth" to lead a religious life. Such a life was peaceful, with good actions, harmlessness and kindness to all creatures. Whereupon Siddhartha vowed to "go forth," abandoning home and family, to become a recluse, meditating in seclusion. "Verily this world has fallen upon trouble...one is born and grows old and dies.... And from this suffering...no one knows of any way of escape, even from decay and death...."[2] Thus he began a way out of his unhappiness associated with the inevitable course of nature.

Disease, old age and death are difficult enough but discontent creates an additional burden. What disturbed Siddhartha most was that he could no longer enjoy life knowing that he would lose what had given him pleasure. He craved what he liked and did not want the things he disliked to happen to him or anyone else. When he looked clearly at the natural course of life, he felt trapped by its inevitable outcomes and by his desires and aversions. This was the discontent from which he sought relief. If I could second-guess the Buddha I might say that, of course, he would begin his journey with disease, old age and death. These were omnipresent in the agricultural world outside his palace and are events of life which erode our innocence. Given my sojourns in the wilderness, as we will see at the beginning of Chapter Three, I was also confronted with the omnipresence of the three signs in nature. Once Siddhartha's eyes were opened, innocence, ignorance and distraction no longer worked for him. In contrast modern societies, tucked in their technological cocoons, seem to cling to these as evidenced by how people spend their time and money.

Leaving society and family, Siddhartha spent six years seeking out different teachers and undergoing various austerities. Even though he applied himself assiduously, he found no release from his suffering. In desperation, he settled under a tree to plumb the depths of the problem or die in the process. That Siddhartha emerged as the Buddha or Awakened One after seven days of meditating is a cornerstone of Buddhist religion. Buddhism claims he conquered the mind's fear of disease, old age and death, that he understood the nature of human greed and aversion and was no longer ruled by them. This is the Buddhist state of enlightenment or *nirvana.* He saw that change is inevitable and that human suffering arises when we hold on to experiences and mold them into what appears to be a solid identity. He concluded that humans do not posses an independent self. It is also said that the Buddha observed the causal chain through which discontent arose. He came to these realizations by means of meditation. With these understandings deeply embedded in his being, it was said that he no longer held harmful intentions for any of his actions and so lived non-harmingly. In addition, the *Sutras* claim he gained what we might now call mystical powers. He could see all the realms of the Buddhist cosmos and the pasts and futures of individuals.

While we may or may not accept Buddhist beliefs as expressed in the *Sutras,* the practice of meditation which emerged out of years of Buddhist investigation is alive. As they have for several thousand years, people still meditate. Whether or not all of the *Sutras* are true, we can use the story of the Buddha along with what is known from meditation practice to explore the origin of discontent. In doing this my focus will be on the interplay between the observations of both meditators and natural historians. Although some of the Buddha's mystical achievements and the idea of *nirvana* as an actual state of being will be mentioned, they are not a cornerstone of my edifice. In this book I take non-harming to be the most important measure of the unraveling of human discontent.

The meditation the Buddha engaged in is one of the many varieties that have been devised over millennia. The Buddha stilled his body and brought his mind into sharp focus so that he could pay attention to the comings and goings of his sensations, perceptions, thoughts, feelings and consciousness itself. He did this without adding another layer of thought to the mix. His meditation was very different from introspection which proceeds using an inner dialogue. It was a kind of a non-cognitive observation. How this is done, what kind of activity of mind it is and its relationship to neurophysiology will be explored in subsequent chapters. We begin at the Siddhartha's starting point.

The impacts of disease, old age and death led Buddha to study his wantings and fears but he was not the only person to have lamented their inevitability. In words reflective of those who see art as redeeming human life, the Irish poet, William Butler Yeats, wrote:

> O sages standing in god's holy fire,
> ...consume my heart away; sick with desire
> And fastened to a dying animal.
> It knows not what it is; and gather me
> Into the artifice of eternity.
>
> Once out of nature I shall never take
> My bodily form from any natural thing,
> But such a form as Grecian goldsmiths make
> Of hammered gold and gold enamelling...[3]

Each of us is a dying animal, sick with desire, who know not what we are. Buddha took this as his problem situation and began to investigate. He came to very different conclusions from Yeats, who sought immortality as an object of artistic beauty. Our animal bodies and minds are a part of the natural world. Discontent has natural historical roots.

My task is ambitious. I seek to understand discontent as a part of nature. As a prerequisite to this, we need to take a brief look at how natural historians describe nature. Many writers, celebrating nature in poems and literature, portray it as a well-integrated organism of beauty and peace. This view is shared by some natural historians. Within such a picture of nature, our discontent strikes a discordant note. In Darwin and some of his successors, there was ambivalence about how well the various traits of species fit together with each other and the environment, and whether nature was peaceful or violent. Sometimes ill-fitting traits helped a species survive but made lots of troubles. In other instances nature appeared rather fearsome, even grim-faced. And sometimes species' adaptations involve characteristics which undermine their comfort. From these perspectives, human discontent is more at home in nature among all its other ill-fitting properties. We will look at the dialogue between both views and touch on the similarities of the demands both Buddha and Darwin make to pay careful attention to life. Although Darwin, the scientist, wrote eloquently about the evolutionary consequences of the struggle for existence, he found it difficult to face its effects when they impinged on his personal life. Although it could have fit easily into his perspective, it never occurred to him to observe his mind as did the Buddha.

Darwin's nature is the setting which we all have as our residence. Long before humans, its other inhabitants experienced disease, old age and death. In looking at how these occur, natural historians carry on the debate about nature's character, be it gentle or harsh. If gentle, then our discontent may seem more out of place; if harsh, then discontent seems natural. We will look at the struggles of other species to see where the animal antecedents of our discontent reside. In animal behavior we find the precursors of our emotions. We will examine raw nature—the context of disease, old age and death—and animals' responses to them. In this exploration, I postulate that old age, which weighs so heavily on the modern world, does not contribute much to evolution. It is an uncomfortable byproduct of the need to secure offsprings' survival in nature.

Next, we move to the human body. First, we look at how it functions in nature without the improvements our genius has provided. On examination, we discover that the human body is as rough-hewn as some other parts of nature. If the blind watchmaker of the eighteenth century, who set the cosmos in motion, were actually planning things, he could have done a better job. Our bodies have all sorts of characteristics one part of which makes us hardy and successful and another part of which burdens life but not enough to undermine survival. For the Buddha, an examination of discontent only made sense in the context of an embodied mind. The mind's reaction to what occurs in the body is an important part of his understanding of discontent. We will look at how humans have responded to the three "signs"—disease, old age and death—both in our ancestors who lived in raw nature and in the civilizations we constructed to protect and sustain our bodies. We will see how nature has built into us attractions and avoidances.

The Buddha's civilization rested on the surpluses of early agricultural production. They made possible growing populations but were accompanied by periodic famine, epidemics and immense mortality. It was a world awash with disease, early old age, and death. It is no wonder the Buddha saw these as the seeds of discontent. The techniques he found to help people come to terms with these tragedies became even more relevant later. During the last hundred years, modern medicine conquered the infections, which so ravaged the Buddha's world, and has set out to correct nature's design flaws. We will look at how our transformed cradle of life affects the discontent which having bodies gives rise to. In contrast to the wilds, advanced civilization has created a safe container in which most *Homo sapiens* will survive into old age, and many people with disabilities will live. These are new demographic phenomena for mammals. Raw nature never before offered a large proportion of a species a third of a lifetime of old age or deaths drawn out by life-sustaining technologies.

After World War II antibiotics, vaccinations, pesticides and increased prosperity promised a Garden of Eden. People began to live much longer with much greater health. Death and disfigurement seemed things of the past. Gradually, as the new environment of changed lifestyles, medicine and synthetic toxins became more pervasive, we emerged from this Eden into a world of cancer, chronic disability, resistant diseases and aging populations. With immense amounts of money and great fanfare, science and industry have tackled each of these, promising again to undo disability and vanquish the ravages of old age. From the perspective of the Buddha, both the success and failure of these endeavors creates an enlarged playground for the discontent set off by our bodies. We will look at some of the ways meditation addresses these new problems.

From nature and the body, we turn to the evolution of mind. First, we need to get a sense of the human mind. What are its roots in nature? How does it operate? Some scientists proclaimed the 1990s the decade of the brain. Neuroscience has made great leaps forward. It has begun to penetrate the secrets beneath the skull and measure how our brains work. Tools, unimagined before, track electrical currents, neural connections, chemical flows and heat movements. We now have brain surgery, sleep clinics, anti-depressants, anti-anxiety drugs, mood elevators, pain suppressors, treatments for attention deficit, truth serums and psychotropics. Despite all of this, much remains unknown about the brain and we know even less of its relationship to the functioning of the mind.

To try to understand the relationships of mind to discontent we will draw on what is relevant in neurophysiology, borrow from disciplines including anthropology and archeology and bring to bear meditation insights. Here we will describe how meditation operates and what it tries to do. With these tools, we will look into the mind and see how its discontent is a product of the way it has evolved. The human mind evolved to accomplish feats available to no other animal. Linked to incredible dexterity, speech and abstract thought, we built a productive environment which has led to our impacting much of the planet. Despite these achievements, we continue to suffer and hurt others. The troublesome parts of our minds and bodies are negative attributes of the capabilities which enhance our survival. We will use the perspectives of meditation and natural history to explore these contradictions.

It is too simple to say that human joys and sorrows are merely the result of the evolution of a few traits. I make no claim that our biology is all-important. There are many characteristics of culture and history that cannot be reduced to natural historical explanations. Cultures vary greatly. Deference and demeanor in traditional Japan do not feel like those of contemporary America. Classic Mayan society whose temples were the scenes of human sacrifice may differ greatly from modern Scandinavia with its investment in individual dignity. Despite the tremendous differences in historical conditions and human cultures which affect societies, it seems more than likely to me that human evolutionary roots can reveal the character of our discontent, how it is embedded in the animal we have become, and how it leads us to affect the world around us.

Since meditation is a process of the mind, we need to use both meditation and science to understand meditation itself. In this exploration, we come across many things built into our brains that contribute to discontent and harmful actions. Central to meditators' concerns are emotions and the chattering mind, as exemplified by the student's meditation journal. Emotions are of ancient origin.

The chatter came later, hand in hand with the development of the brain's complexity. It is one of the many consequences of speech and thought. It is bound to our social development and an intimate part of our relationships.

Without actually looking at your own mind, it may not be possible to demonstrate its role in suffering. The Buddha and other practitioners have observed how much their minds contribute to their discontent. Meditation reprograms the mind so as to modulate the connection between emotions and thought. It gives the chatter space in which to rest so that it has less effect on us and others. Meditation is meant to address this and other troublesome properties of our minds, but meditation, as a function of that same mind, has its limitations. We will look at some of these. Since much of the brain remains a scientific mystery and because meditation has not previously been applied to understanding the origin of discontent in the context of Darwinian nature, any picture of the evolution of the mind will be incomplete. Nonetheless, what we do find is instructive and helps us understand the role meditation can play in life.

My understanding of meditation comes from thirty years of practice, including daily meditation, punctuated from time to time with weeks and months of silent retreat. Sometimes I meditated under the guidance of a teacher and sometimes by myself in meditation centers or out of doors. Although more will be said about the process of meditation in the chapters to follow, one of the main techniques I use goes back to the Buddha. This technique is one of inquiry, a careful examination of mind and body, the same kind I call upon here to make our investigations. Here the inquiry is directed toward an intellectual understanding. In my practice, it is aimed at easing my own discontent.

Because the meditation I learned was handed down by Buddhists through thousands of years, I need to refer to Buddhist ideas and beliefs. The Buddhist tradition that seems most applicable to my endeavor is the earliest, *Theravada,* or The Doctrine of the Elders. I utilize it and some aspects of Zen and Tibetan Buddhism in my explorations. It is my intention to carefully weave my way between what is a matter of belief and what may be observed through meditation practice. Of course, such a line is hard to hold, because in Buddhism most assertions about time and space, cause and effect claim meditation-observation as their proof. So I will draw as little as possible on beliefs, indicate when I am using them, and try to ground my arguments on meditation which can be done independent of the beliefs. First and foremost, this is a book on meditation and nature. Secondarily it is a book about individuals such as Buddha and Darwin, and their world views: Buddhism and Darwinism.

The ultimate goal of this book is to explore attention and wisdom. The answers Buddha found sitting under the Bodhi tree contained two crucial ingredients. The first underlay his method of inquiry and the second characterized the fruits of that inquiry. For Buddha, the development of attention was prerequisite to an examination of discontent and our ability to deal with discontent when it arises. For Buddha, wisdom was the fruit of seeing clearly. It makes it possible to relate to the world in ways that bend toward harmony rather than in ways that create more suffering. If an examination of the origin of discontent has any usefulness, it would be in the ability to live more wisely. I am not sure that the progress of civilization has contributed to wisdom. It seems that the advances of civilization have obscured the sources of our discontent, making it even more challenging to restrain our actions and their consequences.

With an understanding of the evolution of mind, we can look at how attention and wisdom have fared throughout human history. In contrast to commentators who think things are getting better among human beings, other thinkers see our hunter-gather ancestors as having lived more in balance both with nature and each other. That "it takes a village to raise a child" is commonly accepted. The quote conjures up a time when people lived simply, in small groups and much closer to nature. It is often claimed that indigenous peoples' connection to nature gives them deeper, more grounded insights into existence, allowing them to live more harmoniously. It is implied that hunter-gatherers suffered less and cared more for their environment.

Seeing our ancestors in this light, some people look to hunter-gatherers as a model of how to heal the discords of the modern world. Using an understanding of the nature of the mind and body and the changing contexts which civilization has brought, I will look closely at how we have fared in contrast to our ancestors. It may be that their intimate connection to nature gave them a sense of harmony we lack. That connection to nature, so utterly changed with the introduction of agriculture, may have precluded the way of observing life that the Buddha developed in his world of wealth, famine and disease. Nonetheless, there is a shared humanity, and while hunter-gatherers maintained a certain balance with nature, with their stone weapons and tools they were capable of harming others and the world around them because of their discontent. In contrast, the technologically powerful context of our discontent has drawn us into our minds in ways unimaginable in earlier times and this progress may have taken us deeper into the chasms of human suffering. While we have been liberating ourselves materially, we may be torturing ourselves mentally and causing untold damage to nature with our new toys.

Perhaps some hundred-thousand years ago, our preverbal hominid ancestors lived without the life of the mind as we know it. The chapters on mind and on attention will discuss the quality of consciousness they might have possessed. It stood somewhere between what we now experience and what we see in wild animals. Among other things, that consciousness may have possessed a quality of peaceful coexistence. It was peaceful in the sense that there was little mental reaction to events that occurred. The finding of food and shelter, living in an extended family, and aggression and defense solely for survival probably all took place without the added static—the worry and running commentary—of the mind's involvement. There were emotions like fear and pleasure but they occurred without anxiety or regret. There may or may not have been wisdom and compassionate action. Life took place within a kind of profound silence. It is the silence out of which we evolved and towards which meditation is aimed. Such silence is rarely experienced in our civilized lives. Yet, in moments when one is simply present, whether in daily life or in the midst of deep meditation, that silence may be touched. Its presence may connect us to the natural setting with which we have lost contact. This ancestral silence may give us space enough around our thoughts and feelings to quell their more harmful effects.

* * *

One cold winter night in the mid 1980s, a group of staff members from the Insight Meditation Society located in a large old monastery in the central Massachusetts woods came to visit me in the hayloft of a nearby horse barn I was renovating to be my home. We cranked up the wood-burning stove for the first time, soothing my anxiety about whether the place would hold heat, and spent the evening in a ritual of self-exploration. Around midnight, the men wandered outside. One, astride the old mare living below, rode out into the night while the rest spread out across a frozen cornfield. The moon was full and the temperature well below zero. In the stark light, you could see the silhouettes of the men and rider each moving separately. My dog was at my heels, quiet and intensely alert. The reality of nature was unmitigated by human concerns. Life and death made no difference to the moon and stars in the deep black sky or to the intense cold, that grim reaper of winter. In the silence, my mind stopped for a moment and there was an intimation of the world as experienced by hunter-gatherers at their most attentive or a glimpse of what Zen hermits or forest monks may have lived. It is a world that animals without thought experience every day. It stood in radical contrast to the comfort of the fire and the chatter in the reno-

vated hayloft which, in spite of its appearance of voluntary simplicity, was very much a part of the civilization from which its owner sought to differentiate himself. With my mind's stillness came a shiver of fear that I would have to surrender my habitual patterns of living. With it also came a profound sense of peace.

Chapter Two
Darwin's World

I am fascinated by the men who pioneered the theory of evolution. They touched the wild. They were scientists who tried to look at nature as a whole. They shook the complacency of their own society while addressing, directly and indirectly, the deepest issues of family, livelihood and the meaning of life. Starting with careful observation, Charles Darwin, Alfred Russell Wallace, Joseph Hooker and Thomas Huxley created a new understanding of nature. We will examine these issues primarily in relation to Darwin. Like Copernicus, who removed Earth from the center of the universe, the Darwinian revolution decentered us; it linked humans to apes and undermined Christian belief that humans were a special creation of God. The theory of evolution put humans back into nature as one among many animals.

Although Darwin, Wallace, Hooker and Huxley explored untamed lands as few of their compatriots had, they did not pursue the full implications of what they learned. This was because they focused on plants and animals devoting less or, in some cases, no time trying to understand the indigenous peoples they met. As Englishmen living at the height of the British Empire, they believed in reason, progress and comfort. Yet what they glimpsed of nature had the potential to trouble those beliefs. Although Wallace's sojourn among the forest dwellers of the Amazon and the Dutch East Indies was more intimate than the contact any of the others had with natives, the conclusions he drew were an odd mixture of non-evolutionary 19th century spiritualism and English socialism.[1] Darwin was the most inspired and probed deeper than the others. His experience with the hunter-gatherers of Tierra del Fuego continued to challenge his ideas of how humans fit into nature. He came closest to investigating the contradictions between the security of English values and the tentativeness of nature. But he too held back.

In this chapter we will explore some Darwin's evolutionary theory. This theory provides a paradigm for looking at nature. It gives a sense of order to the natural world but also reveals aspects of nature which undermine that order. This duality parallels our struggle to understand how we fit in nature and what life and death mean for us. The overall picture of nature revealed by the theory of evolution was a lot less pleasing than many people, including Darwin, would have liked it to be. As will be elaborated in later chapters, the awkwardness of nature manifested by aging human bodies and unhappy minds is what motivated the Buddha to look inward. When Darwin experiences the harsher effects of nature on his family, he tries to grasp its meaning using his theory. Although this gives him some intellectual satisfaction, it does not address his personal suffering. Darwin's natural history gives us a description of the world we live in. Buddha's looks into how we can face that world and come to accept it.

Charles Darwin was born into the rising bourgeoisie of England in the early 19th century when industry and science were linked with a gentleman's lifestyle. His father and grandfather were both physicians who combined medicine with money-lending. Their growing social and economic class stood in opposition to an entrenched aristocracy with its monopolistic economics, ideas of God and human hierarchy. The bourgeoisie exploited new resources of nature. Their rise to power came from trade and the profits of industry. There was mining, brewing, textiles, pottery, canal building and the railroad. These industries were based on new understanding of materials and forces present in the surrounding countryside. Water, steam and practical science were put to use as never before. Darwin's uncle, owner of the Wedgewood pottery works, was interested in the chemistry of glazes and in geology to locate different kinds of clays. Darwin himself made a fortune from speculating in railroad stocks.

Engulfed in the computer revolution, it is hard for us to imagine how completely industry transformed life in 19th century England. In the hundred years before, agricultural improvements, waterwheels, steam engines and canals precipitated a movement of cottagers to the city. During the 1840s, as Darwin circumnavigated the world on the Beagle, industrial towns were blackened with coal soot and crowded with impoverished workers who were deafened by the roar of the railroad and loom. Later in the century, when the theory of evolution became more acceptable, the grinding misery of industrialism was beginning to be moderated by a combination of the great prosperity it created, riches pouring into the British Empire and social protests. Darwin and his fellow intellectual revolutionaries created ideas which fit well with these transformations and they personally profited from the educational and scientific changes they helped to bring about.

As a student in the universities of the gentry, young Charles encountered controversial ideas. Arguments about geological change that challenged biblical dating intrigued him. Were the seashells found on hillsides or road cuts the result of floods or volcanic upwelling? Besides riding to the hounds, Darwin took field trips to look at rocks and passionately collected beetles. Beetle collecting was a popular pastime of gentlemen, clerics and college students. Given how undistinguished are the three-thousand beetle species of the British Isles, an ability to identify them and find new species was a good training ground in natural history.[2] Along with geology, gardening and agriculture, it attuned the otherwise civilized gentry to the nature around them.

At university, Darwin disliked medicine, the family occupation he had been sent to study. He was drawn instead to the great natural historians of the day who were debating the origins and transformations of the earth and its inhabitants. Soon after graduating, he took a position as a self-supported captain's companion on one of many mapping expeditions by which England was cataloguing its empire. The maps were of strategic importance to the British Navy. The information was also used for commerce and to extract natural riches for the British Empire.[3] England was interested in guano to fertilize its fields, harbors for shipping, sources of wheat and tea and, later, in coal stations to power its fleet. Nature was pickled and dried by naturalists and sent home to museums and rich collectors. For five years, Darwin lay seasick in his bunk, rode the pampas of South America, studied the geology of mountainsides and collected tens of thousands of plants and animals as his ship traveled around the globe.

Because Darwin was the captain's companion and a skilled observer, while his shipmates mapped coastlines, he took extensive terrestrial jaunts winning the position of naturalist traditionally accorded to the ship's doctor.[4] Gathering fossils of giant marsupials in Argentina, he saw immeasurable changes in the rock record of life. Witnessing an enormous earthquake in Chile, he experienced first-hand how much landscape could transform. He saw towns leveled and measured the subsequent rise in coastlines. Collecting a great variety of plants and animals, he was impressed by the diversity of life and how little of that diversity could be explained by biblical creation. It was while exploring the Galapagos that he collected the giant tortoises and subtly varying finches—animals that became a foundation for his theories.

On his return to England, Darwin settled into trying to make sense of what he had seen and collected. His travelogue, *The Voyage of the Beagle,* was an insightful portrayal of the natural history of imperial contact with what we now call the "third world." It was also a best-selling adventure book. With an English gentleman's sense of superiority over the savages he had seen, Darwin tried to figure out the mechanisms by which nature and civilization progressed. How had larger beings come into existence? Why was English society so much more advanced than that of the naked Indians of Tierra del Fuego?[5] In trying to answer these questions, he unintentionally undermined the Christian belief that humans sat at the pinnacle of creation, and came close to divesting himself of the assumption that Englishmen stood at the peak of human development.

Equally important was the fact that he set the stage for the use of natural history to try to understand how human life works and how humans fit into nature. From Darwin on, humans had to take their place among apes, ribbon worms and bacteria. As our knowledge of the diversity of life increases, we find more and more creatures "small and ugly by human standards."[6] The chemical details of some of their lives may be different from ours but they evolved by the same processes that produced us. The Darwinian revolution brings insects and sea slime into our conceptual living rooms whether or not we are comfortable with having them there. As ambivalent as Darwin was about it, the theory of evolution brings the darker side of nature sharply into focus.

I have a friend named Janet. She loves to live close to nature. From being a hippie tree planter in Oregon, to herding sheep in Australia, to picking coffee in Nicaragua, to gathering natural art materials for Cambodian refugee children on the borders of Thailand, she looks at nature and tries to incorporate it into her art. The first question she asks when encountering an animal in a new setting is, "Who does it eat and who eats it?" This alerts her to how the environment works. It also reminds her that she is part of the matrix of life along with the creatures she observes.

This also is one of Darwin's starting points. Along with other 19th century naturalists, he concluded that most plants don't reach maturity and most animals die tragically.[7] Not only is nature harsh, it is arbitrary. Part of nature, our endings are troubling and we too possess awkward traits. However, to best understand the naturalness of the suffering Siddhartha bemoaned, we need to present some of Darwin's ideas and his methods of observation.

Evolutionary Change

In his theory of evolution, Darwin came to a number of conclusions. First, over geological stretches of time species have come and gone. The dinosaurs vanished and large mammals evolved. The origin of newness, he argued, was variation—the heritable change Darwin observed taking place in domestic plants and animals. He did not know how or why mutations occurred, but he carefully traced the processes by which farmers and breeders honed cumulative mutations into new breeds. He then assumed that this process of artificial selection also worked in nature, thus the term "natural selection." Farmers and collectors sustained those changes which were useful or pretty and eliminated the others. Pigeon breeding was a popular lower class hobby in Darwin's England. It was odd to think that some of the fluffed, strangely shaped specialties were related to the wild rock dove with which they can breed. Or that a poodle and Siberian husky are both dogs; they seem almost different species. He suspected that species in nature developed in an analogous way. A change arose and the environment, rather than breeders, selected among those changes.

While breeders strove towards particular changes, nature operated more by chance. Outcomes in the natural world at a particular time and place were the result of the actions of the environment which included weather, food sources, population densities and predators. Those plants and animals, whose random changes helped them to prosper in the world they inhabited, went on to leave many successors and those, whose changes disadvantaged them, left few.

Spontaneous changes were either well adapted or not. The species which were well adapted survived, others didn't, and so there was survival of the fittest. As the 19th century poet, Tennyson, trying to reconcile a belief in a kind God with the terrible death of a friend, described it: "Nature, red in tooth and claw."[8]

In Darwin's natural selection, as opposed to artificial selection, there were no breeders guiding the process. Because the population of a species with abundant food, good weather and little competition continued to increase, there must be checks on their numbers or they would overwhelm their environment. The checks and balances of nature determined the relative numbers of different species at any given time. New forms arose via mutation and competitors, climate and catastrophe pruned them. Species struggled to survive and procreate. It was both beautiful and horrible with nobody in charge. For Darwin, humans were subject to the same process as any other species.

If we are exploring the meaning that nature gives to human life, then we need to be sensitive to the sort of assumptions we build into our world view. The theory of evolution, like any major intellectual paradigm, is subject to many interpretations. Scholarly and popular commentary are peppered with disputes. It is a domain of powerful science, large egos and a reemergence of debates between science and religion. In looking at evolution, we are concerned with the sense of orderliness, progress and efficiency its descriptions create. It is in relationship to those that we can understand how we fit into the world and how nature affects us.

Change is fundamental to evolution. We are a product of that change. How that change comes about is crucial to the kind of beings humans we are and how we fare in the world around us. If nature is orderly, progressive and efficient in ways that support us, then we might hope that our lives will be less difficult and more sensible. If order, progress and efficiency have rough edges, then coping with the outcomes of life becomes more problematic. The Buddha emphasized that the recognition and acceptance of change were crucial for human understanding of life. Discontent for him was the result of our inability to accept life's uncontrollable outcomes. It was in this reaction and how to address it that problems occur.

Evolutionary change arises in two ways: mutation and natural selection. We start with the first. As a result of molecular biology, the character of mutation is much clearer to us. It took eighty years since Darwin's death for the mysteries of DNA and RNA to begin to unravel. Molecular biology now plays a heroic role in our society. Genetics promises new crops, new cures and a new lease on life. Biotechnology is already a significant part of medicine and our economy, and its benefits have only begun.

Mutations happen naturally. Sometimes they arise from routine cellular processes and sometimes are caused by such irritants as radiation. It is likely that mutation creates change much more rapidly than Darwin imagined. From the very incomplete geological record, Darwin got only glimpses of changes taking place over millions of years. He thought change came about slowly, one step at a time. Darwin would later be called an evolutionary gradualist.

Now we know that mutational change can occur much more rapidly. The evolutionary building blocks of newly designed organs lie much closer to already existent parts of the body than Darwin understood. Take the eye. There is no survival value to an eye that can't see. Yet, it is hard to imagine a mutation that would produce a whole eye in one leap. Part of an eye is of no use, so the chance of it being passed on to heirs who might spontaneously build the rest of the eye is unlikely. Because Darwin could not find fossil records of useful intermediate stages, critics held his theory at fault. Even he wondered about it.

Today, however, with advances in molecular biology, we can change genes in one generation. A garden variety bacterium can be changed to inhibit frost formation on tomatoes. The cells of crop plants can be made to produce the toxins of another bacterium which kills caterpillars. Nature also operates more rapidly than Darwin thought. It uses genetic processes to recruit different tissues to achieve the same function or builds different functions starting with the same kind of tissue. The eyes of hummingbirds and shrews have different proteins in them but they have a common genetic origin. Fly, mouse and human eye development are controlled by the same gene. Mouse eye genes when inserted in fruit flies cause perfect eyes to be formed but in the wrong places. The eyes lack optical nerves connecting to the brain so they don't see. Squids have eyes and something called a photophore, a kind of flashlight which they use to illuminate the sea below them so as to hide their shadows. The organ which does this uses similar enzymes as their eye even though it grows out of muscle rather than skin as does the eye. There is a regulatory gene whose role may have been to control structural genes of ancient photoreceptors but later came to create eyes in different ways. The gene is involved in photosensitivity which guides blue-green algae and even the spores of some fungi.[9]

Such changes do not take eons. The three-hundred species of cichlid fishes of Lake Victoria in Africa descended from common ancestors in less than two- hundred thousand years. Some graze algae. Some are vicious carnivores. Others eat insects or mollusks and one group forces mouth brooders to release their young for prey. That is a lot of variation. And changes as a result of symbiosis may happen relatively quickly and account for the lack of what Darwin conceived might be intermediate forms. Examples are the first cells with nuclei, which gave rise to plants, animals and fungi. They derived from the symbiosis of two different kinds of bacteria. And now it has been suggested that larval and adult stages of such things as butterflies and sea urchins were once independent living species.[10]

Although some mutations are useful, most spontaneous ones make little difference or are harmful. Lineages with destructive genes don't survive. In the absence of modern medicine, many human genetic disorders would be painful examples of this. Plants into which a flowering gene has been injected create so many flowers when they are still seedlings that they die because they haven't developed enough leaves to support the flowers.

Mutations occur, and changes are recorded in fossils. Intellectually, this is the easy part of evolution. The more challenging mechanism of change is natural selection. Between Darwin and scientists who objected to his ideas, this was a battleground. Even Thomas Huxley, known as "Darwin's Bulldog" for his championing of evolution, never really accepted natural selection.[11] Huxley, in his battles against the scientific establishment, used the fact that Darwin's *Origin* undercut the fixity of species and made humans similar to other species, but he was an anatomist and embryologist, not a field biologist. He was not steeped in the subtleties of variation and was not convinced that natural selection could hone the exotic and seeming arbitrary forms that abounded or that change occurred bit by bit. His objections were muted and Darwin never confronted Huxley's doubts because of his powerful defense of evolution.[12]

In the *Origin of Species,* Darwin regarded natural selection as paramount in explaining evolutionary change:

> [A]ll organic beings are exposed to severe competition. Nothing is easier than to admit in words the truth of the universal struggle for life, or more difficult...than constantly to bear in mind. Yet unless it be thoroughly ingrained in the mind, the whole economy of nature, with every fact on distribution, rarity, abundance, extinction and variation, will be dimly perceived or quite misunderstood. We behold the face of nature bright with gladness, we often see superabundance of food; we do not see, or we forget that the birds which are idly singing around us mostly live on insects or seeds, or their eggs, or their nestlings

> are destroyed by birds and beasts of prey; we do not always bear in mind, that, though food may now be superabundant it is not always so at all seasons of each recurring year.
>
> ...There is no exception to the rule that every organic being naturally increases at so high a rate, that if not destroyed, the earth would soon be covered by the progeny of a single pair.[13] ...In the case of every species, many different checks, acting at different periods of life, during different seasons or years, probably come into play....[14]

Referring to the succession species in the forest covering the ruins of the Indian mound dwellers in the U.S. whose cities mysterious vanished in the 16th century, Darwin wrote,

> What a struggle must have gone on during the long centuries between the several kinds of trees...what war between insect and insect...between insects, snails, and other animals, with birds and beast of prey...all striving to increase, all feeding on each other, or on the trees, their seeds and seedlings. ...the struggle will almost invariably be most severe between the individuals of the same species....[15]

We usually associate Darwin's language with the Social Darwinists of the late 19th and early 20th centuries. They used the phrases, "struggle for existence" and "survival of the fittest" to justify the success of the rich and powerful capitalists of their times. The struggling masses were portrayed as less fit. In other contexts, these ideas have been used to support claims of racial or gender superiority. We are not interested in these self-serving social doctrines but in understanding the character of natural selection where it has been applied to humans as part of the natural world.

As compelling as are Darwin's and others' examples of natural selection, on closer examination they display discrepancies. Some of the brilliance of Darwin's contribution lies in this apparent limitation. Darwin wanted to establish a theory of evolution, yet most examples of change through natural selection don't quite work when looked at in more detail. The history of Darwinism is replete with supporting evidence needing to be revised.

Science reveres theory.[16] Newton's theories of force and matter sufficed until the new forces explored in the 19th century overwhelmed them. Einstein proposed a general theory of relativity but remained uncomfortable with the ideas of quantum mechanics. At the close of the 20th century, physicists debated whether the discovery of newer, more powerful and subtle forces can ever be included in a comprehensive mathematical explanation such as string theory or quantum gravity. The effectiveness of a theory is often measured by the new knowledge arising from investigation of its anomalies rather than its absolute truth.

Parts of Darwinian theory are evident. Change has indeed taken place over time. Genetics clearly indicates that species are related to each other and have common ancestors from which they evolved. The pace of evolution is subject to discussion. Some biologists argue for continuity while others see evolution happening in fits and starts.[17] The fact that there are no complete examples of one animal species evolving into another has always been a problem for the theory of evolution. Huxley even pointed out that the pigeon breeders, on whose exotic homebred variations Darwin's had relied for one of his major examples of natural selection, had been unable to breed a distinct species.[18] This is, in part, because the definition of species is not completely agreed upon. These two gaps have led to difficulties in applying the idea of natural selection. In trying to explain changes as a result of natural selection, naturalists often have to go back to observing nature more carefully because their explanations fall short. This parallels Buddha's methods of observation. He would not let a person rest with generalized conclusions. He required that individual inquiry be grounded in continuous observation.

For Darwin, "survival of the fittest" was a metaphor, an example of which is Darwin's moth. While studying a night-opening, Madagascan orchid with a long, tubular flower, Darwin speculated that its pollinator was a moth with an eleven-inch tongue. It was not until years after his death that biologists found such a moth pollinating the orchid. In the 1990s another flower, requiring a 16-inch moth tongue was discovered. The matching moth has yet to turn up.[19] These moths "fit" well into a world of such flowers. And if they have no other pollinators, these flowers require the moths in order to survive. Modern evolutionary biologists refer to this as engineering fitness and limit the term fitness itself to reproductive success. The more offspring who survive, the more fit the species. Evidence for fitness is successful reproduction. The word adaptation is now used to describe the moth and flower.

The ideas of competition, fitness, and adaptation are part of a complex of concepts which deeply influenced 19th century views of life. Order, efficiency and progress were heralded in economics, political theory and science. Classical economics viewed competition, free flow of information and pricing as the mechanisms by which goods would be distributed in the most efficient manner. The open marketplace would allow individuals to maximize the satisfaction of their preferences within the constraints imposed by others' preferences. Competition would also encourage innovation and efficient production, making goods better and cheaper so that the economy would grow and society progress. In politics, competition between ideas in the marketplace of free speech would allow, if not the truth, then at least the will of the majority to emerge. Political competition was the foundation of democracy. Nineteenth-century science was regarded as a paradigm of progress. Each new discovery was placed as a building block upon earlier ones, advancing science toward a complete understanding of the animate and inanimate world. Even Karl Marx, who worshipped science but rejected classical economics and free speech, wanted to dedicate his *Capital* to Darwin. Darwin refused the honor on the grounds that he knew little about the subject matter of the book.

Natural selection or "Survival of the Fittest" was evolution's mechanism for creating order, efficiency and progress but the difficulties in applying this concept reveal the crudeness of nature's order and efficiency, thereby bringing into question the idea of progress.[20] Here we encounter a divergence between the way science, economics and politics often view the world and the way Buddhist meditation does. Buddha begins with an individual's concern about his own death whereas the leading ideologies of the 19th century industrial revolution embraced collective progress and glossed over the nature of happiness which was assumed to accompany the efficient satisfaction of preferences or the will of the majority.

This difference of perspectives cropped up during a Ph.D. examination in sociology. The student, who had practiced meditation for many years, asked her examiner, Morrie Schwartz (made famous by the book *Tuesdays with Morrie*) who knew he was dying, whether his own teacher, one of the great mid-20th century social thinkers, ever considered dying or death as part of his ideas. Morrie stopped, surprised because he had not noticed the omission, and said something to the effect that it was never spoken about in his teacher's presentations or elsewhere at the time, but that he would ask his teacher when he met him in the afterworld. In natural history, death plays a crucial role. Death too was one of the Buddha's starting points. So if we want to understand how the roughness of human adaptations contributes to our discontent we can do no better than begin with the ideas of adaptation, natural selection and survival of the fittest, all of which involve life and death.

Adaptation is the cornerstone of a modern school of evolutionary biology. Adaptations are any heritable change which better adjusts an organism to its environment. Adaptationists look for advantageous traits which seem honed by natural selection.[21] For years it was argued that mammals with warm blood and longer gestation produced more flexible species that were better adapted than the dinosaurs they replaced. Warm-bloodedness allows mammals to be more active in cooler environments than would reptiles, and mammalian birth supports intelligence and more complex social behaviors. These traits helped mammals solve problems of survival better than their predecessors. Analogously when the land bridge between North and South America was formed, the placental large mammals of the north overwhelmed their marsupial neighbors to the south leading biologists to regard marsupials as less well-adapted. The latter lost the battle of survival of the fittest.[22] As more has become known about the subtle coexistences of both mammals and dinosaurs, and marsupials and placentals, modern adaptationists no longer see one as simply better adapted than another. These difficulties in applying the idea of natural selection reveal some of the roughness in nature's order.

Nature's Order Is Hard to Pin Down

The most famous example biologists cite of natural selection is the peppered moth, decorated with light and dark patches. It was prevalent in the midlands of England before the industrial revolution. A peppered moth sitting on lichen-covered trees seems to have a protective coloration. It is almost invisible to the human eye.[23] As 19th century factories belched more and more smoke, killing the lichen, observers noticed the number of peppered moths nearby dropped and a rare, black, i.e. melanic, form of it became common. Then, as industrial pollution declined in the 20th century, the reverse occurred. An explanation of the first change was given in 1896, "In our woods...the trunks are pale and the moth has a fair chance to escape, but put the peppered moth....on a black tree trunk.... would be very conspicuous and would fall prey to the first bird that spied it. ...the darker [moths] escape."[24] These changes in peppered moth populations stood as a textbook case of natural selection for several generations.[25] Confirming this common-sense explanation, detailed experiments in the 1950s showed that birds indeed snatched more dead moths of types which stood out when they were placed on exposed parts of either lichen covered or bare trees. (See Illustration 1.)

This was taken as proof of natural selection until 1979 when a researcher threw a monkey wrench into the works by pointing out that the moths do not rest on the open, exposed trunk of a tree. Moths rest either on the underside of branches in the canopy or close to trunk-branch joints, where their color makes little difference. Dead or sluggish live moths placed on an open tree trunk of any color were snatched up by birds. So the moth was subtler than expected. Moreover, moths copulate for twenty hours, leaving them vulnerable to the slightest exposure.

Thus it was back to the drawing board. Analyzing all the known variables suggested factors such as the height on the tree of moss or lichen, January temperatures, the amount of sulfur dioxide in the air, and how far west the moths were of pollution sites.[26] It is conjectured that melanic moths are well-disguised when flying but less camouflaged when resting (which they do not get much chance to do because they are often disturbed by ants).[27] It may be that the physiology, biochemistry, and energetics of the different forms correlate with pollution tolerance. Because not much is known about moth behavior, moths have to be observed much more carefully as they go about their lives. Investigating natural selection as the mechanism for honing better adapted moths has become a program for looking deeper into the life of this species. More recently it is claimed that the experiments of the 1950s were biased to give the desired "proof" of natural selection.[28] It may even be that the peppered moth is not a case of natural selection. Whatever the unfolding of the story eventually reveals, trying to demonstrate selective forces has led to a greater understanding of the behavior of peppered moths.

Large numbers of live peppered moths, *Biston betularia,* and its melanic form were placed on lichen covered trees, uneffected by pollution. Birds indeed took a much greater number of the melanic moths. The reverse happened when the moths were placed on the bark of trees whose lichen was killed by pollution. Birds took many more of the peppered form. Looking at the moths with human eyes, it would seem that being peppered is advantageous in an unpolluted environment whereas being black is protective when the lichen are killed by pollution. Indeed the proportion of each form taken roughly correleates with the degrees of pollution. Yet biologists are not sure what birds see nor where the moths actually hang out when caught by birds. It seems as though natural selection is operating but it is not clear how.

Other examples abound. Why flowering plants, which represent over 90 percent of today's plants, replaced the great ferns, conifers and cycads of the carboniferous era 125 million years ago, Darwin regarded as an "abominable mystery." We may never know because the geological layers containing the "missing links" have been destroyed. So paleobotanists look for the "advantages" flowering plants might have. Insects and flowers seem a more effective arrangement than wind but then problems arise in floral structures which are hopefully less troublesome than the increase in efficiency. The new solutions give rise to new problems, and so on. Progressive levels of relative advantage are investigated. Biologists end up with a tentative conclusion that all these changes led to a selective advantage for flowering plants, but they can't be sure.[29] So biologists are left with a set of questions which are not quite answered and which require closer examination. The picture emerges of a natural world that does not fit together quite so neatly.

As the prominent evolutionary biologist, D. J. Merrell, put it, "The first point to be made about adaptation is that, despite its critics, the phenomena is real and universal. All living organisms are adapted to their environments; if they were not they would die."[30] Natural selection does happen but putting one's finger on exactly how it operates in a given case is more difficult. Better adapted and more fit go well together with common-sense reasons obvious to the observer, but also go hand-in-glove with our deeply held assumptions that nature does its job well. It is our way of talking about genetic variations that seem to contribute to reproduction and survival as if they were the actual cause. It is a useful way of talking.

When DDT was introduced in the 1940s, it proved deadly to insects. Since World War II, where it was used to combat the waves of disease that accompany armies, DDT was literally dumped around the world. After less than twenty years of use, insects began to develop resistance. Experiments with DDT-resistant fruit flies have shown that the genetic variation in natural populations was sufficiently diverse that DDT radically selected for those flies who could survive its application.[31] Pretty soon DDT was not so effective. This illustrates the power of natural selection as an agent of change. Hundreds of species of insects have become resistant since the 1940s. DDT-resistant flies may have a less permeable cuticle, larger fat bodies, more DDT-destroying enzymes, or nerves less sensitive to DDT. Scientists are not sure. The fact that change comes about so quickly when species are under assault shows how opportunistic evolution can be.[32] This story was repeated with bacterial infections such as TB, gonorrhea, staphylococci and antibiotics, and the malarial spirochete and chloroquin, the synthetic version of quinine, a traditional medicine.[33] With HIV, nature's opportunism is terrifying.

While some biologists and philosophers point out that Darwin's use of "survival of the fittest" led to circular reasoning (those who survive are by definition fit, whereas those who are seen to be fit are the ones who survive); others insist that demonstrating natural selection is different from survival of the fittest.[34] The issue in some ways is semantic. The three concepts fitness, selection, and adaptation are interrelated but point at slightly different things. Fitness in modern garb simply means numerical success. Natural selection, on the other hand, points to forces that affect a species. And adaptation or engineering fitness looks at how traits function in a given environment.

In one of the most careful studies of natural selection undertaken, the lives of nineteen-thousand individual finches in twenty-four generations were observed for twenty years on undisturbed islands in the Galapagos, the place made famous by Darwin. The relationships between the beaks of Darwin's finches and environmental changes were mapped in detail.[35] The researchers measured beak size and survival rates of two closely related species. The researchers found that during droughts, because of both the foraging advantages of a larger beak and female preference, the larger beaked species came to dominate. This seems a good example of natural selection influencing evolution. In the harsh circumstances of the Galapagos droughts, most of the different kinds of finches died but the larger beaked species did better. If the trend kept up, the larger beaked species would prevail in a relatively short period of time. This was hailed as proof of natural selection.[36]

In contrast, when El Niño currents brought back abundant rainfall, the balance tipped back again and the two species prospered. They even produced odd hybrids which had a low rate of survival. It is clear that there are mechanisms of selection at work in the waxing and waning of species success, and it is clear that ever more careful observation gives a more complete picture of the complex lives of close finch species. It would seem that if it stayed dry for long enough, we might get two quite distinct species, but that hasn't happened. Also, some other researchers have not conceded that competition is the real cause of change.[37] So given the shortness of the span of observation and variables which may be historically unique, natural selection doesn't quite produce neat adaptations.

Other studies honed to escape the tautological character of survival of the fittest have similar gaps. In a population of frogs, 15 percent had deformed legs making them less agile. After several months, only 4 percent were deformed. From common sense, it would seem that deformed frogs were maladapted. Nevertheless, biologists don't know why they died out in greater numbers. Even though wild populations of animals have much variation, the overwhelming majority of individuals are strikingly similar.[38] When the restraints on species are controlled as in the breeding of show dogs, then change does appear to be directional. When the restraints are eased as happens with pigeons in a city, then variety seems to blossom. Natural selection is playing a role, and it is tempting to speculate how, but to come near to isolating real causes requires complete immersion in the life of a species as the environment changes.[39] The team that studied Darwin's finches comes close to satisfying the criteria of total immersion, though their critics want even more details of finch behavior. One has to respect their dedication to careful observation.

We need to pause to recap. Although it might seem we have strayed far afield, for the author—a meditator and amateur natural historian—the details of attempts to make sense of natural explanations have parallels to meditation. Order, efficiency and progress are values which permeate modern industrial nations. Sitting cross-legged counting one's breaths and trying to observe one's body or how one's mind works seem at odds with those values, except when the activity is called as stress reduction, a part of behavioral medicine used to aid in healing illness or help people work more efficiently. In contrast, Siddhartha took to meditating because of his realization of the inevitability of death. Death shaped meditation practice, and it is death that shapes evolution. When Buddha looked at the effects of death on living humans, he found discontent. When Darwin looked at the effects of death on species, he found natural selection. Darwin's natural selection mandates further exploration and, as we are arguing, the farther scientists proceed in that exploration, the less any simple sense of progress in nature seems to suffice. The only thing the Buddha found that would alleviate his discontent was an investigation of its nature.

We will explore in more detail the character of that investigation in subsequent chapters. What we can use now as a point of comparison, is that with experience, meditators learn to be wary of conclusions about life that they come to in the course of practice. Life has its ways of undermining the conceptual framework we compulsively put on our observations. Like explanations from natural selection, the trick in practice is to keep observing. Otherwise, I have found, the understandings I preserve as concepts often fall short of the realities I am faced with and so I have to observe anew. A metaphorical version of death as a reminder of the need to be constantly attentive is found in the teachings of Carlos Castaneda's character, Don Juan. Don Juan tells Carlos that in order to be fully alert to the meaning of life, he needs to keep death as his ally hovering over his left shoulder.[40] Castaneda may have gotten this idea from Buddhism.

Certainly, Darwin, with his insistence on the importance of the struggle for existence, would say it is true for natural history.

Bound in with the logical difficulties of natural selection, are two more reasons why deductive explanations run into these problems and why returning to observation makes sense. We will see that these also have parallels in meditation. First of all, nature is overwhelmingly complex. Darwin was impressed by this fact. One famous example in the *Origin of Species* links the success of wild pansies and red clover in England to the number of bumble bee pollinators which are preyed on by field mice whose nemesis are cats. Another is the population of feral animals in Paraguay which is reduced by navel flies which are preyed on by other insects which are eaten by birds whose success depends on vegetation, "and so onward in ever-increasing circles of complexity. ...Throw up a handful of feathers, and all fall to the ground according to definite laws; but how simple is [this] problem...compared to that of the action and reaction of innumerable plants and animals...."[41]

Since each situation in nature has more variables than can be knit into an explanation or even known, when a given explanation for natural selection comes up inadequate or is generalized and then applied to a different setting and breaks down, the natural historian has to go back to observation in order to locate even finer structures that seem to govern what is going on. Thus, discovering the habits of the peppered moth as it flies away from bothersome ants, hides for copulation or flutters through lichen-covered trees to elude predatory birds requires more patience and astute observation.

Similarly, the Buddha recognized the complexity of life. In his awakening experience, the *Sutras* report that the Buddha gained the power to see the *karma* of others. *Karma* in Buddhism is the notion that intentions have consequences. Actions growing from good intentions lead to good results. Actions arising from greed, ignorance or delusion bear ill fruit. The Buddha cautioned his followers not to try to trace the karma of a deed or to try to figure out what the trails were that led to a given result. The unraveling was simply too difficult and would lead a person astray. Again, the idea of *karma* can be taken as a matter of religious belief or can be left as a hypothesis to be tested in observation. In meditation, it becomes clear that many times things happen for inexplicable reasons and trying to trace their antecedents frequently feeds discontent. The practical way to proceed is to observe what can be experienced, not all the possible connections to which restless thoughts give rise.

In contrast, the examples which Darwin gave for explaining complex mechanisms relied on common sense. Darwin studied pieces of a phenomenon and generalized from them. His descriptive base was comparative, episodic and experimental, not life historical. Unlike field biologists and ethologists (students of the social life of animals) who spend years in a natural setting observing the life habits of one species, the only species with which Darwin seemed to be intimately familiar was earthworms and they did not display much social behavior. On the *Beagle,* he was a traveler who collected specimens, sampled cross-sections of geological strata and gathered anecdotes from people he encountered. He was able to compare and contrast flora and fauna in distinct settings. His comparisons supported his generalizations but they also had an episodic character. He drew inferences which he then filled in with common sense, as in the case of the wild pansies and feral animals, rather than actual observation of long-term interactions. When Hooker, the botanist to whom he confided his ideas of evolution before taking them public, pointed out that the authority of a generalist was questionable, Darwin decided to master one species and settled into eight painful years studying barnacles.[42] He wrote a two-volume monograph on the evolution of their sexuality which stood as a model for generations to come. Darwin was lucky in his choice of subjects. Barnacles provided unexpected evidence for evolution. They evolved from being hermaphroditic, to hermaphrodites with "complemental" males, and then to minuscule males who were parasites of the females.[43] This was a study of evolutionary change. The factors which selected for increasing sexual dimorphism could only be guessed at.

In addition to common sense, Darwin appealed to experiment. To answer questions about dissemination of plants across oceans he soaked seeds in salt water to see how long they might survive floating about. In clever experiments on the movement of plants conducted in his green houses he manipulated light and had his son play his horn to them to see whether music affected the meanderings of pea tendrils.[44] The results of these observations were then generalized. But as in the case of DDT resistant flies, new causes need to be sought because life often happens differently out in nature.

The second reason for difficulties in applying the idea of natural selection is that nature is historical. The theory of evolution, after all, sets out to explain the history of life. On a macro level, which species follows which is straightforward. In contrast, particular explanations using natural selection are meant to be applicable in many contexts which is how Darwin used them. The problem is that, despite the desire of historians, sociologists and biologists to generalize, history only happens once.

Paul Errington spent years studying the lives of muskrats in the prairie states. His descriptions are insightful and detailed. He even accounts for some of the effects of hunters on the lives of muskrats.[45] But one can't help but wonder whether his generalizations presented as biological verities are influenced by the landscape in which the muskrats lived, a landscape shaped by Indians and then by Europeans. He says little about agriculture fragmentation of swamps or disruptions of the routes of both predators and prey. How much do humans with their pets, farm animals, and pest control create an environment new to the muskrats? Despite the acumen of Erringston's observations, it may take another naturalist to describe current muskrat natural history as a unique result of the impact of humans. As Errington himself said, "...Life selects for what works, irrespective of our human efforts to define and classify."[46] It is a sentiment with which Buddha would heartily agree.

Are Things Getting Better?

The question of progress in nature still needs to be addressed. I have shown that attempts to firmly establish natural selection as nature's mechanism of creating order fall short. Even though Darwin criticized the idea that nature is directed by divine intervention or strives toward a grand design, Darwin's mechanism of explanation gave a sense of direction to evolution. Jean Lamarck, the great French expositor of intention in evolution, put forward his ideas before Darwin. Darwin was embarrassed by Lamarck's evolution and contemporaries who wanted to reinject God into nature. He felt these ideas distracted from natural historical research and that discussions of purpose, divine or otherwise, might overwhelm him. Lamarck claimed that the giraffe which reached higher had offspring with longer necks and the great violinist, virtuosi children. For Darwin the origin of change lay in chance mutation. Darwin and many successors in the 19th century were unable to show that non-genetic lifetime improvements were heritable. Willful directing as a factor in evolution was discredited.[47]

Despite emphasizing the role of chance, Darwin saw the harsh taskmaster, natural selection, imposing a kind of progress on evolution. Since nature continuously selects adaptive advantages and species seem to get more sophisticated in their capabilities, Darwin saw present-day nature as more advanced than it had been eons ago. In his barnacle study he witnessed animals evolving toward separate sexes which, because it allowed for more variation and sexual selection, he felt was a more advanced state.[48] Also, he believed Englishmen were superior to the primitive Tierra del Fuegian Indians, naked and miserable in their southern-most, barren land, sweating when they approached the fires the English invaders needed to ward off the cold.

Several contemporary notions of progress have been expounded. For E. O. Wilson, a prominent expositor of Darwin's ideas, evolution as honed by natural selection is not going in any particular direction except that it inevitably builds complexity.[49] Instead of a God as creator, Wilson praises nature itself as it builds more complexly intertwined beings. The popular philosopher Daniel Dennett takes this a step further and argues that the results of evolution are more wonderful than top-down creation because they are generated bottom-up by the algorithm of natural selection. Bottom-up generation is blind and all-pervasive. The universe is a magnificent self-creation, dazzlingly complex and dazzlingly beautiful. He sees the information encoded in genetic DNA as a move of the organism to defend itself against its own decay. The philosopher's "...job is (to expand the imagination and) to show how the wonders of the world can be the outcome of these processes." And Eric Schneider and Dorion Sagan see energy flows dictated by the second law of thermodynamics as the source of ever-greater complexity in life forms.[50]

Two accomplished paleontologists, Steven J. Gould and Simon Conway Morris, have gone to war over direction in evolution. Gould evokes chance and contingency while Morris claims that evolution lays down systematic building blocks toward a higher order.[51] The arena of contention is how to interpret an outburst of evolutionary diversity which laid down most of the basic body plans for animals. Fossils of the Cambrian era survive in an outcropping of rocks called the Burgess Shale some 530 million years old. Morris, who champions order, sees the possibilities for change constrained by structural demands of life: life tends toward accomplishing things in similar ways. An expert on the Burgess Shale, he feels that the variety of Cambrian body plans is not too much greater than what exists today. He also feels that once complex structures get started, the demands of natural selection and adaptation will push them towards greater capabilities. For him, "intelligence and self-awareness...would surely have evolved...."[52]

Gould, on the other hand, sees the animals of today as the chance survivors of the Cambrian era which produced an explosion of new kinds of life. He interprets the fossils as possessing a great many more types of body plans than animals now have. Much Cambrian body architecture did not survive. There was a kind of selective lottery which could have produced all kinds of very different descendants. Had it gone slightly differently, there would be no being capable of writing on this computer. That intelligence arose was only chance. Such an explosion didn't happen twice because the life that survived the era filled most of the usable niches, and its anatomies constrained the directions it could take.[53]

While the disagreement between the above two may hang on interpretations of the Burgess Shale fossils and the paleontologists' assumptions about the meaning of life, it is hard to show that nature is operating so as to fit all the pieces of the puzzle in an efficient way. Natural selection, adaptation and fitness parallel the language of classical economics. Together they imply that life in nature is orderly and efficient; natural selection ensures that fit species collectively survive and are adapted to the presence of each other and the environment. Such a system automatically regulates itself. It is a kind of moving equilibrium. Change via mutation moves the system a little off center. It then readjusts itself through the action of natural selection. Larger catastrophic environmental events knock the system awry, but it eventually rebalances itself with a different set of species.[54] No other arrangement could take care of all the interrelationships as efficiently. It is like prices in the free marketplace. The very language Darwin used builds order into our conception of nature. In concrete applications, when the language of order, efficiency and progress shows cracks, fresh observation is required to understand how things work and reestablish a sense that it all fits together.

Although nature does have a kind of order, it also fumbles around and can be quite counterproductive. Is the newly built world of the last five-hundred million years any better than its predecessor? Can we lord it over bugs, bacteria, and slime molds or look down on trilobites? Darwin knew better, though he was often ambivalent. As the biologist, Lynn Margulis, has pointed out, "Charles Darwin was correct when he admonished us not to use the terms 'higher' and 'lower' for organisms; all are equally evolved by the very fact of their presence in today's world. Bacteria may be small, but they are just as 'high' as any other life form; they appeared on Earth before any [nucleated one celled ancestors of plants, animals and fungi], and in terms of metabolic diversity they are far more capable than animals and plants."[55] That is, they have been around a lot longer and are very likely to outlast us. They are better adapted.[56] So complexity does not indicate progress. It is merely one of many historical outcomes.

Moreover, evolution can be awkward and sometimes downright inefficient.[57] Humans, giraffes and mice all have seven vertebrae which function in the same way. It is great for us but giraffes are pretty stiff. The only reason they received the same equipment is that they happen to descend from a common ancestor. Seven were there rather than six or eight and so it stuck. All vertebrates risk choking on food because their esophaguses and alimentary canals cross in the throat.[58] This is because the respiratory system evolved from a modification of the anterior alimentary tract. The arrangement was never improved upon although it is a risky set-up. In humans, it is particularly dangerous because the development of speech required an enlarged voice box. Those whose food goes down the wrong way and are saved by the Heimlich maneuver experience this evolved vulnerability.

The form of an organism is the result of a unique history, which may be seen in vestigial organs or unlikely engineering. The evolutionary changes in species may not improve their performance.[59] Both mammalian eyes and external testes have "design flaws" that are linked to the history of their origins. Mammalian eyes have blind spots created by veins in front of the retina, and the exit of the optical nerve through it. This is because things were laid down in the wrong order in evolution. Squids got the engineering right: pupil, lens and retina, then opaque stuff. Mammal scrotums got hung up on the urethra instead of running the shortest distance and because mammal sperm needs to be cooled, the scrotum emerges from the body. Men know how vulnerable this makes them. Such phenomena have been called irreversible evolution or evolutionary black holes. The basic idea is that once crucial structures get put in place, they tend to persist even though the direction they evolve in may create vulnerabilities for species.[60] If things proceed too far, a species may not survive. The human chin has been getting more prominent because front teeth shrank as we ate cooked food rather than raw. As the jaw shrank there was less room for our "wisdom" teeth.[61] Dentistry removes these impediments and plastic surgery can smooth out ever more pointy chins, but with weakened jaws, eating will become difficult if we ever lose access to gas and electricity with which to cook our food.

Humans have other evolutionary tradeoffs. Most small mutations are neutral and most large mutations are harmful. The in-between ones can involve trade-offs. Tropical Africans have a mutation which protects them against malaria. If they inherit one sickle-cell gene they have resistance to malaria, but two genes gives rise to sickle cell anemia, which is usually fatal. E. O. Wilson calls the unfortunate recipients of two genes "Darwinian wreckage."[62] The advantage the mutation confers is enough to keep the species going although it creates enormous collateral damage. The phenomenon of genes conferring useful traits on a species but also having harmful side effects is called in biology antagonistic pleiotropy. It literally means multiple forms working against each other. Humans posses a number of such traits. We will examine thought and its accompanying discontent as one of those.

There are other ways species cope with the environment that don't seem very efficient. Ancient relatives of the Chilean seabass were bottom swimmers who couldn't float. They did not possess an air bladder which other fish use to regulate depth. By evolving fat, lightweight bones, and antifreeze to keep them suspended and alive, descendants of seabass moved into various strata of unoccupied, frigid Antarctic waters. In the Antarctic seas, it is below freezing in the upper strata and becoming warmer as one descends. No reasonable engineer would have chosen such fish as candidates for this kind of environment. Now each species hangs at different depths vulnerable to changes in temperature. And the larvae of the largest of all mosquitoes only feed on the larvae of other mosquitoes. When they have eliminated all other mosquito larvae in a pool, they then devour each other until sometimes only one larva is left.[63] Darwin emphasized such arbitrariness in nature as evidence of evolutionary ancestry.

When Huxley wondered whether the hermaphroditic jellyfish he was studying may have outcrossed by swallowing sperm, Darwin replied, "What a book a Devil's Chaplain might write on the clumsy, wasteful, blundering low & horribly cruel works of nature."[64] Despite occasional upbeat references to nature's vitality, "Darwin's [underlying] view of nature was dark-black." For Darwin "life [was] empty of any divine purpose." He appreciated the wonders of nature but they were a result of harsh struggle. "Heavy destruction was the key to reproductive success."[65]

For Annie Dillard, poet and naturalist, black holes and mortality are as irrational as they are common characteristics of life. "Nature is above all profligate.... Wouldn't it be cheaper to leave [autumn leaves] on the tree in the first pace? This deciduous business...is...the brainchild of a deranged manic depressive with limitless capital.... Nature will try anything once. This is what the sign of insects say. No form is too gruesome, no behavior too grotesque."[66]

The dispute between those who saw nature as evidence of the work of the gentle hand of God or some other reasonable mechanism and those who found nature a lot less sensible became sharper after the publication of *The Origin of Species*. The debate continued in the 20th century, as we will see in the next chapter. The Galapagos Islands were cited as evidence by both sides. Even Annie Dillard thought they were Eden-like because the animals there were unafraid and would cozy up to visitors like no other place on earth. Still the proof that scientists sought there for the honing of species was dependent on the tremendous mortality which resulted form the rawness of the newly-made Galapagos Islands and their devastating swings of weather.[67]

Perhaps Buddha would have pondered such disputes. There are parts of Buddhism which fit nicely into romantic pictures of nature. This is particularly true of Zen and Tibetan Buddhism.[68] But for Buddha himself concerns with efficiency and progress in the natural world would be off the mark. What is investigated by meditation is indeed part of the natural world. Buddha's order came from his assertion that meditation had allowed him to see the operation of the law of karma and from disciplined observation of the mind's reaction to the difficulties that humans, like all other animals, encounter because they resided in a disorderly nature. Leaving karma aside, the inevitability of disease, old age, and death were primary examples of nature's disorder even if they were agents contributing toward honing new species. However novel a species' adaptations, the individuals in it were subject to the same universal constraints. For humans, the peculiarities of the adaptations that made us thinking animals were more grist for the mill, contributing to our success as a species, to our discontent and to the possibility of its surcease.

It is important that the evidence upon which Buddha relied to make his case was intimate observation. There is nothing more immediate than knowledge gained from observing one's own mind and body. Measuring velocity or position in physics and observing animal behavior are one step removed. Although these gain authority because they can be replicated or used to change the physical and biological environment, experience of the mind and body are more primary and unmediated. Darwin's personal struggle with his mind's reaction to disease, old age and death in his family took place during a century which started with God's gentle nature, only to become haunted by the dark side of natural selection, the idea he had introduced. But both the insights of Darwin and Buddha stemmed from observation.

Darwin's Suffering

When Buddha's three signs visited Darwin's family, he was distraught and sought consolation from his theories. As described by his biographers, when his favorite child died, he felt she neither deserved suffering nor death. "'Formed to live a life of happiness' as Charles put it, she had stumbled on ill health and nature's check fell upon her, crushing her remorselessly. The struggle was 'bitter & cruel' enough without the prospect of retribution. Yet against all odds he still longed that she might survive....Annie's cruel death destroyed Charles's tattered belief in a moral, just universe. Later he would say that this period chimed the final death-knell for his Christianity..."[69]

Annie was a casualty of the struggle for existence although Darwin did not know what disease killed her. He speculated that her ill health was due to inbreeding. Because he himself was so often sick and because there was so much ill health among his children, he worried that the cause lay in his marriage to his first cousin. Two of his ten children died young: one, an infant for unknown reasons and another at one-and-one-half years from scarlet fever. His other children were often sick or behaved strangely. In the midst of his family's illnesses, he was writing sections of *The Origin of Species* about the problems which inbreeding caused. He was acutely aware that the red rule of tooth and claw was running its course on his family.[70]

As his biographers describe it,

> The struggle for existence had already set in, and he expected the children's health to fail at any time....he waited for nature to exploit the fatal flaw.... In the [Origin] he belaboured the "evil" effects of inbreeding and the good effects of crossing. As always he was looking for moral meaning: birth, death and chronic illness needed some rationale and Nature provided it. "When relations unite", there is a "decrease in...general vigour" and increased likelihood of "infirmity" among the offspring. The struggle for existence then inevitably takes its toll and the Darwin children were not immune. It is often hard to see the good, but Nature was working for a better world. "The survivors" were the "more vigorous & healthy, & can most enjoy life."[71]

Thus, God was replaced by natural selection producing progressively improved beings, the vigorous among whom living more fulfilled lives.

There have been many different explanations of the illnesses which plagued Darwin much of his adult life after he returned from his voyage on the Beagle. His sickness often coincided with stress from overwork, anxiety about the reception of his ideas, and concern about whether he would get credit for them. Darwin was probably a hypochondriac. The jacket cover of his most popular biography calls him tormented.[72] It is debated whether his illnesses were organic or psychosomatic. He was obsessed with disease and treated his children's illnesses with great concern. He made notes on his children's illnesses as he did for his scientific experiments. He frequently took their temperatures and recorded their symptoms. Although some survived to be healthy adults, they too suffered from undiagnosed symptoms or were overly concerned with their health.[73]

During forty years of illness—from the time of the Beagle until his last ten years of relative good health when his fame was established—Darwin dragged himself and his sick children from cold baths to other promised cures. His own death, at seventy-two-and-one-half, was not easy. In June of 1881, "he was depressed and preoccupied with death... 'I cannot forget my discomfort.... I have not the heart...to begin any investigation.... So I must look forward to a ...grave-yard as the sweetest place on earth.'"[74] Towards the end of the year he had cardiac irregularity and then, "In March, hobbling around the Sandwalk, he had another seizure. He was alone and terrified ....Somehow he picked his way back...Dr. Clark confirmed angina...Charles froze. He felt condemned, a prisoner of his body, an innocent about to be hanged. He gave in to despair." His spirit recovered a bit. On April fourth he,

> had dreadful attacks. With brutal cold-blooded precision, he began taking notes. "Much pain,"...the next two nights were excruciating. "Stomach excessively bad".... And on April 14th, "attack slight pain...." He was running his final experiment. Three days later, the pain came on just before midnight. It was brutal, gripping him like a vice, tightening by the minute.... Charles, in agony, felt that he was dying but unable to cry out.... Seconds later he sputtered and retched; his eyes flickered open....immediately Charles started vomiting. It was violent and prolonged.... An hour passed by and then two. Still he gagged and retched. He was cold clammy...blood spewed out.... "If I could but die," he gasped repeatedly...but the pain was excruciating in any position. Rising he began to faint again.... He lost consciousness.... There was only the deep stentorous breathing that precedes death.[75]

He died in his wife's arms.

For all the optimistic outlook of a successful 19th century bourgeois enjoying the benefits of the most powerful empire the world had known, for all the new explanatory power of science it had put to use, Darwin still had to struggle with the personal meaning of life and death. Encountering the third of the "four signs," death, Prince

Siddhartha bemoans the fact that birth inevitably leads to the point in life when one loses one's loved ones. Siddhartha, as the Buddha, found release from that suffering through meditation. Although Darwin found a rationale for the suffering in the theory of evolution, discontent plagued him to the end. Survival of the fittest had spiritual shortcomings. While pursuing questions about how breeders selected breeds he found in a pamphlet the following: "'A severe winter, or a scarcity of food, by destroying the weak and unhealthy, had all the good effects of the most skillful selection. In cold and barren countries no animal can live to the age of maturity, but those who have strong constitutions....' [Darwin] was beginning to appreciate the darker side of nature."[76] The sacrifice natural selection makes for the progress of evolution is to strike down the weak and unfit. This leaves the more vigorous to pass on their better adapted genes. Yet for all their vigor, they too perish.

Darwin began a great tradition of careful natural historical observation. As will be explored further in subsequent chapters, the way in which his observations of nature related to the explanations he came up with is analogous to the relationship between observation in meditation and conclusions which meditators make about those observations. Both Darwinian explanations and meditation's understandings are merely resting points, often pointing to the fact that one needs to look even deeper.

Although Darwin launched a tradition of careful examination of nature, when it came to his family, he only observed part of the picture. In worrying about life as he did, in trying to reduce his anxiety about family illness, he applied his methods of observation but overlooked the phenomena to which the Buddha would have directed his attention. Darwin was able to observe the container of human life. He noted that, "Man still bears in his bodily frame the indelible stamp of his lowly origins."[77] He carefully studied the expressions of his children in contrast to animals in the zoo, people on the streets, inmates of asylums and the Tierra del Fuegian Indians. He wrote a book entitled *The Expressions of the Emotions in Man and Animals.* Darwin recognized the evolution of animal emotions although he restricted himself to their outer expression. When it came to his own suffering, he never took the Buddha's step inward to dispassionately observe the character of his discontent, how it arose, and its relationship to his mind and his body. In his 19th century world, there was no precedent for this.

Considering Darwin's ample share of life's difficulties, evolution's benefits to the more vigorous were of little consolation to him. The messiness of evolution's footprints became nature's punishment. The darker sides of nature's effects on human beings remained dark. For the Buddha, the fact that the weak perish in abundance, that the vigorous eventually die, and that nature is full of adaptive compromises spurred a different kind of observation. Because Buddha's starting point was disease, old age, and death and his methods were grounded in investigation of the body and mind, to uncover the origin of discontent we need to explore how these occur in nature without humans and then explore how the human body and mind fit into that natural setting.

Chapter Three
Disease, Old Age and Death

For Darwin, there was no understanding of life's processes without "constantly to bear in mind" the universal struggle for life. The same was true for the Buddha. Siddhartha, as is said, "went forth" from the comforts of his palatial home into the forests which surrounded most of the civilized world of his time. The forests put him in direct contact with nature where the realities of disease, old age and death were not obscured by customs, social attachments or entertainment. Dissatisfied with the teachers he encountered in his wanderings, Siddhartha embraced raw nature. He sat under a tree exposed to the elements and the mercy of passing wildlife. He vowed not to move, surrendering to the caprices of nature. Watching his mind struggle with a reality it did not want to accept, he came to understand that there would be no end to discontent until he could look unblinkingly at disease, old age, death and the human mind. Elsewhere I have explored the Buddha's sojourn in the forests and the experiences of monks who followed this path.[1] To explore the problems the Buddha posed in a world with a biological structure as described by Darwin, we need to look in more detail at its Darwinian character. In this pursuit, Buddha will drop into the background for a while as Darwin takes center stage. Buddha informs the questions we ask and we can imagine what he might say about the various natural historians' observations. In the chapters that follow, the perspectives of meditation will take an increasingly more important role in our discussion.

All birth leads to disease, old age and death. How do these occur in the wild? As humans, we are obsessed with our own disease, old age and death. Although we believe it is happening to us in a special way, this is not the case. We will look at the three signs in nature and how some animals' responses to them anticipate our own. This discussion has the same limitations as other observations of nature. It begins with tentative conclusions drawn from observation and can be filled out with further investigation and extension to new settings.

Nineteenth-century Darwinians and romantic naturalists debated whether death in nature was a gentle, purposeful process or cruel and irrational. The controversy continues. We will look at these debates and then put disease, old age and death into a bigger context of how nature functions. In the course of this, we discover that old age contributes little to evolution, and it is experienced by only a small proportion of many species. We will see how the suffering in old age of some other animals presages ours. Because so many humans now live so long, old age has become an important source of suffering in modern societies. If discontent has roots in nature, then understanding them may give us perspective into our own.

With his dying words the Buddha repeated his admonition that his lessons could only be learned through individual experience. No recitation of canon doctrine, no belief in his Four Noble Truths were a substitute for looking at the world and incorporating what is seen into one's being.[2] How investigation unfolds in someone's life is unpredictable. Prince Siddhartha went for a chariot ride and encountered a reality which sent him in search of a way out of suffering. My introduction to the first three of the four signs occurred on an Arctic canoe trip. The canoe was my chariot and my brother, the charioteer.

In 1969, I traveled through the wasteland of the industrial frontier of northern Canada and was deposited in what I thought was a wilderness. The life I left was that of a young, politically active professor in Cambridge, Massachusetts. Cambridge was my castle, the intellectual hub of America, charged with the political issues of the day: race, class and the war in Vietnam. The struggles of the times were reduced to ideas tossed about in the salons of academic opinion makers. Having read all the right history and sociology, these intellectuals believed they understood life and death better than both politicians and the unlearned far beneath them.

Yellowknife, the capitol of the Northwest Territories, with its prospectors, natives, oil drums, permafrost-navigating muskeg-bombardiers, bars and whores jostled the sleeping prince. The Coppermine River, flowing into the Arctic Ocean, and the Barren Grounds, lands north of the tree line, called me to wakefulness. None of my urban youth, teaching or community organizing prepared me for the emptiness, abundant wildlife and the need to rely on my strength and alertness. During the first week of paddling there were a few trees along the river. Then there were none; only the permafrost which turned into puddles in the summer sun. The Barren Grounds stretched as far as the eye could see. When my brother and I stood on land, we were the tallest objects in sight. Down at our calves were bunches of ever-present willow bushes. The landscape was astoundingly stark and beautiful. When my mind took it all in, it was very scary. Little of what I had learned in life was relevant in this wilderness. I was completely dependent on my chariot and charioteer.

Many things happened on that trip. We saw wolves and foxes and unusual birds. We crept up to a caribou herd. Rising after light (the days were twenty hours long), we would fight off blackflies and mosquitoes, have our oats with wild berries and paddle until dusk.

The trip was uneventful and calming until we had to traverse a rapids or an animal appeared. One day while meandering along, we saw two moose grazing by the shore. They were standing knee-deep in water, their heads submerged, eating river plants. They chewed away for moments at a time then lifted their heads for a breath and returned to grazing. From out in the river, we paddled silently toward them. We were quiet except for the slight singing sound an aluminum canoe makes as it glides through water. When we were a paddle's length away, one moose lifted its head.

This was before moose had returned to populous regions of the northern United States. It was before moose, feeling newfound safety, ambled into Vermont farmers' yards to make love to their cows. In 1969, moose were remote, wild, large animals, dangerous during rutting reason. In the Canadian Barren Grounds, I sat looking a half-ton, heavily-antlered animal straight in the face. My heart stopped. The moose looked quizzically at me trying to figure out what I was. I must have been downwind. If it had been able to smell me, it would have quickly departed. My attention was riveted on the moose's face. Its large head was covered with a fungus and its eyes were clouded. Besides the fact that moose rely on smell more than vision, this moose was partially blind. Sitting opposite the moose, my canoe paddle poised for defense, something in me broke and I began laughing at the absurdity of the scene. Instantly both moose jerked to attention, turned and trotted off. My brother and I continued to paddle down stream.

The memory has stayed with me. It was the first of my "signs," occurring years before I began to meditate. Although I had worked in the Chicago stockyards as an adolescent, I didn't understand that there was disease in nature or that the animals I had seen portrayed in Walt Disney's films could suffer. From looking at that moose, I realized that disease and death haunted the wilderness. This was my introduction to suffering in nature. This moose's life seemed uncomfortable and it might perish in the jaws of the Arctic wolves we would subsequently meet. Nature had begun to lose its innocence.

Years later, I learned that other travelers in the Barren Grounds saw a similar reality. "It is often forgotten that wild animals like sheep, elk, moose and deer suffer, sometimes miserably, from disease and injury.... A Canadian biologist once observed a mule deer at the point of death, parasitized by an estimated seven hundred winter ticks. Moose parasitized by winter ticks leave...showers of red [blood] on the white snow where they have shaken. Caribou suffer terribly from warble flies and nostril flies, and moose are sometimes crippled by [golfball-sized] cysts in their lungs."[3]

Death in biological nature occurs in many ways. Different kinds of creatures undergo different kinds of death. When we think of death in nature, we usually think of the encounter between predator and prey. There are those who feel predation is the action of a kindly spirit doing its job as gently as possible and those who see it otherwise. The dialogue begun by Darwin has been carried into the 20th century. An examination of predation not only reaffirms the mean-spiritedness of nature which both Darwin and Buddha urged us to observe, but adds a subtle element of gentleness which can't be ignored. It is interesting how my friends who like nature fall into two camps. Some love romantic nature poetry and change the subject when gory descriptions come up, while others are eager to know the details. My data come from natural historians and nature writers, who variously paint nature with broad lovely strokes or vivid nasty details. The former have deeply influenced conventional views of nature. After this discussion, we will look at predation in comparison to other ways of dying and ask the function death has in nature. Finally, we will examine who dies, non-predatory ways of death, disease, disability and old age. This is the world that Buddha felt gives rise to discontent.

Not Often Seen

The task outlined is not a simple one. In cities we rarely see natural death. Although death was more present in the Buddha's agricultural society, he sought to confront it in the wilds. In my years in the woods, I have only occasionally seen an animal die, and when I did, it was often in the jaws of my dog, Sasha, a formidable hunter. Rarely did I see the actual kill. I usually witnessed the grim aftermath. On the few occasions I have seen natural death, it could be dramatic, such as a paper wasp decapitating a dragonfly or a raptor catching a mouse and tearing it apart. I am not alone in my ignorance. Field biologists in past years infrequently witnessed animals die. Of the hundreds of field studies I have looked at, only a few describe dying.[4] This is changing now with the use of high-tech sensors and video cameras.

Predation is particularly elusive; nature films not withstanding. Other forms of death are also obscure. In E. O. Wilson's monumental work on ants, one finds little about causes of death of any sort. Even less apparent are events leading to death. And it is sometimes difficult to find the *corpus delicti*, especially of insects which are quickly scavenged. But in places like "the edge of the sea," as Rachel Carson calls it, there is more evidence of death. I rarely see signs of death when walking in the New England woods, the mountains of New Mexico, or among the California redwoods, but at the seashore we see bodies of dead crustaceans, bird wings, dead fish or a gull pulling apart a crab.[5]

To find who's eating whom, naturalists use indirect means. They examine droppings, stomach contents and remains. Owl pellets filled with bones and fur have replaced dissection in some high schools. Learning how such meals actually occur takes more work. Here is a paleotaxidermist's description of the death of a Pleistocene bison.

> It is on its four legs, crumpled straight down, because when a bison is killed it doesn't fall over on its side like a moose, it

> drops straight down. These scratches on the hide were done by the lion that attacked from the rear. The lion was no different from a contemporary African lion. You can see the claw marks and then the fang punctures. They are exactly the width of the teeth of the modern lion. There are also the marks on the nose and the claw marks under the jaw and on the neck that show that a second lion held it....[6]

The task we have set ourselves has some built-in limitations: not a lot is known about how death occurs in nature. Yet what is known is revealing.

Predation: Gentle or Violent?

Darwin had mixed feelings about violence in nature. In his youth, formaldehyde and pins marked the end of the beetles in his collection. Darwin, the naturalist, skinned and preserved thousands of animals as specimens. On his voyage, he rode the pampas with gauchos and hunted for fresh meat. Like generations of sailors who came afterwards, the crew of the Beagle ate and discarded the shells of giant Galapagos turtles which turned out to be crucial evidence for the theory of evolution. Live turtles were kept on deck for weeks, secured by ropes through holes made in their flippers, until slaughtered. On a barren island in the south Atlantic, he joined the crew in beating to death hundreds of seabirds. Later he regretted this youthful violence. Nature in *The Origin of Species* was a palpable struggle for each being's very existence, although Darwin tried to give the conflict a gentler interpretation. "When we reflect upon this struggle we may console ourselves with the full belief that the war of nature is not incessant, that no fear is felt, that death is generally prompt...."[7]

In the United States at the end of the 19th century, a number of nature writers began to promote the view that nature was not so violent and that natural death was not tragic. They were responding as much to the Darwinian image of struggle as they were reacting to the carnage they had seen in the American West. It has been estimated that more than two-hundred-million wild animals were killed by 1900.[8] The killing began with the fur trade, driving the beaver to extinction in many locales before 1800. It included the destruction of millions of buffalo by whites and Indians.[9] Scavenging the remains of buffalo, wolves and coyotes multiplied. When there were no more buffalo, they turned on settlers' livestock. Buffalo Bill was a national hero as were hunters, trappers and the men who rid ranches and farmyards of wolves, coyotes and smaller "varmints."

At the end of the century, some who had participated in the carnage became dismayed by what had occurred. They regarded the red rule of tooth and claw as an ideology of needless destruction. So they painted nature as more gentle and cooperative, aiming their arguments at school children. Foremost among these was William Long, outdoorsman and author of many nature books. Amid the reform movements of the turn of the century which sought to ameliorate the cruelties of rampant capitalism and Social Darwinism, Long's books found a large audience and were adopted by schools. Long portrayed nature rhapsodically. He claimed to know what animals thought. For him,

> [A]ll wild creatures are so familiar with their surroundings, and so in tune with them, that it is absurd on our part to think of them as living a life of terror or tragedy. The game birds and animals are all splendidly endowed by nature; their senses are quite as keen as those of their enemies; their speed and their hiding faculties are commonly greater. They can escape from any danger, or think they can and as a natural consequence have no fear of it whatever....therefore do you observe among these hunted creatures plenty of confidence, playfulness, curiosity, but no apparent terror or even anxiety.[10]

Long made a series of points. He believed that animals do not experience pain. He saw the relationship of predator and prey happening without unnecessary suffering. "...the victims of the carnivore have absolutely no experience of pain since the charge and shock of an attacking animal produces a kind of mental paralysis, or stupor which renders the smitten animal insensitive to injury....I have watched many of them...mice, squirrels, rabbits, grouse, deer...when they are gripped by their natural enemies and with few exceptions they gave an astonishing exhibition of indifference."

The "dark Mother," death, protects her wild children. Predators do their work efficiently.

> [I]t is plain that birds and beasts of prey do not torment their game, but kill it instantly.... The bear is not a natural hunter... yet I once saw him do his work as cleanly as ever such work was done.... I was following a buck...when [I came upon]...a big bear crouching low his eyes fixed intently.... As the buck passed...the bear rose and leaped for the shoulders. As he landed, hurling his victim to the earth with the suddenness of a thunderbolt, his heavy paw dealt one smashing blow between the eyes. So the neck was broken and the skull crushed in the same swift instant.... There was no horror, no struggle, no element of tragedy but just a swift merciful ending, such as a healthy animal might pray for when his hour comes.

Long felt that humans misunderstand nature because they attribute to animals human feelings of fear and anxiety. For him, animals do not have consciousness like humans and nature is not a struggle for existence. There, despite eating and being eaten, animals live in peace and harmony. Moreover, Long claims to have lived in the woods in the same harmony. Few others have shared his experiences. On first reading his books, I was envious of how much nature I had missed. Long was joined by a fellow writer Ernest Seton who had been a bounty hunter and then wrote popular animal stories personifying his subjects. One of Seton's heroes was Lobo, the wolf, who escaped from hunters' cruel traps.

There was public opposition to Long and Seton. Their most vociferous critic was John Boroughs, outdoorsman, nature writer and traveling companion of John Muir. He was also a staunch Darwinian. The battle was fought in popular journals of the day. Boroughs attacked Long and Seton for romanticizing nature. He claimed that many of Long's examples were fabricated.[11] Two stories he found improbable were: "an eagle gently dying on the wing high in the air and gliding gracefully to the ground...an amazing aerodynamical feat even if one accepts the idea that birds can die in flight." Nor was "Long's story about the fox that coaxed chickens out of a tree by running around and around the tree in circles until the chickens became dizzy and fell" to be believed.[12] The debate was settled by a heavyweight: Teddy Roosevelt was outraged by the picture of gentle nature. Rough rider, ex-president, big game hunter and Social Darwinist with a reformer's soul, Roosevelt, who, when later injured during an Amazonian expedition asked to be left to die so that others might live, took exception to Long's description of a wolf killing a caribou by piercing its heart with a single bite. He labeled Long and Seton, "nature fakers." Coming from such an authority, the label stuck, and the two faded from the scene, but their vision of nature did not completely die.

There are good reasons to suspect Long of fabrication. Experienced naturalists have not come close to seeing the abundance of interactions he described. As pointed out, observations of dying in nature are infrequent, yet Long saw predation everywhere. His stories are too simplistic, and many things happen in the woods that he never seems to have witnessed. Nonetheless, of the three examples that supposedly discredit him, two possess elements of truth. No one else has observed a wolf killing a caribou as Long described it, but Boroughs is wrong about the eagle. Birds do die in flight. During World War II many common loons were seen to fall onto the deck of a battleship with wings encased in ice.[13] They did not glide gently down as Long claimed to have observed. And as for the fox and the chicken story, I once watched my dog, Sasha—one-quarter husky, three-quarters shepherd—kill a squirrel in the woods in a way that amazed me. Alarmed by the dog, the squirrel ran up a tree. Sasha leapt around the base of the tree, barking. The squirrel (quite safe from my perspective) became agitated and jumped to a nearby tree. Sasha pursued until the squirrel was frantically leaping back and forth between two trees. On one jump, the squirrel missed. It was over in a flash, the dog on top of the squirrel.

Long and Seton's perspectives have been held by later naturalists. In the 1930s, Paul Sears, an ecologist at the University of Oklahoma, raised a clarion cry against the human-induced causes of the dust bowl. In contrast to civilizations which, he argued, ruined things, "[l]eft to herself, nature manages these loans and redemptions [of the substances of life] in not unkindly fashion.... The red rule of tooth and claw is less harsh in fact than in seeming."[14] An example of this is given by Berndt Heinrich, professor, entomologist and contemporary nature writer. He describes one of his captured ravens eating a cowbird nestling being raised by a phoebe couple. Even though the cowbird is a nest parasite who would end up with most of the food intended for the phoebe chicks, the phoebes were upset for a day, then began busily renewing their nest for another litter.[15]

As a graduate student, Richard Nelson went to study the Indians near the Arctic Circle and never returned. On one occasion he watched an eagle attack a gull. Although there was no obvious violence, all his sympathies were with the gull, despite the fact that minutes earlier it was killing fish or crustaceans: "My reactions towards predators and prey are terribly inconsistent, especially considering that I am a predator myself. How can I love the beauty of living things apart from the process that sustains them? Life feeds on life.... when the gull passes over me I strain to see whether it looks terrified, desperate, exhausted or relieved. But it seems like any other seagull, with its expressionless beak and pale blinking eye."[16] I will have more to say about the eyes of wild animals in the next section.

Finally, consider the well-known nature poet, Wendell Berry, who advocates vegetable gardening as a cure for our material and spiritual malaise.

The Peace of Wild Things

When despair for the world grows in me
and I wake in the night at the least sound
in fear of what my life and children's lives may be,
I go and lie down where the wood drake
rests in his beauty on the water, and the great heron feeds.
I come into the peace of wild things
who do not tax their lives with forethought
of grief. I come into the presence of still water.
And I feel above me the day-blind stars
waiting with their light. For a time
I rest in the grace of the world and am free.[17]

Conversation of Death

Among the advocates of gentleness is the claim that there is some violence-transcending rapport to which both predator and prey give knowing assent. This hard-to-pin-down communication may be via eye contact. Although it is clear he made things up, William Long was nonetheless on to something: "The eye of a wild animal is even more wonderful...if you ever look into it in close quarters, you will have a real...lesson in natural history. I have known hunters to throw aside their rifles and to be changed men for life having looked once into the eyes of a dying deer.... The expression of the eye...is...mysterious, reproachful, and so penetrating that it lays your very soul bare....[It] is something that few normal men care to meet a second time."[18]

Aldo Leopold, whose writings helped lay the foundation of the modern environmental movement, went through a similar experience. He had been a hunter and worked for the Forest Service. While timber cruising in New Mexico in 1910, he and his crew saw an old wolf emerge from a stream to be greeted by a pack of grown pups. In the spirit of 19th century slaughter, the men blasted away with their rifles, then approached. The injured, old wolf fastened onto the rifle butt Leopold pushed at it. The dying green fire he saw in the wolf's eyes changed forever Leopold's understanding of wildness, leading him to dedicate his life to its preservation.[19]

Years later, the naturalist Barry Lopez called this exchange a "conversation of death."

> [The wolves] walk on the perimeter of caribou herd seemingly giving warning of their intent to kill. And the prey signals back. The moose trots toward them and the wolves leave. ...a wounded cow stands up to be seen....An ailing moose...who could send wolves on their way by simply standing his ground does what is most likely to draw an attack...he runs. I called this exchange in which the animals appear to lock eyes and make a decision, the conversation of death. ...both animals, not the predator alone, choose for the encounter to end in death. ... We are dealing with a different kind of death from the one that men know. When the wolf "asks" for the life of another animal, he is responding to something in that animal that says, "My life is strong. It is worth asking for." A moose may be biologically constrained to die because he is old or injured but the choice is there. The death is not tragic. It has dignity. In the conversation of death is the striving for a death that is *appropriate.* I have lived a full life says the prey. I am ready to die...I will be dying so that others in this small herd will go on living. I am ready to die because my leg was broken or my lungs are impacted and my time is finished.[20]

Other naturalists have observed both injured and healthy prey give themselves up. Sometimes other members of a herd look on predation of their neighbors without reacting, as if they understood the necessity of death in the wild.[21]

A case can be made that predatory death comes with a kind of informed consent to natural selection. This is seen to redeem nature's violence. For Long, humans feel predation is cruel only because we are anxious and fearful and we project our feelings onto animals. "It appears...that nature knows how to bring to even her violent endings her mercy and anodyne...an exit far more becoming to the animals powers and to his glorious free life than is weakness of old age or...wasting sickness."[22] With a nod to a sentiment reminiscent of Siddhartha's lament, we delve deeper into natural history.

Not so Gentle

The fact is, predation can be crude and very violent. Often predators are cruel but not confident, and prey do not submit. As Darwin pointed out, behind the happy summer day is unremitting death. Civilized people neither see nor want to look at this.

Examples of predation can be upsetting. Giant water bugs grasp frogs from below and suck out their innards. You can see the frogs shrivel like balloons pricked in slow motion. Ants swarm over live bird chicks in the nest and consume them bit by bit. River otters eat fish alive. In the typical suburban backyard, most animals live only a fraction of their life span, dying at the hands of pets, especially cats, and other predators. Killer whales tip ice floes to get Weddell seals. The whales hunt like wolves. They specialize in the young. They ram mother walruses to dislodge the calves riding on their backs. They rush the shore, terrifying sea lions and then seize pups in the ensuing panic.[23] Wild African dogs tear the flesh and eat live prey. They eviscerate their prey to shock it so it won't struggle; until then it fights to get away. Antarctic skua gulls, lacking teeth and the flesh-tearing beak of raptors, prey only on young penguins. They sit on the penguins' backs, peck out their eyes and then eat them. The kills are slow and messy because skuas' beaks are not designed for the job.[24] Spiders and wasps are ancient enemies. They prey on each other. When a wasp gets caught in a spider's web, it will struggle to get free and the spider who otherwise would dash forward, wrap and sting its prey, approaches tentatively running back and forth in the web. Neither animal exudes confidence.

Field biologists report unforgiving violence with scientific dispassion. They warn the untrained to be wary of sentimentalism. As Pierre Pfeffer, the author of *Predators and Predation*, admonishes, "For biologists there is neither good nor evil being in the living world, nothing but honest co-participants, each of whom plays out its assigned role in the natural order of things. There are no such invidious distinctions to be drawn between the peaceful herbivore and the fearsome predator that seeks its life, the delicate insect that feeds only on nectar and the hunting wasp that stalks a victim to be eaten alive by its offspring."[25] Attractive animals whose predation we don't witness or who we somehow forgive, like cats, are well regarded.

As Richard Nelson observed, characterizing an animal as a predator evokes moral condemnation and aversion. But predation is the means by which predators live. Even experienced naturalists have a difficult time watching messy predation. Because of this, it is easier to emphasize predation which reflects the conversation of death than that which seems unnecessarily cruel.[26]

We have seen that there are not many descriptions of actual predation. Although much is written about predators, it is hard to find out what it is like for prey. Some prey lead much safer lives than others. This is illustrated in Pfeffer's book. The skua leads a lucky life. Besides harrying other birds for their catch and killing penguin chicks, it has neither predators nor competitors except other skuas. Similarly for the wasp which temporarily paralyzes a leafhopper and inserts an egg. Its larva then consumes the leafhopper alive. In one season it kills almost half the leafhoppers but it has no known predator.[27] However ants have twenty-eight predators, beetles fifty, crustaceans one-hundred-two, flies twenty-six, frogs sixty-two and so on.[28]

How does this list translate into the realities of an individual's life? A skua might not have to look over its shoulder constantly, but how about the little frogs inhabiting New England ponds in great abundance? They have many predators and are appropriately cautious. If you stand still next to a pond for a long time, occasionally an eye will pop up here and another there. Among the luckier creatures is the fisher, related to the marten and mink. No one knows how fishers die. They have no known predators. And the only predators porcupines have are owls, fishers and mountain lions which may be why porcupines never seem to be in much of a hurry.[29] Another creature without predators is the yellow-bellied sea snake which is very poisonous.[30] The safest of creatures seems to be the bath sponge. It is inedible, so the waterways in its body provide a safe haven for worms and crustaceans.[31] Less lucky are insects which usually have predators for each stage in their complex life cycles.[32]

There is violence in the natural world. Prey are mutilated, eaten alive and often make every effort to escape from their predators. A deer attacked by a coyote defends itself vigorously.[33] A lamprey eel attaches itself to a fish with its suction disk and then begin to eat into the fish. The fish makes every effort to dislodge the lamprey. Even if the fish is successful, the lamprey usually manages to grab it elsewhere and finish the job.[34]

Predators are also at risk in these encounters. When my dog got into porcupines and came home with a mouth full of quills, she would have died if I hadn't pulled them out. The same may be true for mountain lions. Great horned owls sometimes snatch porcupines out of trees and drop them. One owl lived with two quills embedded in its heart.[35] In contrast, fishers have stomachs of iron and pass the quills.[36] A dead gull was found with a mole sticking out of its gut. The gull had swallowed the mole alive and the mole began burrowing its way out of the gull's stomach when the gull crashed, killing them both.[37]

The saliva of the komodo dragon, a macabre, Indonesian alligator–like reptile, is so toxic that it painfully kills within hours. The komodo stalks its bitten victims and eats them when they can no longer resist. It is avoided by prey whenever possible. But if a komodo dragon gets too hot after eating, it risks dying from the undigested meat rotting in its stomach.[38] A shrew will eat two thousand beetles a day or one every ten seconds just to maintain itself. But it risks running into a predatory beetle bigger than itself.[39] And predators often defend hunting territories against competitors, particularly during hard times. Lions kill—but rarely eat—other cats, wild dogs, jackals, hyenas and even people who they feel compete for game.[40]

Some predators are efficient at what they do, while others are inept. Hunting success ranges from 80-96 percent for osprey who always seem to have fish in their beaks to 7.8 percent for wolves, who have to be carefully tracked to find where they have successfully made a kill.[41] Some researchers believe that spiders are hungry most of the time. Thirty percent of their webs, which require a lot of energy to build, catch nothing. Their wait-and-see strategy makes them vulnerable to wasps who take from 12-50 percent of the members of various spider species. For other predators, there may be available prey but they don't seem to recognize it. During a collapse of a hare population, lynxes starved to death except for those few who took to hunting abundant mice.[42] Predators—whose violence makes us uncomfortable—take their chances, as do their prey.

Only for Need?

In comparison to modern human hunters, it is often thought that predators and indigenous people kill only for food. Trophy hunters, buffalo hunters and bounty hunters are criticized for the waste with which they slaughter their prey. For example, in the movie, "Dances with Wolves," the hero (played by Kevin Costner) and his Indian friends look upon a field of dead buffalo carcasses stripped of their hides. This contrasts wasteful Europeans to conservative Native Americans who used everything they take from nature. Leaving aside hunter-gatherers for the moment, let us see if this proposition applies to predators.

Animal studies provide a less idealistic picture. Some predators only eat a portion of their catch. One naturalist observed a chipmunk eating only the brain, a choice part, of a captured mouse.[43] Gulls strip crabs but not the large claws which they cannot break open. Although the chipmunk could have consumed more of its kill, the mouse did not go to waste but became food for a host of other beings, until eventually bacteria and fungi reduced it to its chemical nutrients. Where to draw the line on what constitutes waste is difficult. An uneaten carcass seems wasteful. Sometimes predators do not consume any of their victims. When prey is abundant, fishers and wolves kill more than they can eat. When denning, wolves will try to kill any animal caught off guard, sometimes wiping out whole populations. When a wolf population is small or prey outstrips its food supply, wolves may kill indiscriminately. In general, predators only kill in surplus when swamped with prey under highly vulnerable conditions.

There are examples of killing not done for food, although the prey may be taken as food when other game is scarce.[44] And some cases of self-defense are taken too far. Mocking birds will attack rattlers who approach their nests, pecking out their eyes. They do not stop until they have killed the snake even though it is blind and has withdrawn from their territory. Horses and deer tramp rattlers to death even when the snakes pose no immediate threat.[45] In general, predation is conservative rather than wasteful. Nonetheless, there is killing for the sake of killing and sometimes, as we see later, animals can eat more than is good for them.

Predation not Paramount

Until well into this century, predation was thought to drive natural selection.[46] Although relying on the image of competitive struggle, Darwin used the idea more broadly. "I should premise that I use the term Struggle for Existence in a large and metaphorical sense, including dependence of one being upon another....a plant on the edge of a desert is said to struggle for life against drought...mistletoe with the trees it drapes and seedlings struggle with each other for sufficient sunlight to grow."[47] As biologists did more detailed field observation, they began to think that it was prey that controlled predator populations rather than the reverse.[48] Predators have little effect on squirrel or tit populations although owls take one-third of tit chicks. In the case of voles, their population is so large that it isn't dented by their numerous predators. From a Darwinian perspective, there must be other checks on the growth of populations. Less dramatic forces than predation, such as climate and food supply, take a much larger toll.[49]

Darwin cast weather, particularly the uncertain winters of northern latitudes, as a grim reaper. "Climate...seem[s] to be the most effective of all checks....the winter of 1854-55 destroyed four-fifths of the birds in my own grounds; and this is a tremendous destruction when we remember that ten percent is an extraordinarily severe mortality from epidemics with man."[50] So also the winter of 1994 in New England was a hard one. Because of cold and deep snows, owls took to hunting by day and ducks and geese had only small openings in frozen ponds to feed in. Because New England is the northern range of bobcat and their paws are not designed for walking in deep snow, they took to hunting in backyards. In 1953 two hailstorms killed 150,000 birds in Canada.[51]

During hard times, predation becomes secondary. Prey are already doomed by environmental factors. Predators deliver a *coup de grace* or act as a gory clean-up crew. Fish in drying streams are lunch for birds, coons and snakes. According to Paul Errington, in a musk-rat pond one winter, sixty inhabitants were getting along all right, surviving the occasional mink intrusion, "when cold weather froze the water to the bottom and sealed practically the entire food sup-ply....Three weeks later there was not a muskrat alive on the marsh.... the minks were the final agencies of much of the mortality but their predation made no difference in the end. The muskrats did not have a chance to survive, minks or no minks, once things went drastically wrong for them."[52]

Death and Nature

If there were no death, a housefly laying 120 eggs in seven generations of a leisurely summer would produce approximately 5.6 trillion descendants.[53] When Darwin read Malthus' hypothesis that arithmetic increase in food led to geometric increase in population, he realized that nature required something like selection to keep in balance. Where there is life, there must be death. Besides death's necessity, how we understand its role in nature depends upon how the processes of life are thought to occur. The Gaia hypothesis steps beyond Darwin to include aspects of geology, chemistry and meteo-rology in the organic processes of life.

The basis of Gaia is the observation that without life the Earth's atmosphere would have come to a chemical equilibrium such as exists on Mars or Venus. If only physical events—such as cooling or colli-sion with an asteroid—altered environments, the atmosphere, oceans and geochemistry of the Earth would not be where they are now. According to Gaia theory, life forms collectively act as regulators of each other and of the geology and chemistry of the entire planet. Bacteria changed the atmosphere. Cow flatulence contributes to global warming. Life forms transport minerals, depositing some of the ores we now mine. Gaia views the planet as if it were a kind of giant organism, creating and responding. Although many debate the Gaia hypothesis, it presents us with a broader perspective from which to look at life.

The power of the Gaia hypothesis is that it points to the profound interconnections among life forms and their relationships to the physical environment. Chemistry and geology are no longer fixed underpinnings but an interactive part of what is going on. This adds a new dimension to natural selection. "The 3 to 30 million species... together with their physical surroundings prevent the rampant exponential growth of population...simply put Gaia is Darwin's natural selector."[54] Not only is the environment thrown into the pot, but also within Gaia it is hard to distinguish what an individual organism is, let alone a species.[55] It is now accepted that the sunlight-catching or power-producing parts of plant and animal cells (the chloroplasts and mitochondria) originated as invading or consumed bacteria. So in our evolutionary heritage we are an interlocked component of Gaia. Most of us know that the microbes in our guts are crucial for our health and welfare. To what extent do we wish to regard them as not part of us?[56]

Residing in the bosom of Gaia, life appears more comforting. Gaia emphasizes interdependence and cooperation. As natural selector, Gaia feels not too unkind. But is it? According to one of its originators, Gaia has no sympathy at all.[57] It has no moral imperative not to maim and kill. Whatsoever evolution produces by way of predation or parasitism survives to the extent it works—that is, until there are no more prey or hosts, or conditions change. That the process is painful or torturous matters little. The only thing that counts is that if strategies no longer work, they get modified or their possessors die out.

Gaia includes such things as the eye-eating cichlid fish in Lake Nyasa which pecks out only the eyes of other fishes. It is also the moth ear mite which eats only one ear so that the moth can hear and thus elude approaching bats. A deaf moth is a dead moth, not a useful host. After the mite has finished with one ear of one moth it finds another moth to partially deafen. Indeed if Gaia had sentimentality people would not exist. "Two billion years ago cyanobacteria, newly evolved microorganisms that used the hydrogen of water for photosynthesis, plunged the biosphere into crisis mode. Their 'waste'... the free oxygen that sent thousands of varieties of organisms to early graves...altered the previous habitat forever. From the point of view of an anaerobe [methane breathing bacteria poisoned by oxygen], the global environment was ruined....this ecocide, this destroying of the planetary home, made life (as we know it) possible."[58]

Gaia makes nature seem both more harsh than Darwin's red rule of tooth and claw and also more gentle. Gaia highlights how the very process of evolution encourages mutations which have an advantage to make use of that advantage. An advantage is not restricted to one that is nice to other organisms or one that won't eventually undermine the species possessing it. Novelties in nature are simply experiments; they are not directional. They contend with survival. And survive they may not. The species with a new skill tries it out. It may work a bit, for a while, in a given environment, and then turn out to be a dead end. The damage to others and to the species itself is of no account. In fact it may be that compromises in adaptation like human testes, or the waste of species like anaerobes' oxygen, or natural increases in populations will eventually undo every species. Currently natural selection is not only more difficult to establish because there are more variables than Darwin imagined, but can be a horror show redeemed neither by a moral God nor by Darwin's notion that the vigorous survive to live fully. As Darwin wrote to "the devout Christian evolutionist, Asa Gray....'I cannot persuade myself that a beneficent and omnipotent God would have designed and created the Ichneumonidae [wasp] with the express intent of their feeding within the living bodies of Caterpillars, or that a cat should play with mice.'"[59] Redemption lies not in perfection or even the possibility of perfection but, if at all, in Gaia "herself," a superorganism or quasi-superorganism of pulsating, connected and sometimes self-destructive life, adorned with mutualistic species. If anything, "she" is awkward and filled with violence.

As Annie Dillard saw it, "It is chancy out there. Dog whelks eat rock barnacles, worms invade their shells, shore ice razes them from rocks and grinds them into powder. Can you lay aphid eggs faster than chickadees can eat them? Can you find a caterpillar, can you beat the killing frost?"[60] Evolutionary biologists debate whether the struggle between predators and prey is a kind of arms race. The blind biologist, Geerat Vermeij, found by touch that mollusks developed more rugged defenses as their predators evolved nastier weapons and the development of eyes seems to have led to phantasmagorical defenses among the creatures of the Burgess shale and trilobites.[61]

Predation does play an important role in natural selection even if it is not the major cause of mortality. Predators and prey have an important role in shaping the genetics of future generations by culling the slower, more inept of each. Predators have to become more deadly in the face of more agile prey.[62] Predator-prey relationships also help keep ecosystems healthy. An example is starfish. When otters were hunted out on the West Coast, starfish populations exploded, wiping out mussels and barnacles. When otters were reintroduced, they fed on starfish, reviving the diversity of the system.[63]

The other side of death, which limits populations, is birth and the mechanisms which thrust individuals and whole species into the Gaian churn. Because of the Malthusian imperative, species have the potential of undoing their own successes, not only by running out of food but because either their waste poisons them or it becomes the resource for other species to overwhelm them. Well-adapted traits which promote populations are time bombs except when occasionally balanced by traits which act as limiters. In times of population explosions and diminishing resources, female rabbits reabsorb fetuses. We will see later that human hunter-gatherers sometimes limited reproduction. Although the exact mechanisms by which individual novelties turn into traits of species are shadowy, most species do not seem to take into account their own undermining of their nurturing environment. The gene lines of successful individuals may prosper, but collectively their behavior may contribute to the decline of the species.[64] Humans do the same, as our numbers and global impacts indicate. When we get to Buddha's exploration of the human mind, we will see how he examines the unnoticed beginnings of self-destructive behavior.

Ways of Dying

Estimates of the number of species on earth vary greatly. Entomologists knocking insects out of the tropical rainforest canopy guess there are three to ten times as many insects as previously thought. Marine biologists scraping the ocean floors may have uncovered another ten million species. In addition, there are untold numbers of bacteria, protoctists and fungi to be classified.[65] All of these creatures were born (or hatched, or sporulated etc.) so many of them must die. We have looked at dying as a result of predation or the environment but there are other ways of death.

Although we share ways of dying with many other beings, some organisms have deaths we don't experience. To the Buddha's disease and old age we can add traumatic injury, suicide, cannibalism, rapid degeneration after procreating or giving birth, starvation, infanticide, abandonment and failure of the body to effectively transform. Many species have life histories very different from ours.[66] A Rhine River valley beech may set abundant seed every few years that are consumed by birds. At age 154 it pollinates a tree 0.5 km away, the seed of which flowers and survives. And at age 268 one of the millions of seeds it has produced survives. Aged 316 it rots and falls, leaving only two offspring, one 0.5 km away over 162 years old and one 80 meters away just beginning to flower.[67] In contrast, a small insect may live only a day and have thousands of progeny.

Leaving aside for the moment microbes, plants that clone and animals that die soon after procreating or giving birth, we find, among members of longer-lived species, dramatic deaths. Cannibalism is one of these. From television, we are familiar with female praying mantises decapitating the male during mating. In other cases of sexual cannibalism such as spiders, males try to mate without being consumed while females have other things in mind.[68] In ciliates and cellular slime molds, cannibalism occurs during hard times leading to monster, cannibalistic cells.[69] It is not clear whether ciliates can distinguish between members of their own species and other prey. Predatory slime molds can. Cannibalism occurs among shark embryos who swim about in their mother's uterus eating each other. Both rabbits and some female trees reabsorb their fetuses in times of duress. But it is among a cute, cuddly species that cannibalism and infanticide reach their apex.

Prairie dogs are beloved by environmentalists but the bane of ranchers because cattle break legs stumbling in their holes. In some densely packed colonies after babies are born, many mothers go on an orgy of infanticide, killing and often eating the offspring of others.[70] They take over one-third of the litters, perhaps to reduce pressure on food sources or to add protein to their diet. The killers sneak into a burrow while the mother is unaware. Many of them are caught because they stop to clean themselves at the mouth of the burrow. After chasing a killer away, the mother can be found in friendly interaction with the offender. Some mothers tend to be more active killers or frequent victims. The litters of marauding mothers are vulnerable to attack while they are away doing their grizzly work.

Because killers and victims are equally fit in terms of reproductive success, prairie dog researchers are at a loss to explain why this trait was selected.[71] As we look at the immense variety of life, there is sometimes no reason for a behavior even though survival of the fittest looks like it ought to explain everything. In most infanticide, males are the killers, especially where the mother comes into heat soon after she stops lactating. Natural selection's rationale is that the male wipes out a competitor's offspring so it can generate its own. Female acorn woodpeckers destroy the first laid eggs of other females in the same colony, so competing offspring will not be born before their chicks and be better nurtured.[72] Mothers, too, kill young when resources are scarce. In some burying beetles the parent eats up to half of the egg brood. Flatworms, which can regenerate from any part, have eaten their severed tails which were in the process of regenerating a new head. Cats licking the umbilical cord of young sometimes devour them. In one fly genus, larvae hatched within the mother devour her. And if an ichneumon wasp doesn't find a caterpillar host, it is consumed by its hatched larvae.[73] Some aberrant behaviors get eliminated but others don't.

Besides infanticide, nature includes violent sexual and territorial contexts which sometimes seem like war. Ants continually battle neighbors in what looks like ethnic cleansing. Despite its many predators, few will attack a cornered Norway rat. When the rats abandoned immigrants' ships for the New World, they drove native species out of urban turfs. The most vicious Norway rats are killers who fight to the death any member of a neighboring pack they encounter. Some lions and wolf packs behave similarly. Large ungulates do not use their formidable horns to defend against predators but kill each other in mating disputes.[74] Along with war and intolerance comes fratricide. In eagles, one of two hatched chicks kills the other, and two slightly older great egret siblings attack the third to the indifference of the parents.[75] Suicide also occurs. Some termites abdominally explode when their nest is invaded, exuding foul chemicals, and mother gall midges give their body up to their offspring as food.

Accidents also take their toll. Animals miscalculate. A naturalist saw a pair of eagles hold on too long in their mating dance. They ended up in the water a quarter of a mile from shore and had to swim so they could get out and dry off before taking flight. He also saw a gull swept into the sea by a wave and not reemerge.[76] Some animals have risks built into their life cycle. Crabs die from an inability to molt as they grow older; in other words, they grow too big for their exoskeleton and squeeze themselves to death. Most bees die while out foraging.[77] They get lost, exhausted or caught. For young animals, abandonment often means death. My dog so disliked motherhood that she tried to lose her nursing puppies in the woods. Seal pups, left on a beach by a hunting mother who doesn't return, die. The same is true for bird chicks.

Some animals are doomed when their niche gets filled. Muskrats with established nests drive off excess populations of the young, the less fit and the restless. The home population is virtually immune from predation. Those expelled have poorer food sources and little protection from weather or predators. As Errington relates, the dispossessed, "were young and they were old;...they were characteristically thin and chewed up from encounters with other muskrats; they sat in nests improvised in the shore zone vegetation... They hobbled across open spaces or through rushes and weeds, bleeding, trying to find food and shelter to keep out of trouble. They were dying from hunger or cold their tails feet and eyes freezing...or dying from their fight wounds, or being killed by minks or fox or dog or birds of prey."[78]

In contrast to the animals we have discussed who can breed again, there are plants and animals which breed only once and die soon afterwards. In some insects, the adult stage lasts only a few hours. Many garden ornamentals are annuals and bamboos can spread into huge, impenetrable groves living up to a hundred years before they flower and die. Salmon are the most famous one-time breeders. Born in freshwater streams and rivulets, they migrate to the oceans where they forage for several years, and then skillfully return to their birthplace to reproduce and die. Their heroic journey upstream is awesome to watch. They stop eating when they ascend their natal streams. Using all their stored energy, they throw themselves against the current, jump over rapids and wait in back eddies to gather the strength to push on. As they proceed, their faces become twisted in an expression of grim determination. At the end of the journey, the females are not much more than roe sacks and the males, sperm bags. After the females lay their eggs, the males fight for the right to spread sperm over them, then they both die of exhaustion, becoming food for bears, eagles and other scavengers. Only one mammal shares this behavior, the marsupial mouse. After an orgy of mating, all the males die.[79]

Mortality Rates

Plants and animals have differing life spans—with life span being defined as how long a species will live under ideal conditions. Life expectancy, by contrast, is the average number of years a species will live under natural conditions. Both of these are, of course, of concern to insurance companies that produce mortality tables estimating how long different cohorts of people are expected to live. Biologists have also constructed mortality tables for a few species.

The record for longevity is held by species that clone—that is, reproduce asexually—preserving the same genes. Some grass clones have an estimated age of fifteen-thousand years and some blue-green algae, millions of years. Those aside for a moment, individual redwoods live several thousand years and bristlecone pines even longer. Among more common animals, mollusks live two-hundred-twenty years. Termite queens live up to twelve. Wasp queens live to sixty-four days. Ant workers survive from ten weeks to two years. Mute Swan can live fifty years. Polar bears average fifteen to eighteen years; the oldest found in the wild was thirty-two. There is not much agreement on the life span of mosquitoes, except that the females live longer than males. The average for some mosquito species has been given at four days, with maximums from eight to thirty days, and with hibernation up to six or seven months.

One biologist has put together mortality tables for Dahl sheep:

During the first year	20% died
Prime years(1-5)	8%
Beginning middle age(5-8)	16%
Middle and old age(8-12)	47%
Old age(12-16)	6%

One-fifth die young, another fifth during the middle years, almost one-half die in middle age, leaving 6 percent to survive into old age.[80] And Dahl sheep are hardier than most other species.

Sparrowhawk yearling mortality is 50–69 percent. Afterwards they die at a constant rate of 30 percent a year. Although raccoons are tough and can live till twelve, many die in litters and 56 percent of yearlings starve to death. Sixty-five percent of moles die their first year and then 50 percent of survivors die in each subsequent year.[81] A student of bears states that, "Once polar bears reach maturity, they seem virtually immortal," with survivorship of up to 95 percent. "They have no natural enemy except in rare circumstances," and live on the average fifteen to eighteen years.[82] Mortality in most species falls heavily on the young. For some, like moles and birds, adult mortality is fairly constant. Once the hardiest species, like polar bears and Dahl sheep, reach maturity, they live on into older age. But for most species less than 5 percent reach old age.

Dying Young

Darwin observed that, "Eggs or very young animals seem generally to suffer most, but this is not invariably the case. With plants there is a vast destruction of seeds…Seedlings...suffer most from (being overshadowed by other plants and are) destroyed in vast numbers by various enemies. I marked all the seedlings of our native weeds as they came up, and out of 357 no less than 295 were destroyed, chiefly by slugs and insects."[83] It makes sense that vulnerability and abundance go hand in hand. Sperm and eggs outnumber births, and because of infant mortality, births outnumber survivors. In the biological study of life history, biologists look for the advantages and disadvantages of investments of energy into different stages of life. Some species reproduce once having many, small young who mature rapidly with little or no parental care. Other species have few, large young, slow maturation, intensive parental care and multiple reproduction. Salmon exemplify the first while chimps and humans are the epitome of the latter. In both cases, offspring and mothers are the most vulnerable.

There are many examples of young mammals' mortality. Wolves have four to six pups per litter. Up to 60 percent of the pups die from starvation and cold or are killed by their parents if they are severely wounded in playful fights or exhibit odd behavior, like epilepsy. Occasionally an eagle, lynx, or bear may snatch one. When mature—around two or so—their rate of survival is 80 percent.[84] Mortality among young squirrels is 75-85 percent, giraffes 73 percent, polar bear cubs, 60-80 percent. Malnourished mother polar bears eat their young. Adult males and wolf packs occasionally kill bear cubs. The hardest time is adolescence when inexperienced hunters often have their kill taken by older bears. They then go into the winter with little stored fat. Mountain gorilla mothers often abandon their cubs. Twenty-eight percent of the cubs died of starvation because they were too large to obtain sufficient milk but too small to hunt for themselves. Mortality rate in the first two years is 60 percent or more.

Among species who leave their many progeny to fend for themselves, egg and infant mortality is enormous. Salmon sperm clouds float randomly over eggs which are then buried. Other fish have a feast of the eggs, and hatched salmon fingerlings are prey to birds, reptiles and mammals. Rates of return from the ocean of hatchery bred fingerlings are less than 2 percent.

Mothers too are vulnerable. The only porcupine my dogs managed to kill was pregnant in the winter. Lactating female red deer have higher winter mortality because of low fat reserves. They compete with their weaned fawns for forage. When compared to long-lived clones, sexual reproduction is a costly evolutionary strategy.[85] Mother and child bear the brunt. Because of the risks, the process of egg laying or bearing live young is accomplished as quickly as possible. The young are quick to recover from birth or find protection. Young gazelles are mobile within minutes after birth and hide for the first week of their lives.

If evolution requires replacement and that process is vulnerable, then many of Gaia's checks on growth will center on reproduction. Checks may be gruesome when involving marauding or exploiting others for their own growth. Effects on mothers and their young are not pretty. In the face of such threats, it is no wonder that mothers earn their reputation for ferociously defending their young. A mother giraffe will kick away a marauding lion, and the two mother bears I have encountered were not to be crossed.[86]

Disease

A second of Siddhartha's four signs is disease, a cause of much human suffering. My encounter with the diseased moose in the Arctic taught me that disease exists in the wild as it does among humans.

Parasites have lively and intimate relationships with their hosts. E. O. Wilson characterizes parasitism as "predation in which the predator eats the prey in units of less than one."[87] He gives the example of the cookie cutter shark which takes big bites out of the sides of whales, leaving the scars that earn the species its name. Many diseases are caused by parasites. Viruses are not alive. They capture the machinery inside a cell and use it to make copies of themselves which is how they manage to replicate themselves. Some microbial parasites exist independently of their hosts while others use multiple species as hosts. Some kill their hosts while others need them to go on living. Diseases and parasites are an inseparable part of nature.

Ten percent of all known insect species are parasitic. Sparrowhawks have fungal, viral and bacterial parasites. Some cause diseases but others do not seriously affect their hosts.[88] Wolves get a parasitic brain worm from deer that causes blindness, disorientation, paralysis and death. They get mange, cancers, tumors, rabies and distemper. They suffer from rickets and heartworm in Texas, salmon poisoning in British Columbia, and trichinosis in Spain. Insects also take a toll. Like the arctic moose I saw, lions can become emaciated and desperate from an outbreak of flies.[89]

Just as humans are subject to epidemics that can decimate a population, animals experience epizootics; that is, animal epidemics. When polio struck the chimps Jane Goodall was studying, fifteen chimps got the disease and six died. In 1801 an epizootic among beaver was described by an adopted European living among the Ojibwa. "I found some kind of distemper among the animals which destroyed them in vast numbers. I found them dead and dying in the water, on the ice, and on land; sometimes I found that, having cut a tree half down, they had died at its roots, sometimes one who had drawn a stick of timber half way to his lodge was lying dead by his burden."[90] In New England during the summer of 1995, dead squirrels hit by cars decorated the roadways. Squirrel populations had exploded because a rabies epizootic struck their predators.

You name it and nature has devised ways of getting you. Parasites even have their own parasites. An entomologist found an oak gall parasitized to the fourth degree. The chance experiments which emerge in Gaia are incredibly clever. A nematode of the red grouse lives in its hind end. The nematode's eggs are deposited in feces, and the hatched larvae climb into the flower buds of a heather which another grouse eats.[91] Parasites whose development requires more than one host have evolved means for their passage. Some larvae get their intermediate marine hosts to swim away from protecting weeds, making them more vulnerable to being eaten by dabbling ducks, the parasite's definite host.[92]

This dance between parasites is crucial to evolution. Parasites are usually small, short-lived and capable of rapid change. In contrast, hosts are larger, longer-lived and more set in their ways. To counter the adaptability of parasites, hosts are thought to use genetic variation from reproduction and their immune systems. The immune system is a kind of miniature version of evolution by natural selection. The immune system creates antibodies until one works and, like nature, goes to town producing the effective one. To defend against mosquitoes, blood is supplied with coagulants, platelets, antibodies and sometimes other parasites.[93] Not to be outdone, Gaia as the great natural selector takes a new tack. In AIDS, the virus goes after the immune system itself.

The struggles between species and their parasites wax and wane. Sometimes they evolve to live peacefully with each other and sometimes they revert to war.[94] The symbiotic origins of the cells of plants, animals, and fungi are the peace treaties between host bacteria and the other bacteria which live inside them. This may be one of Gaia's most elegant outcomes. Biologists have a hard time explaining how many-staged parasitism comes about. It is difficult to imagine the intermediate steps leading to co-evolutionary relationships. On the whole, parasitism is more opportunistic than orderly, efficient or progressive. It keeps Gaia off balance. As Annie Dillard puts it,

"Parasitism...is a sort of rent, paid by all creatures who live in the real world.... Teddy bears should come with tiny stuffed bear-lice."[95]

Disability

In the upbeat spirit with which Darwin sometimes held natural selection, one would expect to find nature filled with robust animals living vigorously. That is how nature lovers generally see it. But as I have pointed out, this picture does not bear close scrutiny. Butterflies have bird bites in their wings and become faded. Insects are missing legs. I was surprised once to see a limping coyote try to test a deer. The deer looked on with indifference. In the wild, mountain gorillas appear healthy except for slight physical injuries. Nonetheless they get yaws, leprosy, flukes, appendicitis, hookworm, crab lice, arthritis, small caries, colds and hookworm. Even when diseased, gorillas continue to forage, or they die of starvation. One gorilla with an arm missing above the elbow remained a leader, while others had withered arms or deformed legs and feet caused by accidents or bites.[96] Survival in the wild with a disease or injury is tenuous and takes inordinate effort. One nature writer found the remains of a deer with a partially healed leg fracture, pieces of which stuck out from the leg. It had lived for a long time with a broken, foreshortened leg. It must have been in immense pain and yet it kept going.[97]

Ostracism adds to the difficulties of disability. In litters, adults and siblings attack and kill offspring that are different or behave strangely. You often see disabled gulls standing apart from the others. And an ornithologist claims it is the misfit birds that hang around people.[98] One of Jane Goodall's chimp's legs became paralyzed from polio. He was able to drag himself around, feed and sleep in lower branches of trees. The chimp's sibling wouldn't groom him and unenthusiastically staved off attempts of other males to drive him away. The researchers tried to help him but finally shot him when he dislocated a shoulder and could no longer feed himself.[99] Herds or packs also drive away disabled members, yet a few manage to assert themselves.

As mentioned above, predators seem to know when an animal's behavior is different from what it should be. They test herds to find the weaker, less able and chase them more persistently than the healthy ones. Predators seem to detect erratic motions against the flow of a herd.[100] One day on the range in eastern Oregon, I saw a coyote driving a herd of deer. The coyote moved and the herd responded. It was a graceful dance on the hillside. I assumed the coyote sought a discordant note. On another occasion in coastal California a deer completely ignored a coyote with an injured leg awkwardly attempting to approach it. Disability in nature is difficult to maintain. Starvation, selective predation and discrimination take their toll.

Old Age, Sex and Death

For the Buddha, birth leads to disease, old age and death. Through meditation, he claimed to have discovered a state beyond these "signs." In traditional Buddhist thought the Buddha's awakening, or achieving of *nirvana,* is regarded as a real state of being, in some ways more real than solid material objects. We need not follow the traditions there. For our purposes it suffices to observe that even awakened persons' bodies are part of nature. Yet there are organisms which don't get born in the sense that Buddha understood it and don't really age or die either. It is not that they have not come into existence and it is not that they will not later go out of existence. The solar system is only a few billion years old and life even younger. Both these will only last another few billion years. So all species indeed come into and go out of existence. This jibes with Buddhist cosmology which includes the comings and goings of universes.[101] But there are species which seem exempt from what the Buddha understood as the process of birth and death.

Organisms such as bacteria and some clones which replicate without sex, appear ageless. For bacteria, each of the offspring cells is identical and can divide without end. Accidents happen, and dividing microbes encounter predators, environmental change and limits on food. But they don't seem to age. They pass their cell structure and genetics to offspring which similarly divide. Changes which we associate with aging are absent.

A little more complex than this "immortality" are beings which either do reproduce distinct offspring or clone but do not seem to senesce. Over long spans of time their cell structure does not seem to change. They just get bigger. A list of such beings, their longevity, and size is rather impressive.[102] A quaking aspen has lived ten thousand years, a creosote bush eleven centuries, and a huckleberry more than thirteen thousand. The record may be held by a massive, European pine estimated to be forty thousand years old. It has an extended root structure and multiple trunks.

Given their age, these plants must possess survival mechanisms for a large range of conditions. All the things that normally kill plants are taken care of. Their growing tips do not wear out. They eliminate genetic damages by replacing cells. They manage respiratory burdens and water transport. They maintain large distances from root to shoot. The failing of any of these might be signs of senescence. Although these individuals may not live forever because of changes in geology and weather, they certainly don't age as we know it.

Some animals also live long lives. Lobsters live to one hundred, tortoises to one hundred fifty and quahogs to two centuries. Rock fish, sturgeon and snakes do not seem to senesce. In honeybees, male drones live one to two months, female workers up to twelve months, and queens five to eight years. The queen is killed by workers when her supply of hundreds of thousands of eggs is used up and she starts to smell different.[103] In many birds, after maturity, mortality does not accelerate with age. Except for a decline in productivity, little physiological change is evident. A pair of crowned pigeons nested and laid eggs for forty years but none hatched after twenty. The parenting skills of California gulls increase with age.

The simplest limitation on life occurs when the body of a species is designed to last a given span of time. Some adults can't eat and so die when their stored food runs out. Their only job is to reproduce. Adult mayflies last from a few minutes to a few weeks in a total life span of several years. They hatch almost simultaneously. The males mate and then fall into the water from weakness. Some beetles that can't eat live for a year. Other species have anatomical deficiencies. Male mosquitoes, blackflies and midges eat only nectar sugars and cannot live on them very long.[104] For mammals, aging is synonymous with senescence. The longest-lived mammal is *Homo sapiens.*

Since senescence accompanies human aging, we can ask with biologists, doctors and an agonized public, "Why do we get old? After all, our bodies have built themselves up, why can't they just keep going?" Billions of dollars are spent on research into why people age. Senescence is treated as a disease. I wouldn't mind owning stock in the company that comes up with an elixir of youth. Scientists have proposed four major reasons for aging: programmed cell death, wear and tear, mutational damage and antagonistic pleiotropy.

Programmed cell death has been one of the hottest topics in biomedical research over the last dozen years. Thousands of scientific papers have been written seeking an understanding of its mechanisms and role in disease and aging. It is a complicated subject involving molecular biology, cell physiology and evolution. A simple explanation begins with evolution.

Under ideal circumstances, bacteria are immortal. The genetic makeup of a cell is replicated in each offspring cell. This division appears to be able to go on forever without wearing out. The next, more complicated forms of life are protoctists. They are single or multiple celled beings with a nucleus which bacteria lack. They often divide without sex by a process called mitosis but also can replicate sexually in a process called meiosis. The interesting thing about protoctists is that those that replicate sexually are usually mortal when they are celibate. After a given number of divisions the offspring die. Their death is orderly. Their genes send out signals which lead to the shutdown of their life mechanisms and they shrivel up. It is as if they are programmed to live for only so long without sex before they close shop. When they have sex, there is an exchange of genetic material between the mates, then cell division makes a genetically different being. The old parts are absorbed to make new cell walls and life is renewed. The old cell has effectively died. With or without sex, most protoctists are mortal. From them evolved fungi, plants and animals.[105]

Programmed cell death happens in the development of a mammal. Many more cells are created in the fetus than are used by a mature person. Dendrites in the brain grow outward and if they don't connect with a brain cell they gently decompose themselves to be used as food by neighboring cells. It is a very different process from that of cells damaged by disease or trauma where body defense mechanisms rush in to fight the disease or clean up the mess, causing inflammation and other side effects.[106] Except for sexual (germ) cells in humans, all other cells (somatic) are mortal. They can divide up to fifty times and then experience programmed cell death. The number of divisions is called the Hayflick limit, named after its discoverer.

Why the Hayflick limits exists and its implication for aging are subjects of much research and debate. In exploring these questions, researchers have been able to give youth back to an aging nematode which has a genetic structure analogous to that of humans. By tinkering with cells, they renewed the process of division. The experimental nematodes were reinvigorated and lived longer than their aging contemporaries.[107] This was heralded as a first step in the battle against aging. Some researchers believe that programmed cell death lies at the heart of senescence. Graying hair, stiff joints and failing hearts may be avoidable.

The second explanation for aging is wear and tear. There are no genetic instructions for renewal or repair of some parts of the body or the processes cannot keep up with life's insults. The cells of insect wings and exoskeletons don't replace themselves. Houseflies live one to three months. Their mortality lies on an exponential curve as they degenerate. Housefly wings fray and the male mouth parts get damaged until they can no longer taste. Mammalian brains, heart muscle, and cartilage may be like this or may have some incomplete repair process.

The third reason given for aging is mutational damage. The longer the body of an organism is on Earth, the more it is subject to radiation and other irritants which lead to the breakdown of its basic genetic materials. In addition, every time a cell divides there is risk of error. One reason that bacteria divide successfully and can survive for a long time is their effective repair tools. Repair for sexual beings is much less efficient so they experience more genetic damage. With time, damages accumulate and individuals break down. The long-living pines and birch seem to defy this.

A final cause for senescence is antagonistic or negative pleiotropy. As was mentioned earlier, sickle-cell anemia as a side effect of malarial resistance is an instance of antagonistic pleiotropy. The gene which confers resistance has a downside. As applied to aging, antagonistic pleiotropic traits are those that nature gave you to help you to grow up and procreate but become harmful afterwards. In humans, the capacity to store fats is crucial to growth and survival but increases risk of heart attack later on.[108]

Buddha framed the human conundrum of aging in a non-theistic manner. He never looked for causes of the problem. For Buddha, aging was just something that happened: no reason, no need to ask why. Having lost faith in his god, Darwin, by contrast, found redemption in the natural order where the vigorous survive to lead fulfilled lives. The difficulty with his view is that in most species, most individuals die young and those few who survive to old age suffer senescence. Dying young fits into Darwin's natural selection but what use does evolution have for those who are post-reproductive? Efficiency in natural selection would seem to undermine the need for post-reproductive life. In some social mammals, the post-reproductive do play a survival role. Post-menopausal African elephants can live sixty years, and older females lead elephant packs. Older humans are supposedly wiser. But, there is little research on the evolutionary role of other post-menopausal females nor any about old, sexually uncompetitive males. A quarter of short finned pilot whales are post-menopausal for up to twenty years or more. After menopause opossum can live two years, ringed seals thirty-five and chimps more than forty years. For most species, there is no reason from natural selection for this.

Some biologists claim that the force of natural selection weakens after reproduction because fitness is no longer an issue. Evolution doesn't need to have well adapted oldsters because they do not contribute to population increase. This is the case for animals that procreate, usually have many offspring for whom they do not care and die soon afterwards. But for animals that have few offspring requiring extended care, the situation is different. Because of the need of parents to give care, the risks of survival and the effects of wear and tear, nature comes up with better construction than required by a creature barely making it through reproduction. It is statistical: the better made the beast, the more likely it will bring its offspring to the point of independence. But then there are leftovers. Here is another of nature's tradeoffs. After building hardily adapted bodies to struggle for existence in order to reproduce, natural selection finds itself with some of these bodies left over. They take up resources and no longer contribute to evolution. Moreover, they are not well adapted for their declining years. Here natural selection is a little less orderly and efficient than Darwin and his followers believed.

So nature is like a fine sculptor chiseling out elegantly featured forms but then forgetting functional details. What we see around us are refined organisms which work. They work well enough to insure survival, but the designs do not eliminate all structural problems nor protect organisms from pain or death. This arrangement is not as efficient as it could be for nature as a whole or kind to individuals who have struggled to survive and leave their genetic heritage. Successful offspring are condemned to the same process of aging as their parents.

We'll now look at some of the characteristics of this life sentence, a sentence which greatly upset both Siddhartha and Darwin. The few survivors of violent death who make it to old age await an unkind fate. About the same proportion of mammalian life is spent from birth to puberty with high mortality, as from puberty to mid-life when mortality is low. After mid-life, mammals experience the following age-related changes: menopause, baldness, endocrine-related tumors of reproductive organs, decline of plasma quality, increased fat deposits, beginning of atherosclerotic lesions, decline in antibody production and the start of osteoporosis. This leads to senescence with its associated decreased male fertility, weakening of pulses, loss of weight, impaired response to cold, atherosclerosis, osteoporosis, arthritis, slowed reaction time, diminished vision, hearing loss and decreasing dendritic complexity in the brain.[109] The onset of senescence in cattle is between five and twenty-five years old, dogs ten to fifteen, humans sixty to eighty, mice two to three, and rhesus monkeys twenty to thirty years old. Senescence does not occur simultaneously nor does it progress regularly, so that even though rates increase with age, healthy individuals may not show signs until very advanced ages. After mid-life, age-related mortality grows exponentially. Honeybees suffer a different fate. They literally lose their minds. Worker bees, who survive the risks of foraging, die of senescence by fifty days old. Their brain cells get used up and they wander off confused and empty-headed.

We can document the changes constituting senescence but how do aged animals in the wild come to their end? Remember the polar bear which after childhood "seems virtually immortal." A polar bear is the most formidable predator around. With one paw they catch seals through their breathing holes in the ice. Despite their prowess, polar bears face old age. A student of polar bears saw "a polar bear about two meters away. You don't stop to think at such moments and in an instant I was well out on the sea ice… Carefully...I crept back.... She had a medium-sized frame but appeared small because her body was almost wasted away. Every bone pushed up under her hide, making it appear as if it had been draped over a skeleton... She was a very old adult female in the last stages of starving to death...she could not even stand up... Here in an Arctic snowdrift, this ancient matriarch was ending a long, experience-rich life."[110]

A few other naturalists have observed old, emaciated starving animals. In a matter of months Old B-Ram, a mountain sheep, went from being unchallenged to harassed by younger rams and limpingly feeding in unprotected areas. As he became feebler, he slept and ate in exposed places. He finally died during a winter storm, killed by a wolf who consumed very little of his almost fatless body. Watching old animals decline usually leaves a deep impression on naturalists.[111]

Animal Emotions

If we seek the origin of human discontent in the broadest sense, then finding antecedents in the animal world is a kind of confirmation of Buddha's search. So far we have looked at occurrences in nature of the Buddha's starting point. Before exploring animal emotions, it is necessary to make the obvious assertion that there is no life without attraction and aversion. This is evident for bacteria who move toward food sources and away from toxic substances, in the nervous dance of death between a spider and wasp over who is predator and who is prey, and for the monkey who won't let go of food it can't slip through bars despite an overwhelming fear of the hunter coming to kill it. Animals would not survive without these mechanisms. We will now look at how these are expressed in higher animals as behaviors which in humans we would call emotions.

But how would we know how animals feel? Such inquiries lead us into disputed areas about animal consciousness. Many pet owners swear that their beloved pooch, kitty, horse or parrot has emotions equal to their own. Our friend William Long was a better observer of human feelings than of animal behavior when he claimed animals do not feel pain or get upset about impending fate, both of which torment humans—animals do exhibit anxiety and become afraid.

We begin with the problem of how we know what animals are feeling. Humans have a general repertoire of feeling signals which they use reasonably reliably with each other. When it comes to animals, the closer they are to us on an evolutionary scale, the more likely there is some kind of overlap. Putting humans back in the animal world and therefore animals closer to the human world, Darwin makes a case that basic human expressions are universal.[112] They arise from our animal past and are lodged deeply in our nervous system. In the final study of his life, on lowly earthworms, Darwin found they had continuity with humanity. The worms seemed to take pleasure in eating and liked to touch each other.[113]

Like many ethologists who followed him, Darwin observed domesticated and zoo animals and read all the reports he could find on the expression of emotions in wild animals. He used words for human emotions to describe animal responses. In contemporary ethology textbooks, the word emotion is carefully avoided. Animal behavior is described in as neutral language as possible without reference to human feeling. But some natural historians claim a deep rapport with their subjects, knowing intimately what they are feeling. These latter are often accused of reading too much into behavior while articles in scientific journals may lapse into uninteresting mechanical descriptions.

Early scientific and literary portrayals of animal feelings were naive and romantic. In reaction, ethologists insisted on exact but desiccated descriptions. The pendulum is swinging back again. Currently some researchers try to identify with the behaviors of their subjects and try to see the world as they do. When confronted, a hognose snake "puffs up like a venomous cobra though it lacks any venom. If it does not scare off the enemy, it will go into a flamboyant seizure, writhing, contorting, rolling over, defecating and finally turning over, its breathing stopped and its tongue hanging out."[114] Originally, this was thought to be a fright reaction but it turns out to be playing dead. When the predator looks away, the snake will try to flee.

In the woods in northern California, I was on a field trip when a spotted owl fledgling rose from the side of a trail and landed in a tree overhead. It stayed there staring at us while a couple of horses with riders passed by. After everyone left, I sat with it for a while. It looked at me with great curiosity. I would move to see it from a different angle and it would hop around on the branch to face me. After a time, I began to back slowly away along the trail. When I was about forty feet away, the owl swooped down out of the tree, passing within inches of my face. It had such a curious expression. It wanted to get a closer look. If my hands hadn't been full and I hadn't been a bit frightened at its surprising proximity, I might have reached out and grabbed it. It perched on a tree above me and I again hung out with it a little longer. When it got bored with me, I would hoot and it would look around again.[115]

From this and other interactions with wild animals, I feel a kinship with the ethologists who form strong bonds with their subjects. They remove personalizations when writing scientific reports so as not to impose human-like traits on their species. When Jane Goodall first went to Africa in the early 1960s, she gave her chimps names and personalities. Not having had much formal training, she did not know that "real scientists" numbered their subjects so as not to form attachments or anthropomorphize. In her naiveté, she gave a new slant to ethology.

The issue of animal consciousness is a complex one. There is conflict between people who use animals for experiment, raise them for food and other products or hunters, and animal rights advocates who feel that animals have consciousness and therefore suffer when they are manipulated or slaughtered. In order to place the origin of discontent in our animal nature, we don't have to decide whether or not animals have consciousness similar to ours. We simply need to show that animals exhibit behaviors which would be recognizable as similar emotions in humans.[116] Even if discontent is uniquely human, the foundations of its components such as pain, pleasure, emotions and thought are a product of our evolution. And animals' reactions to the events of life which sent Siddhartha off on his search are antecedents to our reactions. So a natural historical exploration of this is an important part of Buddha's way through Darwin's world.[117]

Although I will address the issue of the neurophysiology of human responses in the chapter on evolution of mind, it is important to note here that when it comes to animal pain, William Long was mostly wrong. He correctly observed that insects don't usually protect damaged body parts and will, for instance, continue mating or eating without seeming to notice they have lost something vital. But there are many counterexamples to his assertion that higher animals do not respond to pain. To varying degrees, mammals, birds, frogs and fish have neurochemicals associated with pain similar to our own.[118] In fact, a lot of what we know about human neurochemical responses to pain comes from experiments on cats and primates. Here we will stick to animals with human-like emotions.

Many predator-prey interactions and intraspecies conflicts show aggression and fear. Predators sometimes hunt with calm dispassion but there can also be a quality of nervousness and rage especially when the prey is formidable or the outcome uncertain. Angry individuals or packs among rats and wolves were mentioned. Among chimps, some groups seem to have a culture of hunting while others don't.[119] On the other side, humans can identify with rabbits who, when threatened, are so overcome by terror that they become immobilized, sometimes even going into fatal shock. Anticipatory fear exists in the animal world and can be out of proportion to the reality of the situation. Whether or not animals worry, they—like rabbits—become anxious. In famous experiments, rats and other animals given frequent, random shocks become anxious and even paranoid, jumping at any stimulus, attacking or cowering in fear. Domestic dogs that have been beaten or otherwise abused will attack without reason or display anxiety or excessive fear. Some dogs are particularly frightened of men. Sometimes I can overcome this reaction by giving them a special kind of non-threatening attention, but other times this does not work.

Other human-like emotions are recognizable in animals. Familiar to us are the various ways animals nurture their young. The more social mammals put a great deal of energy into rearing with a kind of attention and tenderness we recognize and which arouses our feelings. The mothers who vigorously defend their young may be responding to an instinctual Darwinian imperative, and their fierceness seems to correlate to the nurturing energy they have invested. I am not sure how to interpret the behavior surrounding infanticide among prairie dogs. Infanticide certainly existed among hunter-gatherers. Whether the emotions surrounding it were as cool as those exhibited by prairie dogs is not known. It makes little sense from our current cultural understanding of nurturance. In early Buddhism, the unremitting sorrow from the death of a child was an important reason women became monastics.[120]

Animals not only exhibit fear and aggression, they also play. Among lower animals, only bees are known to play. Workers who have been confined to the hive come out and dance around before they take their turn as foragers. Birds play, and play is well developed among social mammals, especially with kin. Play sometimes happens between members of different species despite the risks of predation. There are other emotions that most people don't expect to find among animals. Teenagers seek thrills. Young guenon monkeys tear open stinging ant nests, jump about while they are getting stung and go back for more. Deception is also common among monkeys. A juvenile baboon will cry for help so its mother will chase away another adult eating something the youngster wants. Lower status chimps miscue others as to the whereabouts of food so it won't be taken away from them.[121]

It is often assumed that animals as opposed to humans are sensitive to natural limits, but when it comes to food, animals do familiar things. Even when satiated, animals hoard to deny others the food, and they behave greedily.[122] Ducks and wild turkeys often eat so much that they cannot fly, endangering themselves. I have seen chipmunks, raccoons and skunks become so obese on human garbage that they could hardly move. Hunter-gatherers built traps relying on monkeys' desires overwhelming their fears.

The most extreme case I witnessed were two gray squirrels who hung out in the compost buckets at an out-of-doors retreat center. Totally vulnerable, the squirrels were down in the buckets eating while humans scraped their food remains on top of them. They were obese. Their fur was filthy from the compost they wallowed in. Their eyes bugged as if crazed with greed. They ran around furtively looking for food. They appeared to me like the hungry ghosts in the Buddhist hell realms with gigantic mouths but only tiny bits of food to eat. In some Buddhist traditions there are metaphorical representations of human mindsets. The hellish realms of desire can be portrayed this way. To indicate insatiable desire, they show people with giant mouths and tiny morsels of food—or the opposite, individuals with tiny mouths and big pieces of food. Both sorts of ghosts are always hungry. Looking at these squirrels was like looking into an animal version of a part of one's self: the glutton never gets enough even though the body neither needs nor wants more.

For Buddha, loss of one's love was the most painful of death's insults. We have seen that prey respond in a variety of ways to risks to their own lives, but animals also show a range of behavior when death is present. The phoebes, who got excited when a raven ate the cowbird they were raising as their own, were back the next day renewing their nest as if nothing had happened. Other animals show little interest in death. Wildebeest mothers ignore their dead calves after a few minutes. Chimps sometimes don't recognize death. Jane Goodall describes a recently born chimp who cried out continually from pain. The mother and a sibling groomed it until it died. They continued to carry it around dead for days until it began to rot. The sibling even tried to play with it. They finally abandoned it. Neither the mother nor sibling revealed expressions of mourning.

Other responses to death can be dramatic. Among chimps there are examples of mourning. An adult male chimp became so despondent at his mother's death that he remained with her corpse until he died of starvation and sorrow. And the sibling of a chimp who died and was removed kept returning to the spot looking for his missing brother. If her child dies, a baboon mother will carry the corpse around for a week and appear depressed. Cheetahs apparently mourn as well.[123] One researcher claims that twenty percent of free-ranging rhesus monkeys show signs of depression with the loss of relatives, partner or social status. Depressed monkeys show some of the same changes in brain chemistry observed in humans.[124] Elephants are famous for their reaction to death. An elephant with a dead newborn was observed carrying it into the shade. Elephants bury corpses and become quiet when they walk near a dead elephant.

The most touching example of animal mourning I have found is in the case of a mongoose, Tatu, who had been injured in a fight. As Tatu was less able to forage and the nails of her injured foot grew so that she was unable to walk effectively, her family group fed her, groomed her and hung around her, warming her. When she died in the termite nest in which they were living, they stayed with the body until it rotted. During this period, they did not go out to forage, risking malnourishment, nor did they pay attention to pecking order.[125] While no adaptive reason seems to explain this behavior, it does support those who see nature as more gentle and cooperative, with animals helping each other and responding emotionally to situations of pain and suffering.

Conclusion

We seemed to have strayed far from the Buddha, but not really. We have merely put on the natural historical lenses with which he might have observed had he been a beneficiary of the Darwinian worldview. Granted the Buddha would have focused on what can be personally seen and its impact on the mind rather than the far-ranging observations of natural historians, their fact gathering and the intellectual framework they use to draw conclusions. From what has been presented in this chapter, it makes evolutionary sense that suffering from the loss of a child is one of the greatest of human afflictions. Beside themselves, offspring are the largest biological investment of mammals and the most often lost. Evolution has bestowed an emotional load which fires both the impetus for self-preservation and, through our progeny, the continuation of our species. That the natural world in which we reside looks different from our conventional view of nature is important. Debating nature's beauty, gentility, gruesomeness or violence is really a discussion of what kind of a world Buddha had in mind for us to observe in order to engage our own discontent. As I have argued, life is a curious combination of intricate functional designs and awkward compromises. It is marked both by enchantment and gratuitous violence. That life needs death to recycle materials and make elegant species is something about which Darwin was of two minds. He would have liked the process to be kinder but he knew it wasn't. Buddha, on the other hand, had no doubts. Saying it in many different ways in the oldest of the *Sutras*, he reiterated that suffering arises because humans insist upon embracing the pleasant and avoiding the unpleasant. "Long have you [repeatedly] experienced the death of a mother...the death of a father...the death of a daughter [etc.]...loss with regard to wealth... loss with regard to disease...crying and weeping from being joined with what is displeasing, being separated from what is pleasing—-are greater than the water in the four great oceans."[126]

We want to understand this behavior of humans as part of the natural world. We have seen here that predation can be unpleasant but sometimes improved by a subtle communication between predator and prey which points to how their interaction serves natural selection or nature as a whole. The environment, parasites and species are also interwoven in a web of relationships represented by Gaia. In these relationships the struggle falls heavily upon the young. And oldsters of many species find themselves in the strange position of not giving much service to the new generation, using up valuable resources and having bodies which progressively decline. This is the natural setting we humans have inherited. But we have altered raw nature. We have built civilizations to protect us from the direct effects of the rawness. When we take animals out of the wilds and put them in our houses and zoos, we do the same thing for them.

I noted earlier that figures for longevity of animals in the wild indicate that only a small percentage of most species live into old age. Our pets and animals in zoos and laboratories live longer than their wild counterparts. Polar bears can live to thirty-two years in the wild but forty-one in a zoo. The record longevity for an ant queen was seventeen years in a lab. A zoo elephant with capped teeth lived to fifty-seven. A soft tick survived fifteen years in lab. Horses live to fifty and dogs to twenty, almost twice the longevity of their wild cousins. In 1978, the American Zoo Association tried to discourage zoos from unnaturally prolonging life in order to set world records.[127] Eliminating the hazards of life means that more young animals survive; they live longer and are able to experience old age. This results in a surplus of zoo animals and veterinary geriatrics. If you try to release animals back into the wild, they frequently do not want to go; they become beggars or scavengers where they can encounter the human environment or they don't survive. I once watched seal rescuers, out of public view, release their rehabilitated guests into the ocean. They had to drive the seals away from the docks with rocks and loud noises. These seals often turn up again near human activities. This is analogous to what has happened to *Homo sapiens* following the introduction of agriculture, the development of technology and the growth of civilization. Placing animals in a human-made world of bars and test tubes affects them as it does us. We have made our civilization into a golden cage.

In the next chapter we turn to ourselves as animals living with disease, old age and death. How has our physical relationship to these changed with the development of civilization, and how has this affected our discontent? We have seen that animals express emotions akin to human emotions. So after looking at disease, old age and death in humans, we will look at the human mind and trace the roots of discontent and its animal antecedents there.

Chapter Four
The Human Body

I have argued that Darwin provides tools to examine how the questions raised by the Buddha are embedded in nature. Buddha encountered discontent for the first time when viewing human frailty. In order to discover why failing bodies so distress us he recommends that meditators closely observe their bodies and their minds. To begin a meditator "abides contemplating the body as body both internally and externally," noticing how sensation comes and goes.[1] For the Buddha discontent arises in the complex of the mind and body.

> When touched with a feeling of pain, the run-of-the-mill person sorrows, grieves and laments, beats his breast, becomes distraught. So he feels two pains, physical and mental. Just as if you were to shoot a man with an arrow and, right afterwards, were to shoot him with another one so that he would feel the pain of two arrows.... Now, the well-instructed disciple of the noble ones, when touched with a feeling of pain, doesn't sorrow, grieve or lament, doesn't beat his breast or become distraught. So he feels one pain: physical but not mental... This is the difference...between the well instructed...and the uninstructed....[2]

By observing closely, a meditator can see how the physical and the mental interweave to give rise to discontent. The observations are "independent, not clinging to anything in the world."[3] Dispassionate observation refrains from adding excess mental activity. (We will look more closely at meditation and what it can do in the next chapter.)

Disease, old age and death are the most objectionable parts of our lives, parts we wish we could change. But since we all must experience them, they are a fine place to begin. Some Buddhist meditation teachers claim that one needs no more than contemplation of body as body to alleviate human discontent. It is a beginning and end in itself. However, consistent with my plan of opening Buddhism to Darwin's perspective, I will present how our bodies meet disease, old age and death in raw nature in contrast to how they occur in modern societies. We will look at how different circumstances affect the body which meditation observes.

The Buddha encountered discontent engendered by disease, old age and death in a civilization transformed by herding and agriculture. We, in turn, live in a world permeated by technology. Human bodies once functioned in the wild without the aid of clothing, fire or tools. Technology now impinges upon our physical development and enables our bodies to survive longer. Although we have gained material comfort, longevity and leisure, we pay a price. The discontent the Buddha sought to ease infects our new circumstances.

Dramatic changes in how humans use their bodies accompanied early domestication of plants and animals. As early civilizations developed, stress, repetitive motion damage, and diseases associated with crowding and sedentism increased. Medicine and sanitation later eased some of these. Plagues, famines and infant mortality gave way to exploding populations, chronic conditions and extended old age. Survival made possible by civilization looks different than Darwin's survival of the fittest, brimming with vigorous creatures. What nature would have eliminated, we now co-exist with in new ways. Chronic conditions with some accompanying disability now precede old age for a growing proportion of the population, and the number of infirm elderly in the industrial world is skyrocketing. Neither sickness nor old age is easy to experience. As a friend taking care of his elderly father remarked, "Old age is not for the faint of heart." We will look at the ways Buddha and his followers worked with the discontent coming from the three "signs." Their practices give us insight into how distant we are from raw nature which created us. The brief Conversation of Death we once had has become a drawn out argument.

Early Humans

Although we are not often conscious of it, modern human beings are animals, members of the species *Homo sapiens.* We come in different varieties. There are morphological differences between Australian aborigines, Caucasians, Pygmies and Ainues. Eskimo circulation is adapted to cold, while the Pima Indians can live off desert roots and seeds. Southern Europeans have trouble digesting milk, a staple for the Nuer. Yet all these humans are one species. Ecologist Paul Sears claimed that the earliest human lived in nature, knowing, "little or nothing of nature's laws, yet conform[ing] to them with the perfection over which he [had] no more choice than the oaks and palms, the cats and reptiles around him."[4] How was the human body designed to function, and how did the struggle for existence affect that?

On my first visit to a chiropractor's office, I saw a cartoon which impressed me greatly. It was a sad-looking man hanging by the scruff of his neck from a coat hook, inscribed with the words, "Where are you going to live when your body wears out?" Being young, it had not occurred to me that I needed my body and that it was not guaranteed. What sort of body did we inherit from our ancestors?

Sometime around five million years ago, our hominid forbearers began to change from their monkey-like predecessors. From 3.7

million BP (before the present time), for the next 2.5 million years, australopithecines—like some of their ape cousins—had grasping hands, acute, binocular daylight vision, large brains, expressive faces, upright stance, slow growth, intense mother-child intimacy, peer play groups, age and sex roles, a special juvenile stage and social organization. Evidence for each of these characteristics can be seen in fossils of changing anatomical traits. As hominids continued to evolve, the hand became less ape-like, the body more erect, the jaw changed from herbivorous to omnivorous, the foot became better adapted to walking and the brain got larger. By one million years ago the pelvis, spine and legs had so evolved that *Homo erectus* was standing erect and had begun to walk from its home in Africa through Asia to Indonesia. *H. erectus* has been called an odd creature because of its large skull at the end of an elongated erect body, its weakened jaw and teeth and long thumb. It may have been these simple transformations that enabled a creature once adapted to forest life to survive in the savannas. Dorion Sagan uses the expression "evolutionary lag" for such make-do adaptations.[5]

In spite of hominids' seeming awkwardness, they were hardy and adaptable and so multiplied. A modern human, weak compared to ancestors thirty-five-thousand years ago, can run a marathon, swim the English channel, climb mountains, survive weeks in the open sea and live naked from the desert to Tierra del Fuego. My sixty-year-old landlady used to run my dog to exhaustion. The American Indians captured horses and deer on foot. You may be able to escape a grizzly bear by climbing a tree and a black bear by running down a steep hill. The human nose is as sensitive as a wolf's although it can't discriminate as many smells. A Waorani of the Upper Amazon can identify a species from the smell of its urine forty paces away.[6] How did this relatively hairless, geographically spreading, upright ape fare in the natural world to which it was adapted? First of all, ancient humans were part of a web of eating and being eaten. Although usually not considered, eating, digestion and elimination are some of our most important connections to nature. The Buddha understood this and instructed followers to be aware of these processes.

During silent meditation retreats such processes are often the most stimulating activities one engages in and so one becomes more aware. On one retreat I was carefully observing my way of eating when it dawned on me how animal I am. I was struck by the force with which my jaw came down on the food. It felt violent. Chewing was not violent because of some quality of mind, like gluttony, I was importing into the act. Rather my eating was the action of life living on life. It connected me to the environment which made it possible for me to exist. I could no more be biologically non-violent than I could live without my body. It became obvious, as I ate the cooked, vegetarian food of plants domesticated thousands of years ago, that the meal called upon little of the strength nature had built into my jaw.

Michael Cohen is a founder of the outdoor education movement. He defines "connectors" as those aspects of our being which are lodged in nature. Examples include our senses and such responses as salivating, suffocating and perspiring. All life has responses which connect it to the environment. Microbes, plants, fungi and animals have chemical or physical mechanisms which attract or repel them from stimuli.[7] In his teaching, Cohen gets students to experience connectors in their natural setting. He wants students to feel how their bodies were meant to function in the natural world out of which they evolved. Students spend quiet time out of doors touching grass, spines, hot, cold, and experiencing other natural stimuli.[8]

If we were to dive back into raw nature, without clothes, eating uncooked roots, berries, insects and animals, we would have difficulties. It would push us back in time to an environment the bodies we now possess are not used to. Yet our jaws, taste, digestion and metabolism originally evolved to eat rough fare and survive without protection. Attending to your jaws as they come down on meat or roots will give an idea of how our eating apparatus was made to tear raw meat and grind gathered plants. The way humans obtained food prior to the use of tools was not always pretty. Like chimps, raccoons and wolves we once devoured live insects, creatures squirming in drying ponds, raw bird and reptile eggs, and things rotting by the wayside.[9]

Mortality in the Wild

It is more difficult to construct prehistoric mortality tables for humans than for wild animals. Ancient human skeletons belonged to a wide range of owners throughout a long period of time and are scattered over wide areas of space. One can only estimate mortality rates for a given population. Out of this uncertainty three facts emerge: the majority of skeletons are of children and juveniles; less than five percent of remains are from people over fifty; and the most common cause of death was trauma.[10] In the early debates about what the first hominid, *Australopithecus*, ate, an archeological discovery including human and animal bones was taken as evidence of early hunting prowess. On further examination, the humans were found to be the prey of leopards whose teeth left marks on the hominid bones.[11]

With no evidence to the contrary, we can assume that prehistoric humans died in ways similar to modern wild animals. In the last hundred-thousand years, humans had to be wary of the same kind of animals that threaten us today. There were many more predators of humans then and some of them—such as sabertooth tigers and wild boars—were bigger, more vicious and better hunters than we. Still, their modern versions effectively hunt humans.

In a search for contemporary human predators, one finds a grisly list of beasts and circumstances. One of "the great white hunters" dedicated his book to the "countless....victims of man-killers and man-eaters," including people killed by lions, tigers, leopards, elephants, hyenas, grizzly bears, hippopotamuses, buffaloes, sharks, crocodiles, ostriches, snakes and bees.[12] Because the book is so sensationalistic, it is hard to evaluate the veracity of his stories. Nonetheless, there are enough incidents of humans killed by animals to establish that it does happen and was probably a significant factor in prehistoric human survival.

Contemporary hunter-gatherers and populations living near the few remaining wild areas still consider animals dangerous. The Bushmen of the Kalahari desert in the recent past encountered lions, leopards, cheetahs, hyenas, buffaloes, elephants, mambas, adders, cobras, boomslang, scorpions and centipedes all of which are potentially harmful although they don't usually prey on humans. Because of the risks, the Bushmen do not like to travel at night.[13] The Maasai of Kenya complain about the encroachment of wildlife parks on their traditional homelands. The lions, leopards, buffaloes, and elephants that are entertainment for tourists are a danger to the Maasai. Adding insult to injury, tourists regard the thirty Maasai killed per year by wildlife as part of the balance of nature.[14]

Villagers bordering on jungles in South and Southeast Asia report similar dangers. In the middle part of the last century, an estimated six hundred East Indians were killed by Sundabaran tigers. Most were poor fishermen or jungle foragers who could not afford rifles. The tigers had learned to avoid motorized fishing boats, associating them with the guns with which they had been hunted by the British.[15] India still reports rogue elephants invading villages and attacking people. They are known to rip their victims apart, toss them into a tree, or trample them. And Komodo dragons are a gruesome human predator. Ten feet long, weighing up to 200 pounds, and able to move rapidly, they have attacked children, farm animals and dug freshly buried corpses out of cemeteries.[16] They seemed to have been a predator of early humans in what is now Indonesia. A Hollywood celebrity received a nasty bite from one when he was allowed to enter its zoo cage.

In the United States, successful conservation and residential intrusion into wild areas has brought a recurrence of animal attacks. Joggers and bicyclists have been assaulted by mountain lions leaping on the moving targets from behind. Bears in parks have become dangerous scavengers, learning to rip open vehicles in search of food. Alligators living in canals in Florida neighborhoods snatch pets and small children. Great white sharks, associating swimmers with seals, strike with one blow and then retire to let the victim bleed to death. Sharks also bite but spit out surfers in wet suits upon discovering they are not seals.[17] As pictured vividly in *National Geographic*, humans working near the Arctic Ocean need steel cages to protect them from the curiosity of polar bears who may associate humans with seals but don't often attack. In Medieval Europe, there were stories of giant wolves that invaded towns, stealing children and killing adults; sentries standing watch outside the newly built town of St. Petersburg, Russia in the early 1700s were attacked and devoured by wolves. Except for a resurgence of wolves in Poland when parts of it were depopulated during World War II, most dangerous animals in Europe and America have been wiped out. The vast destruction was often done in the name of safety. As wild populations return, the incidents that created fear are also returning. There is reason to be afraid even if this fear is often out of proportion to the actual danger.

While we cannot ask wild animals how they feel when being preyed upon, observations by fellow humans can tell us the kind of death prey suffer. Since we have no evidence from prehistory, we need to rely on contemporary reports and assume that ancient humans reacted similarly. As for animals, there are two schools of thought: gentle and not-so-gentle. William Long states the case for the first. He cites the famous example of Dr. Livingston, the 19th century adventurer who found himself in the jaws of an African lion. Livingston was badly mauled but remembered a state of dreaminess overtaking him. He felt neither fear nor pain. Long cites similar cases. For Long, "It appears—that nature knows how to bring to even her violent endings her mercy and anodyne."[18] Livingston's story is often repeated, most recently by a doctor writing about modern hospital deaths. He suggests that the measurable rise in the level of endorphins (a brain chemical), causing euphoria, is a protective mechanism originating during the time mammals were prey.[19]

While humans, as some wild animals, can die calmly of predation, such deaths only occur occasionally. Not-so-gentle seems more the order of nature. Death protected in a blanket of soothing endorphins seems unlikely evolutionarily. Human victims of predators are often treated unkindly and they react. As the poet Dylan Thomas wrote, "Do not go gentle into that good night, Rage, rage against the dying of the light." Reports of Sundabaran tiger killings include villagers screaming and fighting as they are carried off in the night. A recommended defense against a black bear attack is to fight back as hard as one can; for a grizzly bear, play dead.[20] In the book *Man as Prey*, there are stories of people being bitten and thrown by large animals. The victims feel fear, not Livingston's calm. During a hyena attack in an African village, "an old woman was dragged screaming from her hut one night by a hyena which had broken through the straw door. The hyena dropped her when a man rushed up." In Nepal in 1907, the "Champawat tigress" killed 436. Victims held on to things while the tiger dragged them away. A man grabbed by a crocodile in Zululand, "twice broke the surface and frantically called to his horrified companions." Crocodiles ate 300 people on the Zambesi River. They took women and children who were washing cloths or drawing water.[21] Although preyed upon rabbits may go into merciful shock, humans in nature who were eaten alive or mutilated experienced pain, fear and desperation—and they put up frantic resistance. A couple in India killed an attacking leopard. The husband wrestled with it while the wife struck its legs with a broom, enabling the man to force it to the ground and then kill it with a club.[22]

Other causes of past human deaths in the wild include infanticide, competition with one's own kind, privation and climate. That social conflict existed from the beginnings of human evolution is not doubted. Archeological bones show crushed skulls, tooth marks, weapon-inflicted wounds and other signs of human-induced violence. (Infanticide will be discussed in greater detail later.) The relative frequencies of various causes of death are difficult to assess. Climate is the major controlling factor for animal populations. We know that hunter-gatherers of the last several hundred years went through periods of great privation, including starvation. Skeletons of hunter-gatherers who immediately preceded the invention of agriculture indicate they survived periods of hunger.[23] That prehistoric weather caused mortality must have been the case, but we cannot establish how prevalent it was. Accidents and the accompanying infections were major causes of death among the young. In contemporary hunter-gatherer societies, few people with serious disabilities survive; you either heal or die. Like wild animals, recuperative powers of hunter-gatherers are great but foragers have many accidents. Today, insects cause the most deaths although snakes take forty to eighty thousand victims a year.[24]

What we know about disease and how it affected ancient human life is fragmentary. Many diseases do not show up in the archeological record. Like muskrats, the weakened and diseased were finished off by predators, leaving little to tell the tale. Perhaps adult diseases did not play a prominent role because trauma and infant mortality took such a large swath of the population.

Having touched upon death and disease, we are left with the third of the Buddha's signs, old age. Like other animals, early humans died very young. Only about five percent of prehistoric humans survived to old age. Elderly Paleolithic people, the immediate predecessors to the first agriculturists, had extensive tooth wear and considerable osteoarthritis.[25] They also exhibited features of aging intrinsic to human biology. As contemporary humans age, some parts of the body begin to diminish: weight at age sixty, height at forty, head circumference and chest breadth at fifty-four, liver at fifty-five, brain at thirty, kidneys at forty-five, spleen at thirty-five, and pancreas at thirty-six. By age seventy the percentage of fat content has doubled. Skin becomes looser, is more likely to wrinkle and is less resistant to temperature change. Cartilage cracks and frays by thirty then gradually calcifies, becomes brittle and interferes with waste exchange. It cushions less and is more subject to inflammation. Bone mass lessens. Muscle mass decreases. The circulatory system has fewer age changes but there is thickening of artery walls and the left ventricular wall. Under stress, the heart pumps less than it used to. There are respiratory changes. The lungs become less elastic and increasingly inefficient under stress. The digestive system decreases its gastric secretions, suffers weaker contractions and a weakened intestinal wall. Absorption of calcium and Vitamin D are reduced. With age the kidney cannot function as well under stress. The brain loses mass and dendrites decrease. The immune system weakens and there is an increase in autoimmune responses. Women experience menopause and men produce less testosterone. Human metabolic rate declines.[26]

I don't know about you, but my response on reading this list over again at age sixty-nine is either, "Let me out of here!" or, "Maybe the Buddha was right but spare me the all-too-close-to-home details." Prehistoric humans, physically more active than modern humans, experienced some of these changes, but most did not live long enough to encounter the progressive effects of senescence. Since soft tissues are not preserved, archaeologists can't tell how our fifty-year-old ancestors aged. We don't know whether old early humans met a fate like that of the aging polar bear or B-ram. Aging humans may have been cared for, consumed or exposed when they became a burden. Evidence of cannibalism and euthanasia is provocative but insufficient to draw conclusions. With the coming of agriculture, however, the situation changed radically.

The Paleolithic

The roots of the transition started 2.5 million years ago. Early humans began to use primitive flaked stone tools and then fire 1.7 million years ago. In the millennia that followed, tools changed, sometimes slowly, sometimes rapidly, until forty thousand BP, at which time the number of different kinds of finely worked stone tools explodes. A series of successive hominid species were toolmakers until one-hundred thousand years ago when two important hominid species existed sometimes side by side. One was *Homo neanderthalensis* and the other *Homo sapiens.* They may have had a common ancestor 300,000 BP. Neanderthals were shorter, stronger, had larger brains and were more primitive-looking. *Homo sapiens* had a richer technology and culture. About 30,000 BP Neanderthals vanished, leaving the field to *Homo sapiens.* Fully modern humans emerged about fifteen thousand BP. In those fifteen thousand years, technology improved rapidly. *Homo sapiens* dressed skins for clothes and made stone awls, boats, slings, snares, animal traps, spear throwers, eyed needles, dyes, art, and pots.

Homo sapiens lived in hunting and gathering groups. Their tools, weapons and social hunting skills far surpassed those of their human and animal ancestors. They employed numerous people to corral prey and could kill at a distance. Once developed, the tool kit of these hunter-gatherers would remain about the same for millennia, saving a few minor improvements such as the bow and arrow and the mortar and pestle.[27] Pleistocene hunting prowess was so impressive that some believe it responsible for the disappearance of species of large animals which vanished soon after the new weapons and strategies began to be deployed.

With fire, tools and clothing, humans crossed a line from the animal world. Technology helped humans adapt to the environment as no other animal could. They began to prosper in environments unlivable without new technologies and their hunting abilities increased greatly. Starting as scavengers, they became mastodon hunters. At this point in prehistory, ancestral human populations began to grow but they lived a chancy existence. Upper Pleistocene humans still walked in fear of predators and the wrath of nature. The next addition to the human repertoire of skills would place our ancestors in an utterly different relationship to nature. The domestication of plants and animals is the most important technology of human existence. With it our ancestors manufactured abundance and created a safer environment in which culture could grow. It gave them a new kind of mastery over life. Domestication and the industry that developed in its wake changed humans' overt relationship to life and death. Significant as this was, it came with important costs. How domestication may have come about remains the subject of much scholarly inquiry.[28]

It seems to have begun in Asia 13,000 or so years BP and gained speed until 10,000 years BP when most of the basics of agriculture were established. Domestication took place independently at different rates and different times in many parts of the world. Before the British colonists landed in North America, the Indians of New England were beginning to domesticate local weeds at just the time when corn from Meso-America arrived, overwhelming their indigenous efforts.[29]

I presented some conjectures about disease, old age and death for early humans who lived as animals. We'll now look at how these three signs changed as hunter-gatherers became sedentary agriculturists and then members of civilizations. With full-blown agriculture, the routines of life changed dramatically. Hunter-gatherers ate a great variety of wild plants and animals. Their diets changed seasonally. Throughout most of their history, they did not preserve food and so moved from place to place as they used up local resources or edibles came into season elsewhere. They lived in units ranging from single families to groups of twenty-five. Because they had such a variety of ways of sustaining themselves, life was one of change. They hunted or gathered for a few days then rested for a few more. They were capable of intense, sustained effort and experienced periods of great privation. Infant mortality was high. Only a small proportion experienced some degree of old age. The major cause of death for adults was trauma or starvation. Their agricultural successors lived a very different life.

Transition to Civilization

A debated issue in anthropology and archeology is whether hunter-gatherers lived easier lives than did early farmers. There were many ways that agriculturists lived better and more securely, but they paid a price. The benefits were straightforward. In ten days, one person with a sickle can harvest a six-month's supply of grain. With a store of grain, you can support a larger population, care for the sick and survive hard times. But farming is hard work. Agriculturists of the Near East were tied to the soil. When grain is ripe, you have a limited time to harvest it before the weather changes, animals get it or it falls on the ground. A farmer and his family need to swing the sickle from dawn to dusk. And harvesting is only part of the process. The most time-consuming agricultural task is weeding a field.

The remains of a community in Syria tell more of the story.[30] Here, between 11,000 and 10,500 BP, just prior to the introduction of agriculture, the inhabitants gathered wild seeds, nuts and berries, and hunted gazelle. Two hundred years later, a completely different people cultivated primitive grains, chickpeas and lentils. Skeletons of seventy-five children and eighty-seven adult farmers showed evidence of degeneration of cervical vertebrae from bearing excessive weight when they were young. The community on the whole was in generally good health, but the women and girls had collapsed vertebrae from continuously grinding grain while kneeling, toes tucked under, bending forward and then rolling the grinder straight out. They had overdeveloped grinding muscles, disk damage, curved femurs and knees with bone spurs. In addition, because of the way they knelt, the large toe had cartilage damage and gross arthritis. The damage occurred when children were still growing which meant there was child labor. In the early periods when they could only coarsely grind the grain, there was little tooth decay. But teeth were broken by stones and worn by the mixture of coarse grain and rock powder from grinding stones.[31] Later, when they began to weave and could thus sift the grain—removing stones and hard grain parts—there was less tooth wear. But weaving meant using teeth to chew plant stems for fiber so there was wear on the particular teeth used, sometimes down to the gums. By 7,300 BP, pottery was used for cooking grain to make porridge. Fine-grain porridge was a life-saver. Among the skeletons from this period was a disabled woman with an unmendable broken jaw who could not have eaten foragers' food nor the earlier course grain. She survived on gruel. This society had the means to support the disabled.

Fine-grained gruel also meant that children could be weaned earlier; thus women could bear more children. In hunter-gatherer societies, population was controlled by late weaning, infertility during harsh conditions and periodically absent males. Such circumstances were ameliorated in agricultural society which both sustained more people and required more labor. Agricultural women began to produce offspring in numbers close to their biological capacity, eleven to thirteen. Populations started to rise. Agriculture meant that a locale could sustain more people and, because fields needed laborers and guards to defend against varmints and other humans, populations grew. Communities of several hundred gave way to larger settlements.

One last factor completes this scenario. It began before domestication was completed. Resources from partially domesticated plants created semi-sedentary, semi-agricultural societies. The southern New England Indians encountered by the Pilgrims planted multiple gardens which they visited periodically as they moved around hunting. In this setting, long distance trade developed. Items traded included tools, specialty foods and, in New England, currency in the form of wampum. With trade came some division of labor where producers of certain products specialized in their manufacture. Thus, by the time agricultural settlements solidified, they were connected to others through trade and could support full-time, specialized craft persons.

We can put some of these changes into Darwinian perspective utilizing the concept of adaptive radiation. Adaptive radiation is exemplified by Darwin's finches. The few who had been blown across the ocean from the mainland alighted upon the Galapagos Islands empty of songbirds. Over thousands of years, the finches changed form and occupied niches that elsewhere were filled by woodpeckers, seed crackers and robins. Darwin did not recognize their common ancestry until an ornithologist in England discovered it. Without competition, finches with mutations which could exploit these niches flourished. Then natural selection shaped changes which allowed better exploitation of the new niches. Evolution to woodpecker-like finches on the Galapagos is an adaptive radiation.

How technology enables humans to live can be viewed through the lens of adaptive radiation. By means of tools, clothing and agriculture, humans have been able to expand their domain and occupy it more productively. That humans are hardy has been noted, but the use of clothing and shelter made it possible to live permanently in the Arctic winter. The use of stone weapons and traps enabled humans to hunt large game effectively, creating an abundant source of protein. And the development of agriculture enabled humans to live more densely and survive in greater numbers. But nature gives no free lunch. Each move humans make which creates a new biological arrangement has its consequences.

Domestication of plants and animals made civilization possible. Although civilization may be a great achievement, it has its darker side. Agriculture is very productive, but it is also unstable. Along with the domestication of plants and animals come weeds, pests and disease. Locusts, which under the right conditions naturally swarm, could devastate farms. In contrast, hunter-gatherers' many sources of food protected them from the massive deprivation that occurred during agricultural eras. Paleopathologists inventory history via skeletons. Hunter-gatherers were stronger, healthier and taller then their successors. Not only did agriculture bring repetitive motion syndrome, but also severe and chronic stress, infectious diseases, iron deficiency anemia and increased mortality rates. The Syrian community mentioned above had infant and child mortality rates of 50 percent, probably greater than among hunter-gatherers.

In other communities bones show infectious diseases and pervasive malnutrition, particularly among the young. These led to the stunting of height.[32] The fine porridge and breads released sugars bound by roots and berries; these cereals, along with their coarser predecessors, led to cemeteries full of toothless people who could eat nothing else.[33] These factors counterbalance increased fertility. In an Amerindian community in Illinois where the transition to agriculture took place around 1200 C.E., life expectancy went from twenty-six to nineteen.[34] While agriculture produced more food, staples like grains have fewer nutrients, except for calories, than what hunter-gatherers ate. This, along with periodic agricultural failure and crowded living conditions, left people malnourished and unhealthy. A number of early civilizations collapsed when agricultural land was destroyed by overfarming, erosion, salination or flooding from deforestation.

The complex of distant trade and undernourished people crowded in settled communities bred the great diseases of civilization. By domesticating animals and bringing the wild into their houses and courtyards, humans encouraged and participated in the epidemics of nature. In evolutionary history, diseases may have been passed down the evolutionary ladder from species to succeeding species. For example, the disease vector of syphilis and yaws may have passed from savanna living non-primates to primates. Then it evolved in the early upright humans who walked into the savannas. Later strains came through cattle to humans. Human diseases are often connected to our food sources. During Pleistocene hunting a half million years ago, humans came in contact with meat and skins. Then came fishing with its associated gastrointestinal parasites. Since hunter-gatherers lived in small, widely scattered groups, a disease would run its course and die out. The conditions of life in civilizations, however, nurtured the parasites by giving them more hosts conveniently close together.

Tuberculosis came to us from birds and cows, then domestic dogs carried it between farms. Houses and granaries sheltered rats and fleas, bringing leprosy and spreading plague. Diseases may have evolved rapidly as humans changed their lifestyles. Measles are related to distemper and may have come from wolves via domesticated dogs. Milk transmits some thirty distinct diseases. Tetanus came with pigs.[35] Smallpox, like AIDS, is a viral disease which crossed the boundary from the wild. The harmless pox of early domesticated cows which were housed with humans for protection mutated into a more deadly form that was passed along by travelers between towns. Because fecal waste and garbage were disposed of casually by most civilizations, they transmitted innumerable gastrointestinal diseases, killers of the young. Thus life went from a few hunter-gatherers living vigorously and dying quickly, often from traumatic injury, to many people with high mortality rates experiencing degeneration, disease and disability. With the coming of civilization, disease, old age and death had a new playground.

It is not my intention to provide an elaborate history of disease, old age and death from hunter-gatherers to the present. What concerns us is how the step from wild nature to civilization transformed the way in which disease, old age and death happen, and how the changes affect discontent. The Buddha lived in an agricultural civilization with its prosperity, diseases and conflicts. Buddha's identification of the nature of human suffering was naturally based on the troubles an agricultural struggle for existence made for the human mind. His methods and insights are useful because they are applicable to understanding both his hunter-gatherer predecessors' discontent and our modern discontents. Disease, death, aging—three of the four signs of Buddha—are affected by how humans evolved. To understand our suffering we must understand our evolution as we moved to new habitats and our bodies did not always adjust in a timely fashion. First we attempt to tackle the body in the modern world. Later we will look more closely at the mind of hunter-gatherers.

Diseases of Civilization

Aside from unsubstantiated longevity mentioned in classical myths and the Bible, prior to agriculture, few lived beyond fifty years. In Mediaeval cemeteries, one finds people who lived into their sixties. By the Renaissance, five-to-ten percent of the population was over sixty, but it was still a world racked with plagues and death of the young.[36]

By the end of the 19th century, mortality tables rapidly changed in Western Europe. Recall that life span is how long a species can live under ideal conditions and life expectancy the average amount of time people actually live. It takes into account infant mortality. Life expectancy had been around thirty for millennia; by 1900 it had increased in the West to almost fifty. In other parts of the world it was a different story. Statistics from a mid-19th century estate in Russia revealed that 45 percent of the population was dead by five and life expectancy was twenty. In contrast by 1980 in the U.S., 97.8 percent of those born reached the age of twenty and 31 percent reached eighty-five. In 1994, life expectancy at birth was almost seventy-six, and the U.S. is not the healthiest place in the world. People in Scandinavia and parts of the Commonwealth live longer and healthier lives. Disease, old age and death continue to change.

Of those progressive states in the U.S. which kept statistics in 1900, 16 percent of babies died within their first year of life. Half of those deaths occurred within the first seven days. About 1 percent of mothers died in childbirth. These figures did not include Southern or Black Americans. Other major causes of death were 17 percent heart-related, 12.5 percent pneumonia, 12 percent diarrhea and 9 percent tuberculosis.[37] By 1980, the situation had changed—30 percent of deaths were cardiovascular, 14 percent cancer and 3 percent accidents. Infant mortality and infectious disease had almost vanished.[38]

Infectious diseases which so plagued civilization have been reduced in industrialized countries. The specters which hung over the Renaissance and still troubled Darwin's family included tuberculosis, cholera, typhus, typhoid, virulent pneumonia, syphilis, malaria and childhood diseases. They are no longer of much concern. Smallpox, the only disease to be completely eradicated, took its last public victim in the 1960s. This has been counterbalanced somewhat in the 1990s by AIDS, resistant infections, and the feared recurrence of a 1919-like flu epidemic.

Except for cancer, most deaths in industrialized nations occur in old age from chronic degenerative conditions like heart disease. As a result, mortality tables have taken a new shape. In nature, the graph of human death rates is similar to that of many other animals. It starts out high in the first years, declines during adult years and then climbs exponentially in midyears. With the coming of agriculture, this graph did not change much. There were more births, but infectious disease kept child mortality even higher. If you survived childhood, there was a slightly better chance of surviving to old age because of the support of the economy.

In industrial countries, public sanitation improved in the late 19th century followed by medical knowledge, which grew rapidly during the 20th century. As a result, the mortality curve took a new shape. Almost everyone survives through middle age when mortality begins to rise slowly until a point toward the end of one's sixties, when an exponential death rate comes into play. Remember, zoo and laboratory animals live longer than do their wild cousins. Indeed, the shape of the curves for humans and laboratory fruit flies and other laboratory animals is almost the same.[39] This is an important observation. It means that when it comes to life and death, the support technology provides affects other species in the same way. For zoo-type animals in the wild, natural old age was infrequent. Life was short and was ended quickly by predators. But for humans, lab rats and animals living in zoos, age is greatly extended. The extension of life—and of death—involves technologies of maintenance, rejuvenation and transition to death. Sanitation and medicine are examples of civilization's adaptive radiations which have transformed the struggle for existence. But precisely these biological benefits create new mental challenges.

Except for cancer, chronic degenerative conditions are what eventually do us in. We saw that the human body is in decline by age fifty. It suffers worn teeth, arthritis, diminished physical size and ability of organs to function, wear and tear on cartilages and reduced muscle mass, particularly the heart. In nature, with nothing to buffer bodies from the environment, these changes greatly reduce chances of human survival. In civilized middle-age, we see evidence of these changes. People have dental fillings, receding gums, eye glasses or contact lenses, tricky backs and knees, incipient prostate problems, a bit of forgetfulness, hearing loss, tennis elbow, computer wrists, high cholesterol and sometimes heart attacks. Such conditions were set in place in earlier years.

Clogged arteries and elevated cholesterol may occur at twenty. By forty larger plaques have been deposited on artery walls leading to leg pain on exercising by fifty. Angina sets in by sixty and strokes from mild to worse by seventy. For cancer, there may be cell damage by forty, becoming cancer *in situ* by fifty, not clinically observed for twenty years or more. Diabetes evolves gradually, blood sugar imbalance increasing until the sixties when it may require medication. And osteoarthritis, part of mammal physiology millennia before the advent of civilization, runs its normal course but is exacerbated by a civilized lifestyle. At forty bone spurs occur, joint pains increasing by fifty and some disability by seventy.[40]

Our civilized lifestyle contributes to the development of these chronic conditions and cancer. They are a new generation of "diseases of civilization." And there is plenty of time for them to develop. Thirty or forty percent of middle-aged people will live until eighty-five or ninety. Although these "diseases" are lodged in changes that occur naturally, our gilded zoo bars encourage them. A list of things we can't resist for a starter: saturated fats, sugar, salt, tobacco, alcohol, processed foods, chairs, physical inactivity, passive stimulation and loud music. An immediate response to this list is to say, "OK, I know I'm weak willed. I'll clean up my act." Such a response misses the point. There is nothing on this list that isn't part of our make up as *Homo sapiens*. Every known hunter-gatherer society has been seduced by them! We mentioned Michael Cohen's connectors, the attraction-repulsion mechanisms that make it possible for all of life to operate in its environment. Items on the list appeal to our connectors.

As a psychologist points out, "We can decide to avoid refined sugar, but we can't decide to experience a sensation other than sweetness when sugar is on our tongue."[41] Sugars, fats and salt were scarce in the world in which we evolved. We have built-in attractions to getting as much of them as we can. Nature had no reason to limit our intake or revise our metabolism so that an overabundance was not harmful. There were not enough of them over long enough spans of time to let mutation and natural selection weed out their harmful effects. Things which functioned to promote our survival in the wilds have picked up harmful side-effects in civilization. This dynamic is a kind of antagonistic pleiotropy; that is, a trait or traits, one aspect of which promotes welfare, another aspect of which undermines it.

Sugar craving helps survival, but as a taste for soft drinks and "carbs" (carbohydrates), it can contribute to diabetes. Without fluorides and dentistry, we would be as toothless as the early gruel eaters. And Big Macs, ice cream, clogged arteries, high blood pressure and cancer are a package. As are chairs, computers, blue collar labor, TVs and rock music with slipped disks, carpal tunnel, myopia and hearing loss. Cars and sedentary living go with tendonitis, aches and pains and osteoporosis. Chimps and humans have been called avaricious snackers. Like greedy squirrels, when pleasing food is available, humans eat increasingly too much of it. As opposed to the malnutrition of early agriculturists, a complete diet, but in greatly reduced amounts, has been shown to contribute to health and longevity. This is consistent with what it was like for early humans in nature. Nature had little need to select out the negative effects of abundance. Our problematic sweet tooth was of great ancestral advantage, rewarding organisms for locating scarce rich sources of energy—sugars—in their environment.

Before 300,000 years ago, the size of anterior human teeth had been stable for eons. As *Homo erectus* moved into temperate climates, they used fire to cook their food. This was a necessity for frozen meat during the ice ages. Teeth began to shrink at the rate of .1 percent every 2,000 years. Because cooked food was easier to eat, both the millennia that followed and the development of grain grinding, led to a doubling of that rate of decrease.[42]

As my eating meditation made clear, what we now eat employs little of the power of which our jaws are capable. In early agricultural villages, filtered ground grain cooked into soft gruel destroyed people's teeth, which were no longer needed to survive. The food we eat today is not as tough as fifty years ago. Factory-farm, fast-food chicken is edible, bones and all. Dentistry and prepared foods means that hardy jaws contribute less to survival. This may contribute to even faster shrinking of our chins. Teeth of many wild mammals are worn down by use as they age. We rely on technology to repair the damages of our lifestyle and to provide false teeth when ours wear out. It is a dental challenge to fit teeth into a jaw that is getting smaller as age diminishes our head size.

Other examples of suffering caused by the lag between the ancient body and the present environment abound. The way hunter-gatherers lived was almost guaranteed to produce hardy adults. Natural selection weeded out the less fit. Pleistocene hunting involved the stalking of animals faster than us. As recent hunter-gathers in Africa demonstrate, endurance enhanced by mostly hairless bodies cooled by sweat glands allows hunters to keep up with faster wounded animals. Early hunters and the Tarahumara Indians for whom distance running was almost a way of life as well as today's marathon runners suggest the human body was designed to run. Yet fifty percent of runners injure their knees, ankles, feet, tendons, muscles or cartilage. Hitting unforgiving pavement after a sedentary childhood is brutal punishment for a body designed to move barefoot on uneven, variously textured surfaces. The body was designed to engage in various activities rather than train every day. Running shoes, stretching routines, exercise equipment and sports medicine are technological attempts to plug our ancient adaptation into the new, built-up environment. Their success has not yet been mapped, nor do we know if such new fitness regimes have an impact on senescence. It has been said that one would not want the old age of a professional athlete.[43]

Other ways of keeping fit encounter the same dilemma. A friend and colleague has been pumping vigorously on a stationary bike to keep himself in shape as he ages. In talking with him about senescence, I pointed out that without the use the body was designed for, it suffers ailments like those mentioned. My friend asserted that his daily hour-and-a-half of biking was a prophylactic against degeneration. I replied that the healthy bodies of our ancestors were damaged while being used. By biking he was increasing his chances of accident, of just those injuries that affected people before the diseases of civilization. If he didn't bike, he then faced the more gradual type of degeneration brought on by civilization.

When Europeans first landed in New England, they thought the Native Americans were lazy. Yet the Indians were taller, physically stronger and able to withstand greater privations. Understanding how hard it was to survive, they did little more than they had to. In that they were like the rest of life which paces itself, aware of the costs of expending energy. Children have a reservoir of energy, but when there is no need to use it, or it is subverted by television or imprisoned by schools, the same "laziness" exhibited by Indians predominates. Unless we are disciplined by an environment that makes us use our bodies, we tend to let them go. And without the robustness that hunter-gatherer lifestyle gives the body from childhood on, even hunter-gatherer activities may become harmful. The atlatl or spear thrower, which made the hunting of large animals easier twenty-thousand years ago, has experienced a comeback with its own association and journal. The only problem is that it also has its own disease, "atlatl elbow," a tendonitis caused by practicing heaving spears with the spear thrower.[44] In a single generation with schoolbooks and audio-visual aids, Inuit Eskimo children became myopic. Our bodies and our civilization are deeply intertwined. There is no retreat. New ways of living lead to new ways of dying.

Cancer is the second leading cause of death. Some cancers occur because of faulty cell instruction, others from external agents like radiation. Body defense mechanisms eliminate most of these before they do any damage. But some get away.[45] Why has cancer become so significant in modern societies? Most cancer seems related to modern lifestyle. Much breast cancer can be viewed as a form of antagonistic pleiotropy. It is connected to the production of hormones—hormones which would not be so prevalent if women lived harsher lives, bore children earlier and nursed them longer. Other cancers come from radiation, synthetic chemicals, smoking, diet and other human-caused changes in the environment. Holes in the ozone layer may contribute to skin cancer. If you build an airtight house in areas of Maine where there is radon in ground rock and heat your house with a wood stove, you increase your chances of lung cancer. Smoke particles capture the radiation which is then embedded in the lungs. Energy-efficient houses may pose a risk similar to that of the Abnaki Indians of Maine, who warmed themselves with an open wood fire which filled the tipi with carcinogenic smoke. But the Abnaki may not have lived long enough for the cancers to appear.

Finally, although they are not yet significant killers in industrial societies, some familiar infectious diseases have mutated around the technologies which kept them under control and now test the weak places in medicine and public health. Resistant microbes haunt hospitals and mass produced foods. Tuberculosis and malaria, which once threatened New York City, kill millions in the third world and can be found again in the U.S. The shrinking of the globe by airplanes and the widespread movement of people means that communities are no longer isolated. Flu originating in Chinese chickens or pigs infects millions of Americans each winter. AIDS, probably mutated from the wild in Africa, is a disease of lifestyle and communication. We have created new niches which parasites are adept at occupying. Though infectious diseases are frightening, it is living with cancer and chronic conditions which now provide fodder for the Buddha's methods.

Technology has been able to prevent accidents which killed or damaged our ancestors and medicine is able to ameliorate the damage when they do occur. Medicine and public health have constrained the infectious diseases that accompanied human crowding and filth. This left us surviving in a new environment to which our bodies were not well adapted. We now experience cancer and chronic conditions, diseases of civilization that can leave their victims lingering. These are addressed by additional applications of technology and medicine, which in turn create new consequences. This is where the human struggle for existence now resides. (See Illustration 2 for how scientists would redesign the body to better adapt it to our world.)

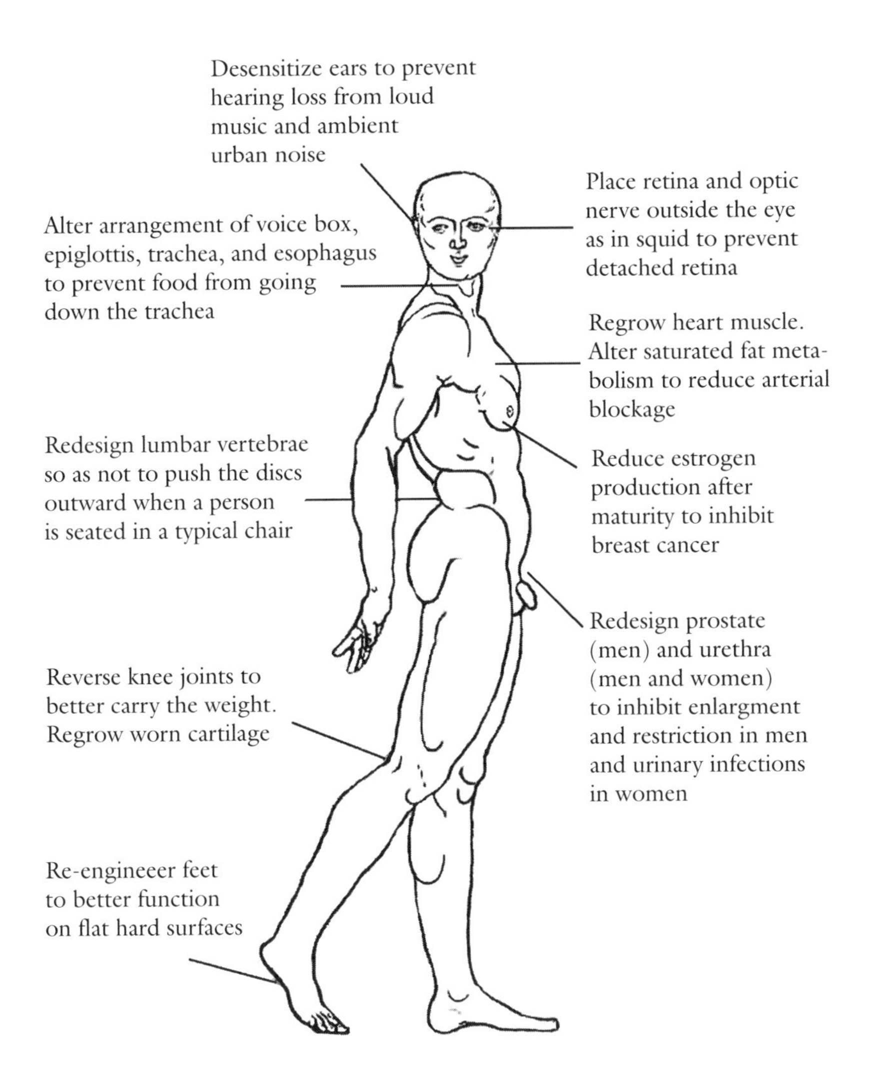

Evolution has left humans with design flaws as it has some species. An engineer could better adapt us to the physcial environment of the civilization in which we now live. See Olshansky, S. et. al. 2001, "If Humans Were Built to Last," *Scientific American.*

Senescence

In modern society the first three of Buddha's signs often collapse into one, old age. Extended old age for a large proportion of the population is not evident in nature and new in human history. Since infectious diseases are neither chronic nor deadly, we are subject to maladaptations or wear and tear until our bodies finally give out. Where the average dividing line falls between relative vigor and disability is hard to calculate. Somewhere in middle age medical problems increase. Health insurance rates increase progressively from age forty to sixty-five, at which time in the U.S., Medicare takes over. National health insurance for the elderly is the biggest budgetary problem of industrial nations because of the cost of caring for an ever-increasing number of elderly. It the U.S. costs are heading toward one fifth of the gross domestic product.

While most elderly report good health and mobility, the title of an editorial in a scholarly journal tells a different story: "Aging is riskier than it looks."[46] Although gerontologists generally portray aging optimistically, this view is not held by the general public who may have a more accurate assessment of the situation. Old age is unlike starvation among animals. Starvation is both likely and deadly whereas old age is more like car driving, not so risky but at some point likely to incur real damage. Gerontologists' statistics indicate that 5 percent of the aged become demented and 70 percent need help before the end. Thirty percent—twice as many as before World War II—spend their last days in an institution of some kind. While only 3-4 percent of those between sixty-five and eighty are demented, 11 percent at sixty-seven will have severe dementia by eighty-three or before they die and 20-25 percent less severe dementia. A similar number will have other mental conditions as severe as dementia. In total from 40 percent to 50 percent of the elderly suffer some kind of clinical mental difficulty before death. Nineteen percent of seventy years old increasing to 32 percent of those eighty-five feel lonely rather often or nearly always. Forty-five percent of the whole group felt lonely sooner or later. Seventy percent of men and 30 percent of women live with someone when they die. Most women are living alone before they die.[47] Women get sicker than men do, but men die earlier.[48] It is not clear whether the causes are biological or social.

Modern medicine often treats old age as if it were a disease. A review of a recent book on senile dementia in a scholarly journal states that in the 1980s dementia was regarded as a disease and there was much optimism it could be cured. The reviewer is not so sanguine now because, as the number of elderly grow, it is harder to tell the difference between the "diseased" and "normals" used in research control groups.[49] We noted before that old age is an evolutionary anomaly. We live three to four times longer than required for survival of our species.[50] Humans in wild nature give birth at age fifteen or sixteen. Females can bear young until their early forties. By about twelve, hunter-gatherer children can care for themselves, so all that is required for population growth is for parents to live to fifty or so. By fifty, most early humans have fulfilled their evolutionary function. Nature and evolution have done a good job; we begin to wear out when we are no longer needed. The 5 percent who survive are surplus to the process. It is not a surprise then that senescence occurs. As Siddhartha's story illustrates, civilized humans are uncomfortable with this. The Buddha proposed one solution. Medicine and the public seek another. They want to know how medical aging works in order to reverse its effects.

In the last chapter four theories of aging were presented: programmed cell death, wear and tear, accumulated genetic damage and antagonistic pleiotropy. The proponents of programmed cell death claim it is the governing reason for senescence. It may be that the graying and wrinkling of humans can be reversed. It is not clear, because our cells still can divide for a hundred or more years after we have begun to senesce. Or it may be that by transplanting stem cells (the cells which, like the first cells in the fetus, generate various body part or organs), cartilage or heart muscle can be rebuilt. This promise of science has yet to be delivered. At the present time, all four theories seem to describe aspects of aging.

There is much literature on persons in the United State who need significant help because they are physically unable to care for themselves. The immense wealth of industrial society allows it to maintain a sizable disabled population. In fact the wealth is so great that, in addition to paid care arrangements, society can afford to have 10 percent of the population give eighteen hours a week of unpaid labor to caretaking the sick and disabled.[51] In the U.S. in 1991-92 almost 20 percent of the population had some disability, about half of it severe. About 11 percent over age fifteen have difficulty seeing or hearing, and about the same percentage cannot lift and carry ten pounds or walk three city blocks.[52] While some early disability comes from survival of children who would have died in the wild and some comes from people not using their bodies, it is among the elderly that percentages rise rapidly. Senescence includes debilitating chronic conditions. About 55 percent of those over seventy-five have chronic arthritis, 22 percent cataracts, 35 percent hearing difficulties, 19 percent orthopedic impairment or deformities, 12 percent diabetes, 40 percent heart disease, 42 percent high blood pressure and 14 percent chronic sinus problems. For males, cancer rates skyrocket after age fifty-five, doubling every ten years. For females, the rise is more gradual. In autopsies done on people over eighty-five, 50 percent have evidence of cancer, although it is not what kills them.[53] If experience in nursing homes is born out, after eighty-five the combination of chronic conditions, dementia and disease leave more and more of the elderly disoriented or unable to care for themselves. It is not unusual for residents to swap daily reports of their multiple ailments.

In gerontology there are debates about quality of life as the aged population gets larger and older. The predominant but unpopular view is that women will be disabled for one-quarter and men one-fifth of their lives. Increased health and longevity correlates with wealth. The percentage of life lived disabled is growing faster than the percentage lived healthily. This translates into longer vigor in youth and middle age at the cost of an even faster increasing segment of the population moving into difficult old age.[54]

The data are unclear. More recent research suggests annual declines of 1-2 percent in disability of people in the U.S. sixty-five and older since 1982. The percentage with high blood pressure fell from 46 percent to 40 percent, arthritis 71 percent to 63 percent, and emphysema 9 percent to 6 percent. In 1982 27 percent of men aged sixty-seven to sixty-nine were unable to work. This dropped in 1990 to 20 percent. The number of people over sixty-five with diseases declined 11 percent since 1982. Because the current generation of elders is better educated than the preceding one, they may take better care of themselves. There are also new medical technologies which allow the body to function for a longer period of time. These include hip replacements, knee replacements, lens replacements, organ transplants and drugs to slow osteoporosis. The computer has also given rise to less strenuous work, allowing people to take part in the work force longer. The age at which one is defined as elderly has increased. Sixty-year-olds are no longer as frail as they used to be.[55]

The two versions of reality expressed here may not be so contradictory. The experience of my friends with elderly parents indicates although the age of frailty has advanced, aging parents have more diseases and chronic conditions, and they live longer while frail. When one becomes old now is hard to judge but a growing number of old people are often uncomfortable and unhappy. When my mother died at eighty-seven, the hip replacement done in the last months of her life was one of the better functioning parts of her body. Unfortunately, the anesthesia used for the operation added paranoia to her senility.

The literature on nursing homes and elder care vividly describes the pain and suffering of a large percentage of the elderly who, as the scholars above report, realistically describe their lives as difficult.

We will return to this subject after looking at Siddhartha's third sign, death.

Medical Death

How does dying in old age compare to dying in nature? One physician, Sherman Nuland, reacting to romantic portrayals of humans' last moments, wrote a book with vivid descriptions of dying. "An octogenarian who dies...is the victim of an insidious progression that involves all of him, and that progression is called aging."[56] Most autopsies reveal three to nine major pathologies, some not known at the time of death. Fifty percent of the elderly are found to have a cancer without symptoms. The cause of death is just one of the many factors which have become more or less advanced in the body. Infection is most prevalent for people over eighty-five.[57]

"By and large, [normal] dying is a messy business," writes Hayflick. Only one in five die easily or quickly. To achieve this goal, "...treatment decisions are sometimes made near the end of life that propel a dying person willy-nilly into a series of worsening miseries from which there is no extrication...surgery...chemotherapy... intensive care...."[58] Most medical expenses in a person's life accrue in their last few weeks. "[H]aving lost the major battle, the doctor may maintain a bit of authority by exerting his influence over the dying process, which he does by controlling its duration and determining the moment which he allows it to end....With the vast increase in scientific knowledge has come a vast decrease in the acknowledgment that we still have control over far less than we would like."[59] And so physicians try to control the process of dying. " Though the hour of death itself is commonly tranquil and often preceded by blissful unawareness, the serenity is usually bought at a fearful price...and the price is the process by which we reach that point."[60]

Death for animals is usually quick, violent and painful. Even in old age, except in rare cases such as the disabled deer, death happens within weeks. For most, death is likely to be like that of a muskrat, capitulating to a hungry predator, than it is to be drawn out like the last ditch efforts of a starving polar bear. Darwinian death was the tortuous attrition Nuland describes. Because the dying and their relatives and their caretakers want to avoid such an experience, a whole armament of medical technique is brought into play, one which, paradoxically, extends the process. Still the person dies in pretty much the same way. To die of old age as my Uncle Sam did, over ninety, relatively healthy, falling asleep at his granddaughter's graduation party, is a gift. Only 20 percent of deaths are as peaceful. The drawn out death of the remaining 80 percent creates the conditions the Buddha identified as a major source of suffering.[61]

At the turn of the century the great physician William Osler said, "Pneumonia may well be called the friend of the aged. Taken off by it in an acute, short not often painful illness, the old man escapes those 'cold gradations of decay' so distressing to himself and to his friends." Although most autopsies reveal infection, the continuous use of antibiotics in the elderly keeps those infections at bay allowing the patient to live ever longer. Death frequently comes when the weakest parts of the human body no longer function. Most people now die because blood flow is restricted to their heart or brain. These are particular hominid weaknesses. Other animals cannot be made to die this way. That is, if you experimentally try to cause their arteries to clog as they age, something weaker in them fails first. The conditions for rats in a pathogen-free lab are similar to humans in an aseptic hospital, yet old lab rats die of kidney failure. Our heart or brains go first, oxygen starved by clogged arteries. Our kidneys are over-designed by a factor of five and so last longer. On the other hand, old fruit flies succumb to gastrointestinal senescence, constipated and immobile.[62]

Discontent

Our next task is to look at how humans manage this new lease on life which technological civilization affords our bodies. How do humans relate to disease, old age and death as they occur in industrial society? Again William Long offers some insights. If it weren't the case that he drew his ideas from a completely different context, his could be taken as a version of Buddha's discourse on the "two arrows" (i.e., the original pain and the secondary pain of dwelling on it): "Pain has apparently played a very important part in the development of human consciousness... The degree of suffering is not in proportion to the physical injury, but to the quality of mind which reflects upon the injury.... Without...self-consciousness...there can be no real joy or sorrow."[63] Nor would there be our fear of pain and death which he feels we improperly attribute to animals. For Long it is anticipation of pain or death which distorts reality and causes humans so much anguish. Once more, Long's human psychology is more reliable than his natural history.

Annie Dillard, a much more observant naturalist, writes:

> Evolution loves death more than it loves you or me. This is easy to write, easy to read, and hard to believe. The words are simple, clear...but you don't believe it, do you? Nor do I. How could I, when we are both so lovable? ...Cock Robin may die the most gruesome of deaths, and nature is no less pleased; the sun comes up, the creek rolls on, the survivors still sing.... We value the individual supremely, and nature values him not a whit.... Nature loves the idea of the individual, if not the individual himself....Our excessive emotions are so patently painful and harmful to us as a species that I can hardly believe they evolved....The terms are clear: if you want to live, you have to die; you cannot have mountains and creeks without space, and space is a beauty married to a blind man...(who) does not go anywhere without his great dog Death.[64]

I have tried to show that civilization has lengthened Death's reign for many years. Civilization brings technology, industry, science and medicine to hold the great dog at bay so that it gets in only a few bites at a time until little remains to hold an individual together. Because of our civilized blinders, we don't see the nature behind this process. In everyday life we are not even alert to Darwin's observations of struggle and change. Without an understanding of evolution or how civilization has contained nature, it feels like Buddha's three signs of worldly suffering are assaulting each one of us personally, and we expect medicine and technology to heal their effects. This, and the fact that disease, old age and dying are so drawn out, create a greater opportunity for discontent. The domain originally explored by Buddha has expanded with the "progress" of civilization.

Buddhist ontology barely intersects with Darwinian evolution. The Sutras claim that, as part of the Buddha's awakening experience, he could see the process of rebirth in the physical world and many of the other realms of the Buddhist cosmology. One interpretation of Buddhist rebirth is that when a being is born it inherits the predominant wholesome or unwholesome intentions of some other being (itself?) in a preexistent life which will then influence it. What the being does with its given karma will, on its death, be the inheritance of another, new being. And the process continues. The Buddha is said to have had many past lives of which he gained knowledge. The idea may be taken as a metaphor or a literal truth (many Buddhists believe the latter). In Buddhism, beings are thought to be reborn in six different realms: those of animal existence, hell, hungry ghosts, humans, gods, and fighting demons or jealous gods who are always at war with the gods.[65] Because Buddhism asserts that humans have had many, many rebirths, it is said that all beings have been each other's father, mother, brother, and sister in many lives. This belief is a foundation of a sense of compassion and interconnectedness.

That human existence is only one of infinitely many possible incarnations is emphasized in one of four reflections used in meditation practice. It is the reflection on the specialness of human birth.[66] One is encouraged to remember its rarity in the great chain of beings. Being human presents a special opportunity to become awakened. Humans balance uniquely between realms of bliss and suffering. The lower incarnations are places of greater suffering, while the devas and gods reside in bliss. It is only when balanced between the two that there is the possibility of transcendence. According to the Sutras, devas and gods actually have further to go than humans. Because humans are so interrelated with all beings, they have an obligation to be compassionate. If actions are not done with care, then harm comes to one's self and others and the being reborn will encounter more suffering. Meditation is a tool for examining that part of our being which is discontented because of our embeddedness in nature.

Right now we are concerned with the Buddha's attention to the body. In the next chapter, we will look at the mind. What is relevant is that for Buddhists the ending of discontent via meditation is something only humans (and possibly devas or gods) can do: in the long chain of rebirths, beings only rarely get a chance to be human. While human rebirth may be rare in Buddhist cosmology, in natural history birth is quite common. What is rare is survival.

In the Buddha's time, most humans had little chance of meditating because they were not around long enough to do so. Childbirth and infancy were the most dangerous periods of life. Twenty-five-hundred-years ago, the Buddha's farming society had so progressed that hunter-gatherer reality was fading. There were many more births than among hunter-gatherers but also tremendous infant mortality and famine, widespread infectious diseases and epidemics. Agriculture made possible a growing population and some survival with disability into old age. In contrast to hunter-gatherers where there were many fewer births and much less death, early agricultural civilizations were immersed in death. Living in this environment, the Buddha must have been acutely aware of the sorrow of infant mortality, the loss of the young, the ravages of epidemics, the difficulties of injuries from repetitive agricultural labor and the suffering of disability. It was not easy for a woman or her mate to lose half of the eleven children she bore.

In the "Poems of the Elder Nuns," some of the earliest Buddhist texts, women relate the unbearable feelings of losing children and loved ones. This was the reason they renounced the world to practice meditation. Although we do not have pictures of the ordinary people of the Buddha's time, one might guess many would have had bad teeth, unhealed infections and the scars of smallpox or other injuries. Peter Bruegel's scenes of 16th century Flanders offer striking examples of how common were deformities. A vigorous old person was a rarity. It is no wonder that the Buddha emphasized the suffering that arises from disease, old age and death.[67]

In the industrialized world things have so changed that, except perhaps for the remote poor, we can no more imagine the Buddha's era or even the struggle our grandparents had with disease than Buddha's contemporaries could imagine their hunter-gatherer predecessors' problems. Over half of the 100 percent increase in U.S. population since 1900 is due to the elimination of infant mortality.[68] The specialness of human birth as part of the Buddhist cosmology has transmuted into the biological commonness of human aging with its accompanying chronic conditions. This is a profound challenge for meditation.

While trying to write the first version of this chapter I heard a National Public Radio report on the one million elderly Americans who are moving into assisted-living housing. It included an interview with an eighty-five-year-old woman, Mary, who had decided to move out of her home of thirty years because she could no longer care for herself. Mary is symbolic of the situation faced by many aged Americans. A combination of chronic conditions, a manufactured environment designed for the physically able and collapsing social networks lead to undesirable living arrangements and isolation. At least Mary is not poor, which would make matters still worse.

Before I describe Mary's circumstances, let me say something about images of successful aging now prevalent in our society. Morrie Schwartz was probably the best known dying American of the last ten years. The book describing his last year, *Tuesdays with Morrie*, sold millions of copies. Morrie died of ALS, popularly known as "Lou Gehrig's" disease, a progressively debilitating nerve disorder which breaks down one's muscular control. Morrie, as likely as not, chose to die when he could no longer swallow his own phlegm and so progressively risked suffocating. I knew Morrie well, having been a colleague and friend for almost thirty years. During his last year, he was a font of wisdom and love. In the few years before his disease began, he worked on a project studying the lives of extraordinary old people. The people he interviewed were vital and engaged, but they were also physically able. In contrast to them and Morrie (whose physical body went quickly but whose mind remained sharp and whose social network surrounded him), most elderly face increasing mental and physical limitations in an environment of diminishing social support. In such circumstances, otherwise bright and constructive people may pull inward, and some elderly question whether it is worth living.

Mary is such a case. Her husband at eighty-nine had died a few years earlier and she could no longer manage the walkway up to her cul-de-sac, southern California house. The interview was heart-rending. Mary cried because of her lonely evenings. She missed her husband. She remembered how the house had been filled with her children and their friends. She felt lucky that she had her last child when she was forty, because her parenting then lasted until almost sixty. Now her children were far away and she felt it was better that she move into a place for assisted living. Mary had trouble walking, had poor vision in one eye and a cataract in the other. When she mentioned how lucky she felt that she could still see, she cried again in fear of going blind. A friend had argued against her moving from her home and the things she knew and loved. Mary, uncertain, hoped the move would be better. She could return home if needed. Her children were not going to sell the house for a year.

Here is old age as Prince Siddhartha saw it. At eighty-five, modern medicine has kept the more graphic symptoms of senescence invisible. If she was "weltering in her own water," it was not described on the radio. There was no mention of the soaring sales of adult diapers. But Mary certainly is "bent as a roof gable, decrepit, leaning on a staff, tottering as [she] walks, afflicted and long past [her] prime." Siddhartha asked, "What has [she] done, that [her] hair is not like that of other people, nor [her] body?" To which his charioteer answers, "[She] is what is called an aged [woman], my lord." One can see Mary, as my mother did at that age, asking, "How did I get to be old? How did it happen?" as if she never imagined that such a thing would ever come about. One evening, as I sat at dinner with a woman in her late eighties who had recently lost her husband, she burst into tears when she broke one of her few remaining teeth. "Now I am old and ugly. I never liked old women. Now I am one of them." The Buddha asked, "And what, friends, is aging? ...brokenness of teeth, grayness of hair, wrinkling of skin...weakness of faculties...this is called aging."[69]

Mary brings in Siddhartha's third sign too, death. She is not statistically near dying. She has a 50 percent chance of living five more years and a 25 percent chance of surviving ten more.[70] Yet, she is experiencing what Siddhartha saw as the suffering in dying. "Then he saw a dead person who would not be able to relate to his loved ones." For Siddhartha, brought up in the bosom of a doting family, this was the most painful suffering of all. Death robbed one of one's relationships. The beauty of Morrie Schwartz's dying was that circumstance catapulted him into an intensification of his relationships. A few days prior to his death, he engaged in a loving interchange with a famous interviewer on national television.

As was pointed out before, current mortality tables and modern medicine collapse the three signs into one—old age. Even one's dying moments are a sideshow to the last years of life now. Mary is physically alive but she is suffering the losses of death. Her house is empty and her children, though concerned, are all gone. What awaits her is an institution. She is experiencing social death. It may be the most painful part of the death the Buddha describes.

An irony of the interview is that while the interviewer is sympathetic, she does not respond to Mary's reaching out. Mary is asking for loved ones to be around to help in her extended dying. Her children are unavailable. They possibly have fractured families, small dwellings, their own growing children and busy lives. The structural arrangements of contemporary life provide only nursing homes and occasional sympathetic listeners. One wonders what the reporter felt as she left Mary's house. Perhaps it was, "I have so much already going on in my life, I can't take care of her too." And I, listening to the report, go back to my writing. That is contemporary "reality." The modern world prolongs and intensifies the discontent of aging; thinking, removed from the immediacy of day-to-day survival, turns to a macabre and trembling attention to death. If one wanted to be mired in the three classical conditions which generate suffering, then one could hardly do better than to be old now. Chronic disorders and the indignities of old age occupy a greater proportion of our lives than ever before in human history.

There is a voluminous literature on aging. The *Journal of Gerontology* is published in four separately bound sections: biology, medicine, psychology and sociology. A recent issue points out that the coming generation of elderly with fewer children, more broken families and less savings will have fewer caretakers. There will be increasing numbers of widows and older women having to fend for themselves. Research indicates that almost no family member cares for the fewer surviving elderly men who have not married or those whose families break up. Children care for their mothers but not their fathers. Although external social conditions add additional twists to the difficulty of life's situations, the Buddha lived in what we now consider to have been a more functionally integrated society. The family was intact and there were viable local institutions such as small villages and people living in extended families and local kin networks. Nonetheless, he was compelled to address the discontent coming from disease, old age and death. They were as real then as they are now. But without family or helping neighbors and a lingering dying process, the burdens accumulate.

Acceptance

I spend time with the elderly. Three of my friends just had hysterectomies, a close friend is wasted by chemotherapy, another is debilitated by AIDS and I have my own medical complaints. Nonetheless I have a hard time reading about the travails of disease, old age and death in the contemporary world. The Buddha proposed remedies for this discontent engendered by our bodies. He talked about what he called mindfulness of the mind and body. Leaving the topic of mind for the next chapter, let us look at some of his meditations on the body. The Buddha formulated them as a kind of reality check, so that human hopes and fears would not disconnect an individual from what the Buddha observed in his own meditation as the nature of life. These meditations may seem to focus only on the negative because they are meant to cure the discontents arising from "the second arrow"—our unwillingness to accept the inevitability of disease, old age and death.

The first of the meditations focuses on sensations as they occur in our bodies. Buddhist meditation starts with the breath and then extends to sensations wherever they are noticeable. In the next chapter, we will discuss the mechanics of meditation in more detail. The Sutra on Mindfulness is the most fundamental of all the Buddha's sayings. In it he recommends that practitioners cultivate mindfulness of the body while sitting, standing, lying, walking and sleeping. Sitting quietly while observing or moving about, one is instructed to pay close attention to how the body feels in all of its parts.

An example from my meditation practice will illustrate. Middle-age has made me more concerned with the tenuousness of my body than when in the bloom of youth. After years of practice, I have enough concentration to hold my attention on the parts of my body which worry me. One day while sitting and paying attention to transient sensations, I settled in on a pain in my foot or leg. As I watched it, for the first time I realized that the pain was not in my head, but actually down there in my limb. I had heretofore unknowingly experienced pain in my head. It really was not in my head, but felt in the hurting body part itself. The realization shifted something in me. Pain, for that moment, became less personal and the reaction of my mind to it—that is, the suffering that it caused—diminished. There was a feeling of relief. This pain became another body sensation.

The Sutra further instructs the observer to "contemplat[e] the body as body both internally and externally."[71] By this it means to hang out with whatever sensations occur until the meditator sees that whenever sensation arises, it is merely sensation, arrow one, and that arrow two, the mind's reaction to the sensation, is something extra. Pain isn't personal, although it may be a result of a foolish action. It is just pain. Such observation creates a kind of natural history of the body. Like Darwin's discontent when natural selection pressed in on his family, humans are often unhappy that pain or decay is happening to them. The Buddha's remedy is cool observation until the fact that the event did not pick you out personally is accepted. Natural Selection and Death care not a whit. They happen.

Eating meditation, as mentioned before, can be a part of formal body observation. While eating meditators put all their attention on the process: bringing food to the mouth, the feel of the food in the mouth, the details of chewing and swallowing, and the feel of the food as it is ingested into the body. This is done slowly with each morsel of food. This crucial part of life is brought into attention so that meditators become aware of how their bodies are individually connected to the world. The biological question raised by my friend Janet to assess the ecology of a natural setting is relevant here. What we eat is one side of the equation. We too are eaten, whether by microbes or, as an advertisement for calcium supplements asserts, "Osteoporosis, the disease which eats away at your bones as rust eats away at metal." Included in the Mindfulness Sutra is careful observation of the other end of the process, elimination of waste products. The Sutras are not shy about this. Zen monks were said to become enlightened while shitting. It is an extraordinary activity if you really attend to it, and crucial to our existence.

The Mindfulness Sutra also recommends contemplation of the various parts of the body. Here investigation is more akin to anatomy and physiology. From what they learned by doing body meditations, Buddhist monks in early India laid the foundations for Ayurvedic medicine, a tradition of treatment still widely practiced in Asia. "And again, a monk reflects precisely on this body itself, encased as it is in skin and full of various impurities... There is connection with this body hair of head...flesh, sinews, bones, marrow, kidneys...blood, seat, fat,...saliva, mucus...urine...."[72] In an era lacking clinical dissection, monks doing charnel ground and slaughtering place meditation observed the anatomy of animals or humans. Although hard to read for denizens of the modern world where blood and gore are mostly experienced in movies or on television, the Buddha gave a kind of poetry to guts and decay.

> Monks, it is as if a skilled cattle-butcher or his apprentice, having slaughtered a cow, might sit at the cross roads displaying its carcass.... It is as if a monk might see, thrown aside in a cemetery a body that had been dead for one day or for two days or for three days, swollen, discolored, decomposing; so he focuses on this body itself thinking, "this body too is of similar

> nature...." It is as if a monk might see thrown in a cemetery a body which was being devoured by crows or ravens or vultures or wild dogs or jackals or by various small creatures; so he focuses on this body itself...it is as if a monk might see thrown aside in a cemetery a body which was a skeleton but with [some] flesh and blood, sinew-bound...a skeleton which was fleshless but blood bespattered...a skeleton which was without flesh, sinew-bound; or the bones scattered here and there no longer held to together: here a bone of the hand, there a foot bone, here a leg bone, there a rib, here a hip bone, there a back-bone, here the skull...[then] a body the bones of which were white and something like sea-shell...a heap of dried up bones more than a year old...the bones gone rotten and reduced to powder....
>
> Even so monks, does a monk drench, saturate, permeate, suffuse this very body with the rapture and joy that are born of aloofness.... [Among the many benefits of this practice are]: He is one who overcomes dislike and liking...He is one who bears cold, heat, hunger, thirst, the touch of the gadfly, mosquito, wind and sun, creeping things, ways of speech that are irksome, unwelcome; he is of a character to bear bodily feelings which, arising, are painful, acute, sharp, shooting disagreeable miserable, deadly....[73]

In a society where the dead were thrown into charnel grounds near villages to be consumed by dogs and varmints and eventually rot away, a vivid picture of decomposition made clear the contents of a body and its inevitable destiny. The flesh and innards of an individual's life are the food of other life. As the ecologist Paul Sears put it two-and-half millennia later, "The face of the earth is a graveyard and so it has always been."[74] At the end of the process, our body dissolves into dust which blows away in the wind. The Sutra's reason for meditations on the foulness of the body is that by facing squarely the fact that everybody will experience the pain of disease, old age and death, the fear which the prospect and then the reality of this creates can be overcome. This directly addresses William Long's assessment of human anxiety and Mary's discontent. The Buddha claimed that a meditator could find a place of peace in the midst of these inevitable transformations.

In ancient Buddhism, the foul nature of the body was a constant theme. "The Poems of the Elder Monks and Nuns" are some of the earliest Buddhist texts we have. Kappa, the monk, says:

> Full of stains of different sorts, a great producer of excrement, like a stagnant pool, a great tumor, a great wound, full of pus and blood, immersed in privy, trickling with water, the body always oozes foully....the body is covered with ignorance...caught in the web of latent tendencies.... Joined with five hindrances, afflicted with thought, followed by root craving.... Those who thus avoid this body like a dung-smeared snake, having spurned the root of existence will be quenched without [defilements].

And Kulla:

> I Kulla going to the burial ground saw a woman cast away, thrown away in the cemetery, being eaten, full of worms. See the body, Kulla, diseased, impure, rotten, oozing, trickling, the delight of fools. Taking the doctrine as a mirror for the attainment of knowledge and insight, I consider this body empty inside and out.[75]

These sentiments may be more accessible to us in the words of a contemporary disabled meditator who has experienced the indignities of the body.

> In silent meditations and daily activities, I looked closer. I experienced the desire for control, the mental grabbing at plans or solutions to work everything out just so. Without undertaking the schemes my mind conceived, I stayed close to the fears always lying underneath. It arose, passed and arose again, changing texture and shape over minutes and months. Guided by the Buddha's clear perception, I saw how frightened I was of the

> vulnerability and impermanence accompanying my human body; this sack of sinew, bones, and blood came without a warranty. My final prognosis...like everyone else's...was certain. I could not dictate the time of death's arrival or the pains and frailties encountered before its Day. In touching this utter unpredictability, I let go.[76]

This practice is by no means easy.

Even the Buddha found the discomfort of degeneration a challenge. Aged and ill, there was some relief from pain through disciplined meditation, but death is inevitable.

> Ananda, I am old and worn out...I have reached the term of life which is eighty. Just as an old cart is made to go by being held together with straps, so [the Buddha's] body is kept going by being strapped up. It is only when the [the Buddha] withdraws his attention from outward signs, and by cessation of certain feelings, enters into the signless concentration of mind, that his body knows comfort.[77]

His final words were "all conditioned things are of a nature to decay...strive on untiringly." The fruits of that striving are serenity in the face of disease, old age and death.

Buddhist traditions boast many distinct meditations upon death and dying. Zen masters are known for the poems they write just before they die. There are stories of their sitting crossed legged calmly writing their death poem, and then simply expiring. Tibetan Buddhism is famous for the "Tibetan Book of the Dead" which includes Tibetan beliefs about the boundaries between life and death. One elaboration of meditations on the body comes from the 10th century Buddhist saint Atisha, who recommended that meditators keep the following facts in mind.

1. Everyone has to die.

2. The life span of an individual is always decreasing.

3. Death happens with or without meditation practice.
4. The point at which you die is uncertain.
5. There are many causes of death.
6. The human body is very fragile.
7. Wealth cannot help at your time of death.
8. Friends and family neither.
9. Even one's own body cannot help.[78]

Some of these contemplations point to the reality of death, others to its social circumstances. Framed in the context of mortality within pre-industrial society, dying is not thought of as following years of chronic disability. From Darwin's perspective, 1, 2, 4 and 5 are obvious but somewhat misleading about life processes as a whole. They don't say very much about the way in which death comes for the majority of people. By Atisha's time Buddhism was a world religion with millions of adherents and large monasteries and universities but societies were still agricultural with high rates of birth and infant mortality. Since neither mothers nor infants were part of the community of monks and nuns, the model of an adult dying of disease and old age was the center of focus. Number 6, pointing to the fragility of the body, is true for the individual; that is, the person who has the problem of dying. But it is not so true, as we have seen, for the species. As far as animals go, human beings are relatively tough. Numbers 7 and 8 speak of social circumstances. Wealth doesn't make any difference at the moment of death, but it may make a great deal of difference when it comes to a long senescence. It provides for care and medical assistance to lessen disability and pain. And absence of friends and family, as we saw in the case of Mary, is one of the most painful aspects of extended dying.

Finally for number 9, the physician we cited who described scenarios of medical death pointed out that the very end of life is a place where, although the body does not save a person, it often gives them peace in their last hour. The difficulty with respect to number 9 becomes painfully evident in industrial society with its advanced medicine. Many years before a final lucky few moments of serenity, each of us has to deal with the physical effects of living in ways which are out of balance with how the human body is adapted to function. When looked at soberly, the amount of chronic disability caused by our routine lifestyles means not only that "one's own body cannot help," but that medicine, which promises to make things right, is limited. Conversations focused on current ills are no longer confined to old people. Thus, while civilization has conferred great material benefits, it has left its citizens living with bits of death's reality throughout their lives.

There is fertile ground in Mary's and all our lives for the Buddha and Attisha's practices. They are particularly aimed at the fears to which our bodily changes give rise. From the perspective of the Sutra on Mindfulness, these are natural and going to happen to everyone in one form or another. What is not dealt with in the Sutra is the ways they might occur. According to statistics, for one quarter of Mary's life her old age will give her the opportunity to apply the Buddha's meditation and Attisha's contemplations. Mary is not alone in this. A growing proportion of the population in modern society faces the same prospect. While Buddhist practice encourages confronting directly the decay of the flesh, the civilized world spends immense resources trying to stave off the effects of Gaia with her sword, natural selection, or keeping the effects hidden in hospitals or nursing homes. Moreover, most of the society's symbols direct attention as far away from the charnel ground as possible. They glorify youth, sexuality and vigorous consumption—the very things Darwin held as a saving grace of the devastation wrought by natural selection. They also romanticize aging, speaking of the golden years. Some elderly satirize this. The aged mother of a friend once said, "I will tell you what I think of the golden years," and then she made the sound popularly known as a Bronx cheer.[79] And another elderly friend when sick and alone remarked, "What is the point?" From the Sutra's perspective, the popular images encourage greater delusion about the real character of life and death. If chronic conditions, disability and old age are inventions of modern civilization, the discontent they give rise to represents a pressing challenge we have hardly begun to meet.

To engage this situation, the Buddha set a task of careful observation. That observation creates a natural history of one's body. This personal natural history is a kind of ongoing Conversation of Death. Death no longer strikes directly, allowing only minutes for an exchange of messages. "I have lived a full life. My time has come. I die so my species may live full and vigorous lives." The way that nature is currently expressed in industrial civilization relegates Death to the shadows where we hardly acknowledge it until it takes over so much of our lives that we find ourselves in an interminable, nagging argument. The Buddha's recommendations are intended to bring our natural history into the open. We are to look squarely at Death; the Conversation of Death becomes direct. This may take the edge off the increased discontent that civilization creates by extending the process of dying.

Whether or not a person is able to engage in that observation and act upon what is seen has a lot to do with how the human mind works. It is to that realm we turn next. The mind, as much as the body, is a creation of nature. What can we understand about its natural roots? How does the natural history of mind relate to meditation? And finally, given that the body had a particular set of problems to contend with as humans went from hunter-gatherers to civilized moderns, how has that transition affected our minds?

Chapter Five
Evolution, Mind and Meditation

The human mind is a product of evolution. Meditation adds a new dimension to evolutionary biologists' and neurophysiologists' descriptions of the relationship between the mind and the brain. Since there is an overlap between these scientists' observations and what meditators have known for years, we will examine how each perspective illuminates the other. How is discontent rooted in our brains and minds, and by what biological processes does meditation address it?

Over the years I've been practicing meditation, there were periods when I did so intensively. During much of the fall of 1990 I was in solo silent retreat on an island in the Puget Sound, spending most daylight hours doing sitting or walking meditation out of doors. Then in the winter, while teaching university and living near the Insight Meditation Society (IMS), I spent three days a week working and the other four meditating at IMS. This went on into the spring. During one calm and peaceful sitting in the meditation hall, it seemed like something was slowly taking possession of me. As I was meditating, I felt my face twist into that of a growling demon from a Tibetan Buddhist painting. Eventually the muscles in my face relaxed. I found the incident amusing. Since I have had many odd sensations while meditating, I did not think it was important. As I continued to sit that winter, the mask occasionally reappeared.

Then one spring evening during a therapeutic session with friends who act as guides for each other, I went to the bathroom and glanced in the mirror as I passed it. Instead of my face, I saw the demon. Like a frightened child, I scampered back to my guide and said, "There's a ghost in the bathroom!" She took me back into the bathroom and wisely suggested that I get to know what I was seeing in the mirror. Thus began a terrifying and fascinating adventure which lasted well into the night. As I stared, the reflection melted into a second and then a third being. I knew that the images were somehow part of me and yet more than my particular personality. I don't remember much of what went through my mind. I had thoughts such as, "Where does this come from?" and, "This is scary!" I occasionally came up with explanations like, "It's just my past lives," or, "This is my rage." But most of the hours in front of the mirror were without mental chatter. I just silently looked at the images.

There are many ways one can explain such phenomena. They can be seen as imagination, hidden parts of the psyche or esoteric metaphysical states. Whatever the psychological meaning, relevant here is the final image that appeared. Some time in the early morning, I saw a simian-looking creature facing me. It was me but not me. It looked like an early hominid, standing somewhere between ape and human. It was muscular with an intelligent face. It did not have language as I knew language, but understood much about existence, which it comprehended without words. I was in awe of its intimate relation with reality, seemingly not mediated by thought. The being had great dignity.

Two things came to mind. One is the phrase often used by the Korean Zen master, Soen sa Nim: "before-thinking-mind." This is the understanding of existence which comes without thought or prior to thought becoming prominent. It is an important kind of awareness which meditation practice aims to develop. Soen sa Nim cajoled students to "Only use before-thinking-mind!" It seemed to me that the mirror creature existed in an era before the thinking part of the human mind had completely evolved, and that it possessed this consciousness of "before-thinking-mind." My second thought was that the being represented an ancestor from an evolutionary past and that I was connecting with shared aspects of my own makeup. Somewhere in my brain and body, it seemed, was the ability to experience the world as our ancestors did before thought and social life as we know them had evolved.

I was profoundly moved by the experience. I felt I had touched some ancient part of my humanity buried under a chattering knowledge-filled mind influenced by a civilization steeped in complex technology. The image was a reminder of the dignity and directness which so often evades me as I struggle with my problems and the world around me. I am grateful for the experience.

There are a number of possible explanations; the most obvious is a psychological interpretation of the first image. Tibetan Buddhism is replete with pictures of Wrathful Deities and saint-like beings. These representations play complex roles in Tibetan beliefs and meditation practice. The Tibetan practice I've done has not included the use of these representations but the images do resonate. One day, several years before my mirror experience, I was driving back to Santa Fe from a frustrating trip to the airport. I was trapped in one of those no-win, irrational, bureaucratic disputes with an airline. In order to change a ticket before a certain date, the airline insisted that I drive a hundred miles to the airport to speak to an agent. They hadn't told me about the airport's restricted hours so when I got there the airport was closed. Talking with them on the phone only made matters worse. I was completely out of sorts when I noticed a Tibetan Stupa or shrine among the used car lots and junkyards on the road into town. I stopped and entered the low door into the shrine room. I thought that it would be a good place to meditate until I calmed down. I settled on to the floor and sat for a few minutes, but I was too restless. My mind was bouncing all over the place. I opened my eyes and noticed the pictures painted on the walls. Among them were different wrathful deities. I stood up and walked around the room looking at each. When I got to a red painted deity, I stopped. There was my state of mind vividly displayed. While looking at it, I gave a sigh of relief. My anger abated.

The other image in my mirror experience, the early hominid, can be elucidated in a number of interrelated ways. One is found in the research of neurophysiologists, psychologists and anthropologists who try to understand the nature of consciousness and how it has evolved to its present state. These scientists use their own experiences and match them with 1) sophisticated scientific probings into the workings of the brain, 2) an increasingly detailed reconstruction of ancient human anatomies and 3) clever psychological testing of contemporary emotional and intellectual responses. They study the evolution of the brain and how it relates to the working of the mind. My mirror experience seems to illuminate the relationships between speech, thought and nonverbal knowing that these scientists are researching.

A second way to explore the mirror image is found in the work of ethnologists, biologists and historians who study indigenous peoples. With proper appreciation of the differences, the life of recent hunter-gatherers may be the closest we can get to learning how our human ancestors experienced their world. Being much closer to nature than we are and necessarily more reliant on gut responses, hunter-gatherers openly use parts of the mind that rely less on speech and thought. Before-thinking-mind may have been more directly accessible to them than it is to us. So the study of hunter-gatherers reveals something about the character of the human mind. In addition, my vision would be familiar to many tribal people; for example to shamans who ritually navigate between a hunter-gatherer's experience of nature and the ideas they have of the unseen forces which hold their world together. The inarticulate and the world of visions was and is often more important to them than to us.

Finally, meditation is concerned with the relationship between the mind of thoughts and the mind of silence. For years, meditators have explored how their minds work and have done this without creating more mental activity. Meditation is no stranger to my mirror vision. Such things are not often talked about and sometimes regarded as irrelevant or even dangerous. Nonetheless, there are some meditation teachers who know a lot about the strange things minds do. After my experience of feeling possessed, I consulted teachers I thought might be able to give me guidance. When asked about demons, one teacher who had studied all the major Buddhist traditions, answered me with a kind of fearfulness in his voice, that he had heard of exorcisms but knew little about them. Eventually I went to see a renowned Tibetan Lama. As I was relating my experiences, his translator became more and more agitated, finally shouting at me, "Who taught you such evil practices?" When he calmed down I was able to explain that it wasn't something I was purposely doing, it was just happening. The Lama turned to me and said that if it happened again, I was to shout, "Go away demon," which he bellowed out.

The Origin of Mind

We have seen that nature endowed humanity with a strong and durable body. Evolving over millions of years, this body demonstrated an extraordinary ability to survive in a number of different environments. Parallel to this physical evolution has been a series of transformations of the hominid brain, endowing it with an even more extraordinary ability to use the body to manipulate the physical environment and to create social life as a means of controlling nature. For Darwin, human consciousness was a product of evolution, governed by the same laws that apply to a beak or a wing. It differed from animal consciousness only in the matter of degree.

Scientific reconstruction of the evolution of the brain and mental functions is mostly speculative. It is based on brain size, skull shape and archeological evidence of behavior. About five million years BP (before the present), chimpanzee and hominid lines split. As noted, around four million years BP, our australopithecine ancestors began walking erect, lived in nuclear families which shared food, and had a sexual division of labor with smaller numbers of children dependent on their parents for longer periods of time. Perhaps this was the point when we began to behave differently than other animals. By two million years BP, *Homo habilis* had a larger brain and used crude stone tools. *Homo habilis* may have engaged in cooperative living and group hunting. Half a million years later, *Homo erectus* appeared. Its brain was still larger and its tools more elaborate and better finished. It made hand axes and choppers. Groups of *Homo erectus* migrated seasonally from one source of food to another, setting up base camps from which to hunt and gather. The use of fire spread. In the next million years, the brain size increased and skin shelters, clothing and cooking implements came into use.

About three hundred thousand years BP, *Homo neanderthalensis* appeared with another increase in brain size and with a vocal tract closer to what modern humans possess. About fifty thousand years BP, the immediate predecessors of truly modern humans are found and, although brain size does not increase, their ability to manipulate the environment and engage in complex social activities accelerates dramatically. Neanderthals who dominated northern latitudes until thirty thousand years BP, formed societies, produced ceremonies and engaged in warfare.

These various hominids did not follow a direct line of descent. Many lineages died out, and who is whose ancestor is not always clear. One of the great puzzles of archeology is the relationship between Neanderthals and *Homo sapiens.* They are usually thought to be a distinct species that had a common ancestor. What is clear is that as *Homo sapiens* spread out from the Near East, Neanderthals became extinct about thirty thousand years BP. Scientists speculate the Neanderthal extinction may have come from competitive pressures, disease or interbreeding.

Although Neanderthals were stronger and had larger brains than *Homo sapiens*, their technical and cultural skills were less sophisticated and less effective. Neanderthals left none of the elaborate cave paintings and pictograms that modern humans did, nor did their culture change as rapidly. Both these omissions are taken to be indications that the Neanderthals were less intelligent than *Homo sapiens.* Some paintings are found in such inaccessible recesses that intricate rituals may have been connected to their creation. Their creators had to squeeze through narrow passages in the dark for long distances in order to paint the pictures in a location where it would have been very difficult to carry on the activities of daily life.[1] By thirty thousand BP what we think of as the modern human mind was thoroughly in place.

Even though we have some physical evidence, we do not have crucial ingredients for knowing what the minds of our ancestors were like. One would love to be able to watch various hominids going about their daily activities or to find records of their exchange of information which might give us insight into what their activities meant to them. Without this, all we can do is infer their mental capacities from their anatomies, the objects of their cultures, the thinking abilities of animals and from analogies with modern hunter-gatherers.

Nonverbal Communication

We saw in Chapter 3 that some animals have expressive emotional repertoires which we can understand. They exhibit fear, pleasure, aggression, greed, sadness and play. Not only do animals communicate emotions but individual animals have personalities that may not contribute to survival. In contrast to our image of the noble, wild hunter are the cowardly wolves who hang back during a kill until the contest is decided. Afterwards the braver hunters allow them to eat. Not all of the squirrels who scavenged the compost buckets ate excessively, although there was certainly enough garbage for all. One day on a walk, I came across a young badger who seemed confused. It did not scamper off, but alternated between hissing aggressively and going into submission. The hissing was typical behavior but his submission seemed off. I thought it might be a dispossessed teenager, driven from home and not yet fully attuned to danger. Its chances of survival were probably not great. Animals have feelings, make thought-like discriminations and seem to be able to communicate both to others. Still it is not clear that they think as we understand it.

Animals also utilize tools to do things they otherwise could not have done. Chimps and birds use sticks to extract grubs and insects. Having learned how to navigate a maze, animals can teach others to do so. They can give directions as in the famous dance of bees. They communicate subtly. Vervet monkeys and whales signal, making refined distinctions. Animals assess the capabilities of opponents. Red deer stags competing for mates roar at each other to see who gets worn out first before the challenger engages in a fight. White-throated sparrows recognize the voices of neighbors in their right place and only challenge wandering males and territorial males who are in the wrong place.

Animals remember where they've hidden food. Rats learn to be wary of food on the breath of friends who become ill, and teach their children about the safety of food which has affected kin the children never knew. Vampire bats who have eaten, share the blood with those who haven't found a victim. They feed their kin first and then known others who fed them in the past. They do not feed strangers.[2] Some animals engage in a chain of acts without having to remember the specific stages. Responses to friends and foes in songbirds are of this nature. Finding hidden food is more complex, requiring the stringing together of individual episodes. Scrub jays even seem to understand dissembling. Some jays watch other jays bury their food and then come back later and steal it. When they in turn bury their food in the presence of another jay, they quickly return after the other jay has left and rebury it. Non-robbing jays never rebury their food.[3]

Chimps and apes, our nearest relatives, have a more extensive repertoire of thought-like behavior. Apes exhibit a capacity for complex internal representation. They seem to understand the relationships among a sequence of present situations which they can string together. Chimps can remember a large number of relationships. They engage in situational analysis but cannot re-present situations; that is, teach by means other than demonstration. For all of the marvelous behaviors we can teach chimps and apes—such as forming some sentences, expressing their feelings and innovating new grammatical forms—their overall vocabularies remain small and they are unable to teach each other how to do these behaviors.[4]

The gap between the mental abilities of humans and their animal relatives is so great that a co-author of the theory of evolution by natural selection, Alfred Wallace, among others, felt that the human mind could not have developed merely from the workings of evolution. Viewing evolution as gradual, he thought natural selection could have only given rise to a human brain a little better than that of apes. In contrast, Darwin and later Darwinians felt that human intelligence fits more continuously into the animal world. These scientists also locate a contributory series of evolutionary developments involving the thumb, voice box and brain. Several million years ago hominids became right-handed. This is associated with brain lateralization, i.e. different functions being controlled by different hemispheres. Improvements in tool making and increase in social life appear related to this change in the brain.

A theory of the evolution of mind is presented by Merlin Donald in his *Origins of the Modern Mind*. He proposes four stages:

1) procedural mind as in animal instincts, 2) episodic mentality, including stringing together instinctual responses, 3) mimetic mind and 4) semantic mind. The latter two, Donald argues, are unique to human evolution. In his presentation, Donald draws heavily on Philip Lieberman's studies of hominid vocal anatomy and Darwin's theory of vestigial function.[5] Lieberman traces the modern larynx to its beginning about two hundred and fifty thousand years BP. He argues that the anatomy of Neanderthals' voices limited the rapidity and complexity with which they could communicate. Neanderthals probably spoke more slowly and probably could not make the range of sounds which would enable them to communicate as clearly or rapidly as *Homo sapiens.* The latter could say, "'There are two lions, one behind the rock and the other in the ravine' in the same time as [Neanderthals could say] 'lliiooonn rooockkk.'"[6] Neanderthals had neither fully encoded speech nor the neural apparatus it requires. This would have been a severe selective disadvantage. Neanderthals wouldn't have been capable of the degree of abstraction of *Homo sapiens* nor could they have engaged in the coordination that underlay the profound rapid changes our immediate ancestors evidenced in their complex culture. Following Donald, I assume that speech, mental abilities and brain capability are interwoven, and that the mind of modern humans along with speech as we know it is evolutionarily recent.

Before *Homo sapiens*, some hominids possessed culture in the form of games, rituals and collective activities which distinguished them from apes. They must have had cognitive skills intermediate between the two. Darwin drew evolutionary inferences by looking for surviving vestiges of earlier adaptations which don't have significant function in later species. An example is the human appendix. For Donald, "human emotional expression" is a cognitive vestige. Our expressions have roots in the expressions of our ape ancestors. Ape expressions are programmed and don't have much variety. Hair rising at the back of the neck when frightened is a good case in point. Later we will look at the part of the brain, the amygdala, which processes these primitive uncontrollable responses. For Donald, australopithecines were more ape-like whereas *Homo erectus*' mind was mimetic but not yet fully human.[7] Evidence for this is indirect. *Erectus*' tools imply development far beyond that of apes. Since these tools did not progress much over the million years in which they were in existence, it is unlikely *erectus* had the symbolic language necessary for the more rapid change typical of modern human populations.

It is important to understand the nature of mimetic thought: it is genuine thinking, but it is not linguistic. The impressive range of behavior of prelinguistic children, illiterate deaf mutes studied extensively in the 19th century, and Brother John—a monk, who had epileptic fits which left him functional but nonverbal—show there is more to consciousness than the verbal. Arguing from Darwin's notion that vestiges are indicators of precursors, Donald believes mimetic mind may be seen by examining the preverbal capacities of human thought—these have, presumably, been passed down to us from our evolutionary ancestors. "The cognitive style of people stripped of language is familiar to us," Donald writes, "because a significant part of normal human culture functions without much involvement of symbolic language."[8] Examples are found in trades, crafts, games, athletics, arts, theater and ritual. Early jazz was, and ancient crafts such as pottery and weaving still are often taught and learned without language.

When I worked as a mechanic, I was impressed by some of my almost illiterate shopmates who could do things with their hands that I could not and moreover had no language to convey how they did what they did. When the car or truck manual failed me, I would go to them for help. I learned not so much by instruction as by unthinking imitation. I can not explain how they did what they did with their hands, but now that my eyesight is getting worse, I find myself working on cars by touch as they did. Years ago, when my brother and I built log cabins in the wilderness of Northern Canada, we would work silently with ax and Swede saw for hours. Tools would be offered just when needed, and we would proceed to the next step without speaking as if the work were a dance. This immersion in the realm of mimesis was some of the most satisfying work I have done in my life.

Donald argues that mimesis not only sets the stage for speech but is essential to it.[9] Mimesis is representational. Its foundation may be well-developed eye-hand coordination. Without well developed language, *erectus* improved tool making and fabricated them distant from where they were used. This was not easy. Archeologists take months to learn how to find appropriate materials and make *erectus'* simple, chipped tools. When given stone-aged axes, modern lumberjacks cannot cut down trees with them.[10] They break the axes. These activities require skill not determined by immediate trial and error. They were taught. It may be that *Homo erectus'* anatomical changes allowed for more expressive faces and that, combined with improved vocal anatomy, led to richer communications.

Human expressions are much more varied and subject to conscious control than are those of apes. The expressions of squirrel monkeys are emotional both in tone and content, but it is like the babbling we use to satirize them, a bit out of control. Phonetic speech is different and uses more recently evolved areas of the brain. *Erectus* may have gone from primate prosody to the use of expressions in representation. From Donald we get a picture of a nonverbal society acting out through mime. They gave directions for the chase, expressed conflicts and resolved them. They showed love and sadness and taught tool making. *Homo erectus* was still anatomically incapable of language. Australopithecines had differed from their primate predecessors in terms of female reproduction, diet, hand and foot use, brain size and emotional behavior. To this *Erectus* added communicative and neural complexity. *Erectus'* mimesis stands midway, as it were, between an involuntary alarm call with a look in a given direction and a "by the way old chap, that lion on the rock above is looking at you hungrily." The mimetic building block was automatic emotional expression which it transformed and used in voluntary representational communication. Mimetic communication of a lion threat might include a toned down alarm signal, a facial expression as to the degree of the threat, and the use of sound, hands, eyes and face to indicate not only the nature of the threat but where the danger lay and an avenue of mutually assisted escape.

As suggested by Donald, we can find other evidence for the preverbal primitiveness of expression and mimetic representation in our own experiences. The human smile is half automatic behavior and half voluntary. In the past, I felt that smiles displayed in some peer therapy processes were phony. I could not put my finger on why until I learned that there is voluntary control over the lower half of the face in smiling but not over the upper.[11] Actors have to work very hard to learn how to smile authentically. Many achieve it only by the Stanislavski method of conjuring up happy thoughts. Then their smile is real.

We can explore further what it is that we know without words. The Indian guru Bhagwan Shree Rajneesh taught a mirror meditation from which one gets some sense of the depths from which expression directly communicates. Set up a mirror in a dark, candle-lit room and look directly at your face in the flickering light; look at your face without blinking for as long as you can. With a little practice you can stare without blinking for an hour. Look especially in your eyes.[12] What people experience staring in the mirror is that the appearance of their face changes, sometimes radically. What you see, you cannot quite classify: fear, anger, joy, and other emotions. But there is something else for which words are not adequate. Some find it difficult to continue looking. Others become frightened. I caution students who want to try this exercise to stop if they feel like they are in over their heads. It is important not to blink, because when you blink you fall back into thinking rather than just looking. Rajneesh claims that if you stare long enough, your head will disappear. What is seen in the mirror illustrates some of the shared content of life that gets communicated without words.

This can also be seen in others' faces. Strange how little we consciously notice when we look at another person. Identifications in courts are notoriously unreliable. We treat communication as if it were mostly verbal although even the cadence of sentence structure comes from the separation calls of young mammals.[13] Nonverbal communications frame many of our interactions. In my notes I have the following: "Dinner with six-year-old Elizabeth this evening where, when I really looked at her face, I could not figure out what was the meaning of the acting out she had been engaged in. She was demanding, noisy, and did not answer me when I asked if she had had a rough day. But when I said I had had a rough day, her expressions changed and she calmed down for a moment." My large reservoir of psychological concepts did not possess words for what I had seen in her face. Her category-defying expressions seemed to come from the vestigial parts of my brain "seeing" what I had no language to name. I knew something quite real from a more ancient place.

I have noticed that when I drop into a space of meditation while listening to my students in my office, a door opens in my mind and I see their faces as I saw little Elizabeth's. I sometimes see them as ancient preverbal beings. Often when I open up that way, the students become calm, although speech may continue as if on automatic pilot. The talking no longer seems very important. Communication is taking place elsewhere. In such moments, I have seen extraordinary things in my students' faces —great heroism, suffering, the dream body image of an abused student or the death scene of a baby brother. I also see qualities to which I cannot put words. I can subvert this by again becoming the "professor." As in my mirror vision, these experiences fall into the realm of before-thinking-mind. I have often heard Zen Master Soen sa Nim say, "In thinking mind you and I, different. In before-thinking-mind, no difference." It may be that the hominid in my mirror vision was symbolic of the *Homo erectus* in me, and at their proto-human best they communicated with before-thinking-mind.

Speech, Consciousness and Thought

For Donald, narrative is the next step beyond mimesis: "Narrative is so fundamental that it appears to have been fully developed, at least in its pattern of daily use, in the Upper Paleolithic. A gathering of modern post-industrial Westerners around the family table, exchanging anecdotes and accounts of recent events does not look much different from a similar gathering in a Stone Age setting. Talk flows freely, almost entirely in the narrative mode. Stories are told and disputed; and a collective version of recent events is gradually hammered out as the meal progresses. The narrative mode is basic, perhaps the basic product of language."[14]

Language depends on a number of physiological changes in the brain. The incredible facility with which children acquire language and the large amount of space in the brain that is used for speech are evidence of its importance in evolution. Lieberman regards the vocal apparatus and its related areas of the brain as a self-contained, specialized adaptation, although other scientists disagree. Speech in infants develops before the cortex is mature. Children's ability to learn language diminishes as they grow. They can make all the sounds of all the languages then lose those they do not use. Mimetic communication uses other parts of the brain.[15]

Before considering human's ability to think and its ramifications, we need to touch on the thorny issue of human consciousness. The question of consciousness has plagued religion, philosophy, psychology and science for a long time. The earliest reference I have found to the difference between animal awareness and human consciousness is in a translation of the pre-Buddhist *Aranyakas*, or forest books, of the Brahmins in India twenty-five-hundred years ago. These forest ascetics transformed ancient *Vedic* cattle sacrifices into jungle fire rituals and tried to understand what it is to be human rather than simply appeasing angry gods. The authors of the *Aranyakas* felt that animals did not understand the meanings of their acts as humans do. For humans, rituals were deemed necessary to purify behavior. Since animals were not aware of the meaning of their acts, they could neither correct nor redeem their behavior. Self-consciousness was thought to be uniquely human. When it is unbalanced, it troubles its owner and, for the Brahmins, it is also out of accord with the natural order.

We have seen that animals have various repertoires of emotions that are shared by humans. We have also seen that some animals display what we can recognize as discontent. They seem disappointed when denied pleasure; they can be traumatized by fear; they mourn and get depressed. They even die of sadness. That animals suffer is undeniable, but the character of that suffering in their consciousness is not clear to us.[16]

In the view of Darwinian adaptationists all traits are the result of variation and natural selection. Even human consciousness is the product of these processes. But it is difficult to pin down what consciousness is. As a biologist who defends animal rights puts it, "... whether we see consciousness as so unique and mysterious that it will forever be beyond the reach of science, or as a temporarily baffling phenomenon that will one day be brought into science, conscious awareness is one of the biggest problems that our conscious minds can allow themselves to contemplate. At present, nobody understands it."[17]

Putting aside the perennial questions, "What is consciousness and how can we know what it is?" we can don adaptationist robes and ask how consciousness might influence hominid selection. A list of useful functions includes a) connecting actions to consequences, b) selecting similar aspects of different ongoing events, c) keeping steps in order in complex learning, d) correcting for inappropriate automatic responses, d) using memory for long-range planning, e) imagining possible outcomes, f) comparing outcomes to past experience, g) reorganizing memories and plans and h) making speech possible.[18] Animals possess various aspects of these but, taken together, they confer tremendous selective advantages on humans. They enable us to theorize about, plan and test actions. Human power over the present world is based on these.

But there are downsides. The mind that can accomplish all these things for survival also attaches emotions to free-time modeling, cannot stop comparing outcomes, engages in endless planning beyond any reality of possible control, battles continuously to change automatic reactions and talks to itself incessantly. As one of my meditation students asked when I offhandedly said that the busy mind probably came in the evolution of humans along with the development of speech, "Why would nature design a feature that so tortured its possessor?" The student felt that to make something with so many liabilities was regressive evolution, if not the punishment of a cruel god. Being caught off guard, I did not really answer the question, but now I would use the concept antagonistic pleiotropy. This is a technical way of saying that nature is not so efficient in its design endeavors; its useful features can also have adverse aspects.

Rajneesh liked to point out that, "Nothing fails like success." Anthropologists, psychologists and neurophysiologists have noticed that the successes and failures of consciousness are intertwined. For some the mind evolved in a simple set of circumstances, but the features it came up with were too complex for the situations it faced.[19] Evolution overdid its job. The very success of the mind in terms of fitness is now our liability, in part because communication evolved out of emotion. While the distress cries of monkeys efficiently tie emotion to action, a flood of human feelings may not contribute to humane solving of complex technical or social problems.

For animal bodies, we have seen that traits which enhance survival in one circumstance have features which are liabilities in another. Humans, for example, have paid a high physiological price for speech: unlike most primitive air-breathing animals which swim, our larynx no longer keeps water out because structures for speech-making have become dominant.[20] We are also more likely to choke when eating or drinking than other terrestrial mammals, and we can't chew as efficiently as they. Our teeth have become so crowded that we are subject to infections from impacted wisdom teeth, often fatal until the invention of modern anesthesia in 19th century dentistry and of antibiotics in the 1940s. The only thing the vocal tract of humans is better suited to do than other animals is to speak (and sing)! In fact newborns, whose risk of choking is great, have vocal tracts similar to non-human animals because at their age not choking is more important than making complex sounds. This antagonistic pleiotropy is outweighed by the incredible signaling power of speech. Fifty sounds can be spoken and easily understood in two seconds. At five sounds per second, one-fifth the rate of speech, Morse code operators are so busy transcribing that they cannot understand the message. The brain can decode speech even more rapidly, and think so frenetically that we sometimes pray it would stop.

For a period of time in the 20th century, behavioral psychologists equated inner speech with thought and looked for such physical evidence as subvocalization, a non-noisy moving of the vocal chords while thinking. When, as a graduate student, one behavioral psychologist first heard the theory of inner speech, he felt, "Oh, thinking is nothing but talking to yourself."[21] As I watch my mind chatter while sitting silently in the meditation hall, I too have tried to discern whether or not my vocal chords were subtly moving as I talked with myself in my mind. I never felt anything when I brought my attention to my vocal apparatus. After several days of silent meditation, I am often quite hoarse and have difficulty speaking. I wonder whether this was the result of straining my vocal chords from the nonstop chatter of my mind.

Research shows my meditation observations were accurate but my speculations off base. There is no correlation between inner speech and subvocalization. Inner speech either involves muscles not used for speech or no muscles at all. Mentally talking to oneself is faster than talking out loud, and is sometimes difficult to control. Some researchers think the rapidity comes from deleting parts of words during inner speech. Inner speech is mostly monotone and sexless. Evidence that there is a separate speaker and listener inside your head when you are talking to yourself (so the listener can self-correct the speaker) is highly questionable. Yet when people mentally talk to themselves, the brain centers for vocal activity and analyzing speech are working, but not the brain areas that stimulate the vocal cords.[22]

If you closely observe the inner dialogue during meditation, it does feel like you are talking with someone. But looking even closer, that someone is you. "Should I continue sitting here in the sun typing?" "Well if you move, your hosts will notice." "I guess I will stay, but the sun has gone behind the clouds and it is getting cooler." The inner dialogue is really a monologue with the thinker renaming the voice in accord with grammar. Rather than "I am talking to myself," a person is simply thinking. One thought follows another. There is merely the appearance of dialogue. It may be that because children hone their biological speech capabilities through imitating and being corrected by adults that this model is embedded in our minds. We habitually recreate our mentors. They are either given our own voice when talking to ourselves or given the voice of others.

We know that the brain is not fully developed at birth and that interaction with parents and others is crucial to complete its growth. When children are not given the appropriate feedback, they suffer developmental disabilities. Children who are speech-deprived have great difficulties communicating. So-called feral children, recovered from the wild, never learn to speak properly. People whose internal other is flawed have trouble interacting normally. Such abnormalities can be created in experimental animals, but since human social life is so much more important, the damage to effective living is greater.

In humans, there is an obvious interrelationship between thinking and speaking, although their evolution may not be identical. In addition to the verbal, we have thoughts which are abstract, visual or tactile. Years ago, while driving through the endless expanses of northern British Columbia with my artist friend, Janet, I was chattering away to myself in my head and I turned to Janet and asked what she was thinking. Her response stopped me. She said, "Oh nothing, I am just seeing colors and textures." Thinking wordlessly in images and color also evolved, but its location in the brain is different from speech.

Meditation

How is the brain put together and how does that relate to the structure of the mind? Neither scientists nor meditators can more than guess at the neurophysiology of the brain, the structure of mind or how they are linked. Buddhist meditation analyzes human experience into five aggregates. These are lumps or heaps of stuff, so-called *skandhas*. The five aggregates or *skandhas* are: form or body, feeling, perception, mind and consciousness. This analysis comes from experiences that meditators have while they are meditating. The Buddha observed the aggregates during his own meditation and recommended that others investigate the accuracy of his description.

There are many different kinds of meditation. The technique recommended in the core Buddhist text, "The Greater Sutra on the Foundations of Mindfulness," is quite simple: to bring one's attention continuously to one or another of the aspects of experience.[23] Although the instruction is straight-forward, it is not easy to figure out how to do what Buddha describes. The breath and the body are the most ancient anchors for attention. In the pre-Buddhist practices of the Brahmins and shamans, the breath was often seen as the symbol of life and the measure of its quality. By paying close attention to the physical sensations of the breath in some part of the body, yoga practitioners were able to concentrate their attention and their energy. This concentration was used to achieve extraordinary physical control, like slowing the breath, standing for days, not sleeping or reversing peristalsis. The Buddha was less interested in extraordinary bodily control than in using concentration to look at life and its processes. Careful attention to the body is the first foundation of mindfulness practice.

As those who try this practice know, keeping one's attention on a changing point of sensation —like the breath as it passes in and out of the nostrils —is no easy task. It is no accident that the mind has been compared to a chattering monkey, swinging nervously from branch to branch, grabbing a leaf or berry here, batting a fly there, all the while babbling away. When you try to pay attention to sensations, the mind stays put for a few seconds, and then meditators find themselves far away, thinking about this or that. At the beginning of practice, it might take five or ten or even twenty minutes before you remember the task of attending you had set for yourself. When you remember, it is similar to waking up. "OOPS, I forgot what I was doing. I had better go back to looking at my breath." It takes effort to bring your attention back to the sensations as they occur at a chosen part of your body.

A second component of meditation is found in the name of the Sutra; that is, mindfulness. What is meant by this is a little tricky to grasp without actually doing the practice, because mindfulness seems to defy our common-sense ideas of how we know what we know. Once the mind settles down a bit and can actually pay attention for a while, then one begins to observe what the breath is like: hot-cold, regular-irregular, fast-slow, shallow-deep. A concentrated attention can be focused on different experiences, other sensations or perceptions such as the registering in the mind of simple sensation. The feel of breath flowing by a point is sensation. Noting the sensation is hot or cold is a perception. "Boy my nose itches. I wonder when this meditation session is going to end?" is part of what the Buddha called mind. Attention can be pointed at each part of this process. Instead of the nostrils, one can attend to the coming and going of thought.

If you have never meditated, it might be a little puzzling to understand how you can watch these things without adding more thought. It seems circular. When one notices one's mind chattering away, it is like the fox guarding the chicken coop. "Oh, I'm thinking." But that is another thought. Here we come to one of the aspects of Buddhist meditation which is hard to grasp. It penetrates mental dialogue. The easiest way I have found to describe it is in its simplest instance, the breath. A meditator can become so completely attentive to the breath that nothing else exists. There is simply in and out, felt in great detail, no mind saying, "Oh, I am calm now watching my breath;" merely breathing occupies one's entire being. While this is difficult to describe conceptually, it has, nonetheless, been experienced by meditators for thousands of years. When the mind becomes somewhat concentrated and thinking is not in the center of the stage, a meditator can actually observe the comings and goings of sensation, perception and thought without creating lots of additional mind chatter. This is sometimes called "bare attention." With bare attention, a meditator can observe how experience presents itself. It is as if there is a kind of knowing in the mind that is independent of thought, possibly the "before-thinking-mind" mentioned earlier.

In the above paragraphs we have been using Buddhist and psychological terms rather loosely. Many of us, if asked, would categorize our experiences in the language of modern psychology. There are external physical stimuli. There is raw sensation. There is the cognition of sensation. Then there is thinking about experiences.[24] These are the categories of mind I tend to use when I file what I experience, and I use these categories when I talk with others about it.

An example illustrates these divisions. As a professor, I have attended hundreds of scholarly presentations and university meetings. As a meditator, I have listened to as many spiritual teachings. After years of being gripped by the content of the talk, either agreeing, disagreeing or being bored, I began to meditate by concentrating my attention on sound. I carefully listen to each word as it is spoken. From time to time, my mind makes sense of what is being said, but I don't pay much attention to it. I recognize that there are words and voices but I let go of thinking about what is being said. I restrain internal comments. At the end of talks, I have a keen sense of the sound of the speaker's voice and the feeling which it carries, but I often cannot recount the content. Such a strategy has gotten me through many an academic meeting. A liability can be that I pick up the speaker's quality of mind, its mimetic communication. It might often be less painful to simply have listened to what was being said rather than to tune into the person speaking. I "know" that the sounds are physical stimuli. I experience this strategy as a kind of hanging out somewhere between sensation and cognition. I hear the sounds and recognize them as coming from a speaking voice. But I hold myself from thinking about the content.

Some meditation masters use analogous situations as a teaching device. Teachers from Southeast Asia have been known to talk for hours. One teacher was famous for keeping his monks sitting in meditation while he chatted away with his visitors seated at the front of the meditation hall. If the monks could not find calm inside and instead used their mind to listen to the endless small talk, sitting there would be torturous. Once when I was meditating with a distinguished Sri Lankan teacher, he talked on interminably. I was squirming with boredom. After an hour or so, something clicked. I let go and rested back into his presence. I became peaceful. I could feel what a profound human being he was. It mattered little that he was trained to present the driest of religious texts.

Stimuli, sensation, cognition and thought are western psychological categories in contrast to the Buddhist aggregates of form, feeling, perception, mind and consciousness. The latter are components of experience which, the Buddha observed, made up the ego or self. The Buddha's reason for meditating was to understand suffering and bring it to an end. In that pursuit, he claimed to discover that grasping onto experiences and freezing them into a fixed self is the root of discontent. Further exploration led to the observation that experience and the self in which it is lodged are constantly changing. Nothing we know is permanent. The Buddha's examination of how experience happens uses the aggregates as categories to direct attention to the process by which an illusion of a solid self occurs. Here is how a Tibetan teacher characterizes the aggregates as they can be observed in meditation:

> The process...takes place in...something like a five-hundredth of a second. First you have an impression of something. It is blank, nothing definite. Then you try to relate it as something and all the names you have been taught come back to you and you put a label on that thing. You brand it with that label and then you know your relationship to it. You like it, or you dislike it, depending on your association of it with the past.
>
> Now the very, very first blank which may last a millionth of a second is the meditative experience of the primordial ground. Then the next instant there is a question...you do not know who and what and where you are. The next moment is a faint idea of finding some relationship...then memory...then strategy...liking, disliking...It just flashes in place.... Primordial consciousness flashes out, the unconscious flashes out, which creates tremendous open space.... Ego tries to posses that open space,... It's like water freezing into ice....that kind of freezing of the space starts

at the level of form, continues with feeling and now manifests fully with perception.[25]

As purely intellectual categories, the five aggregates and Buddhist psychology do not mesh with contemporary ways of describing the world. They do not assume a materialistic distinction of mind and body, with scientific phenomena considered real and out there. Form or body includes both the material substrate, an awareness of the presence of a body in which sensation is happening, and also the sensation itself. Unlike the physical stimuli underlying Western psychological sensation (e.g., frequency and amplitude for sound) there is an element of mind in the first aggregate. Form and body may contain a primitive recognition of the sensation. Form provides rudimentary names, rudimentary concepts such as good feeling and criteria such as hot and cold. Since the whole meditator is the one experiencing things, there can be no complete separation into distinct aggregates. "If sight comes first or sound or smell...it just happens to you. You are just insensate, just crawling along."[26] The very beginning of having what we call an experience is where the ego or self begins to solidify. Each of the remaining aggregates builds the self more solidly. This is an analysis of components, not necessarily a time sequence of how experience unfolds. And, as for Buddhist categories, each implies the other. They do not stand alone.

The second aggregate is feeling. Feeling for the Buddha is the habit of mind which categorizes primitive experiences. Buddha thought it important to notice as subtly as possible how each sensory event (sight, sound, taste, touch, smell, and mind which is regarded as a sense) is perceived as pleasant, unpleasant or neutral. This habitual categorizing adds another layer to the self. Perception, the third aggregate, involves recognition and identification and includes memory. It "is the factor responsible for noting the qualities of things."[27] While meditating, perception might be a subliminal, not an articulated reaction: "My knee hurts. I will never walk again." Here the sensation which set off my reactions has begun to become "MY

PAIN" along with a hint of the fear of future consequences. With perception, the self (me, my, mine) coalesces even more.

The fourth aggregate, mind, includes what we experience when we are talking to ourselves. "I" becomes the words I say to myself as I sit here writing. But the Buddhist term mind also encompasses emotions, dreams and the subconscious.[28] There is no good translation for the fourth aggregate, *samskhara*. It is a collection of the above-mentioned mental/emotional phenomena as a kind of territory. With the fourth aggregate, the "self" becomes more solid. Our feelings, thoughts, and unconscious psychological reactions —the stuff of our daily world —all belong to the fourth aggregate.

The fifth aggregate, consciousness (*vinanna*) is variously regarded as the awareness without thought which observes the other aggregates and how they function or as the field in which all the other aggregates occur, similar to our notion of consciousness, or something that only exists when the other aggregates arise. Consciousness may be thought to underlie all the other aggregates or to be the most sophisticated. "Consciousness is that sort of fundamental creepy quality that runs behind the actual living thoughts...."[29] Since consciousness is necessary for the awareness of form, sensation, perception and mind, the aggregates are interwoven with each other. None is considered to be prior.

An understanding of Buddhist categories is meant to be practical rather than theoretical, so abstract descriptions are bound to be inadequate. From our conventional way of looking at things, form or body exists 'out there,' almost as physical stimuli. They take existence as something that can be experienced through the senses only when consciousness arises. And it is in consciousness that cognition and thought occur. For some meditation teachers, sensation is the key to understanding the makeup of experience. Sensation is our connection to the physical world. We recognize sensation and react to it with liking and disliking. Thus, the mind is affected. But the source of stimuli can be the mind as well as the external world. "Every thought, every emotion, every mental action is accompanied by a corresponding sensation in the body.... Sensation is the crossroads where mind and body meet."[30] These teachers assert that by focusing on physical sensations in the body, one is able to see exactly how the external world sets the mind going and to observe each movement of mind as soon as it arises as a sensation in the body. In so meditating, one is able to study how selves are constructed.[31]

When the mind becomes very quiet in meditation, consciousness itself can be observed with an awareness that is prior to consciousness. How that is can only be pointed to. Its description is a metaphor, words in the mind, elements of the fourth aggregate. It may be that before-thinking-mind can only be experienced. The contradictory sayings of Zen are a way of pointing to and demonstrating what may defy description.

What has been touched on here is the Buddha's method of observation and some of the components of being he observed. These can be thought of as part of a natural history of mind. Meditators with patience, persistence and careful observation have spent many more generations observing humanity through the Buddha's lenses than have the followers of Darwin, who have been looking at nature in terms of evolutionary biology for only the last one-hundred-fifty years. Meditators are not superior to natural historians; they offer a different perspective into mind than science. This kind of observation is what the Buddha meant by mindfulness. Its may be a crucial ingredient in understanding the nature of discontent. It is the Buddha's remedy for humans' resistance to accepting the inevitability of disease, old age and death.

All sorts of bookstores, even airport newsstands, sell books touting meditation. Meditation is offered as a way of overcoming jet lag, improving business acumen, developing psychic powers or healing the world's wounds. However, despite the claims of some self-help books, it is difficult to learn Buddhist meditation on one's own.

That is why meditation has so often been transmitted from teacher to student in a committed relationship like monasticism. I have presented a simple model of Buddhist meditation, drawn mainly from Theravadan Buddhism. I recognize that there are many different practices within Buddhism. One can meditate on the "Pure Land," chant the Buddha's name, visualize a tantric deity, gaze at the moon, surrender to a guru, label each little piece of experience as it arises, or walk bowing every third step. Looked at superficially, some of these practices seem silly. Nevertheless, each has substance, the understanding of which requires practice. In what follows, we will rely on my description of mindfulness meditation and fill in other aspects of meditation as they are needed. Some references on how to do meditation can be found in the footnote.[32]

The Brain and the Mind

Ajaan Cha, a Thai forest monk who died in the early 1990s, described meditation in the following: "Try to be mindful and let things take their course. Then your mind will become still in any surroundings, like a still forest-pool. All kinds of wonderful animals will come to drink at the pool, and you will clearly see the nature of all things. You will see many strange and wonderful things come and go, but you will be still. This is the happiness of the Buddha."[33]

What are the strange and wonderful things which appear in the mind? Where do these animals come from? The Greeks thought the essence of a human resided in the belly. We, on the other hand, experience the world from our head. As it turns out, both of us are correct. Recent research has shown that both the brain and the gut possess cerebral materials which communicate with each other semi-independently. Neither rules the other. The neural materials in the gut are ancient, arising in ancestors whose heads were only mouths. In situations of danger, the brain tells the gut to shut down for action. Terror overstimulates the gut and it lets go.[34] When your gut is upset, you become unhappy. These observations support the Greeks. On the other hand, what we see when we look out at the world is actually represented on the back part of the brain, and can be mapped there. The visual world is experienced in our head. It is interesting that no one knows why we feel that what we see is "out there" and not inside the brain where the "seeing" is actually taking place.[35]

As evidence of our evolutionary superiority, the human brain is claimed to be the most complex object in the universe.[36] The brain processes 10^{27} bits of information a second.[37] There are not enough genes to specify the brain. The brain's physical structure of neurons, axons and soma does not resemble any other physical signaling system. Its development in a growing human gives it an individual structure so that its detailed elaboration in genetically identical beings is different. The specific outcome depends on a developing child's interactions with the environment.

As a well-known brain researcher, Gerard Edleman, describes it, "We see that the development of brains is enormously dynamic and [random]." Genes set cells migrating with instructions such as: if you make certain connections, then complete the structure; if not, then self-destruct. "Indeed the circuits of the brain look like no others we have seen before.... Their signaling is not like that in a computer or a telephone exchange; it is more like the vast interactive events in a jungle.... And yet despite this, brains give rise to maps and circuits that automatically adapt to changing signals. Brains contain multiple maps interacting without any supervisors, yet they bring unity and cohesiveness to perceptual scenes. And let their possessors...categorize as similar a large if not endless set of diverse objects...."[38]

Darwin would have liked the jungle analogy, as might Ajaan Cha. Although there are schools of research that create models of mind in the image of a computer, Edelman's contemporary "neural Darwinism" seems to me a more productive approach. There is interesting research indicating that during early human development other cells in addition to brain cells experience survival of the fittest.[39] For reasons that do not seem programmed by DNA, some cells survive and others die off. In the embryo, certain cells, indistinguishable from others, develop into specific organs apparently because of where they are located. Brain cell groupings seem to respond to environment and interact with each other. The tremendous language ability of children gets hardwired into a particular configuration by experience. Of the two Kissinger brothers, Henry, the elder, learned English later and speaks with an accent, while his younger sibling came to America at an early age and speaks like a native.

Research increasingly indicates the importance of early experience in forming the structure of the brain which then affects later behavior. There is evidence of a relationship between offsprings' schizophrenia and flu or malnutrition among pregnant woman. Brutalized children become violent adults. This is associated with malfunctioning neuroreceptors. Young hamsters placed with aggressive adults become bullies when they grow up. Regularly neglected children become depressed, isolated adults. As adolescents, violent people can be reprogrammed by being given lots of loving support. They can be taught to recognize when violent episodes are going to arise and be prepared to control them.[40] There seems to be a correlation between brain chemistry and behavior, which is sometimes alterable.[41]

There are long-standing debates about whether unhappiness or deviant behavior is the result of nature or nurture. Important social policies and rancorous political discussions have been predicated on one or another of these views. In some cases, the distinction breaks down because early neglect can become physical structure, although that structure may still be plastic enough to be changed or overridden using other facets of the brain. Until recently, most psychotherapy and counseling focused on inadequacies of childhood nurturance. This approach has been increasingly overtaken by the modern, yet crude, use of psychoactive chemicals. Pathology is seen in a somatic framework.

Dr. Peter Kramer, author of *Listening to Prozac,* says depression is usually accompanied by low self-esteem. With the administration of Prozac, depression lifts and there is higher esteem. After finishing with Prozac, a patient often experiences no more depression but may still have low self-esteem. Kramer believes that manipulating brain chemistry will lead us to understand more about how our humanity derives from our being animals. Young chimps' neurochemistry is altered when they are taken from their family and traumatized. When returned they continue to be dysfunctional.[42]

Contemporary research into neurophysiology and neurochemistry is progressing at a gallop. Researchers are using powerful new tools. Positron emission tomography (PET) is an imaging technique which traces areas of activity in the brain, as does functional magnetic resonance (fMRI). Micro-surgical sampling of brain chemicals in laboratory animals gives insight into neurochemical interactions. All these research techniques are being used to link behavior with brain states. Prozac and other anti-depressants affect brain chemistry. They are taken by millions of people on the advice of a doctor because patients want relief from suffering. Even though anti-depressants may be a modern version of Freud's treatment of hysteria or a substitute for crucial social support in a pathological society, patients testify to changes in their moods and behavior.

It was the odd behavior of people with brain damage that led to preliminary understanding of brain physiology and how it relates to mind. Phinias Gage, a 19th century dynamiter who survived a blasting rod shot through his brain, is a classic case. He functioned but his social behavior degenerated. Many other cases of brain damage illustrate how parts of mind and personality are located in specific regions of the brain. An example is a man with brain damage who had above-average performance on every psychological test and yet was unable to make judgments which involved feeling as to appropriateness. He suffered a loss of planning and moral considerations. In the condition called prosopagnosia which affects stroke patients, a person has no awareness of a face and yet can discriminate known ones. Some brain lesions cause "blindsight" where the person can locate objects but has no awareness of what is being seen. Recently neurophysiologists have studied people who experienced localized damage to their prefrontal cortex when they were very young. They develop above-average intelligence but are incapable of articulating and consistently adhering to rules about not harming others.[43] In these examples, the sense of "I" that plays a large role in how we function has been impaired. The list of adaptive benefits of human consciousness is replete with recognition of the person doing those things. There is an "I" who is thinking, planning, comparing and remembering.

"Primary consciousness" is the name some psychologists give to the ability to match events to memory and act thereon, without awareness of the process. Humans, chimps, most mammals and some birds have primary consciousness. An example is pulling your hand from a fire, a learned behavior done automatically. Animals without cortices don't exhibit primary consciousness; you cannot teach them through repetition something which then becomes an automatic response. Snakes at the right body temperature have it.[44] In humans, the much more elaborate frontal cortex has repertoires that can categorize primary consciousness. We learn about a whole series of hot things and can warn ourselves beforehand about how to behave toward them. This is established by symbolic means using comparison and is socially reinforced. The ability to distinguish symbolic models from ongoing perceptions allows the concept a perception of a past.

This is a higher order consciousness with a sense of past, present and future. Language and speech in an interacting community allow an inner life which ties experiences in the world to concepts about the world. Inner life affects most of what humans do.[45] Although much less developed species can be regarded as having an inner life in the sense that they can discriminate between self and other, we have little evidence they posses higher consciousness. This evolutionary development happened in a relatively short time. Although Neanderthals had mimetic consciousness, full-fledged speech and thought developed in *Homo Sapiens* in the seventy or so thousand years before thirty thousand BP, when cultural evidence becomes undeniable. As in the evolution of the natural flashlights of squids called photophores, the source of change were regulatory genes which built new structure out of tissues already used elsewhere. The rapidity of the development of speech and thought may contribute to their various troublesome qualities, their antagonistic pleiotropy. We may still be in a shakedown stage of natural selection where it is unclear how much the benefit of consciousness and speech outweighs the destructive.

Using PET, fMRI and microsurgery, areas of the brain activated by behavior can be studied. The cortex of humans is highly developed. It is where much of our thinking occurs. Although they had larger brains, Neanderthals had flat foreheads, leading some to speculate their cortex was much less highly developed. We have prominent foreheads making room for bigger cortices and theoretically much more thinking capacity.

When we see something, the neural patterns on the visual cortex resemble the stimuli. When we remember or conjure up a mental picture, there is a reconstruction of the original in response to internal brain stimuli rather than the original sensations. The package is not just visual, but auditory, including those things associated with the image. Memory and perception are relative, affected by intervening events and current experience. The chattering mind has auditory and visual components.[46] They occur as patterns on the early sensory cortices. Except for non-complex sensory information, there is no anatomical way of getting information into the thinking part of the cortex without going through the early sensory cortex. So even something as abstract as mathematics is done by images. Internally generated images are "faint" compared to "lively" ones which come from direct experience.

Earlier evolutionary developments underlie and are interwoven into our higher-level consciousness. Reptiles developed the basal ganglia (controlling organs and movement) and the midbrain structure (sorting signals and responses). Early mammals added the cingulate cortex and later mammals the neocortex. Each of the earlier structures continued to develop as new parts of the brain came into existence, so that the basal ganglia or brain stem of modern humans is considerably larger and more complex than our ancestors.[47] The innate circuits which regulate basic survival systems such as endocrine, immune, viscera and instincts all interact without necessarily involving the higher-level more plastic circuits. So, underlying our mind are brain functions which keep our bodies going.[48] (See Illustration 3.)

Our mental reconstructions of the world are adequate enough for present survival but they are interwoven with other stored material and constantly affected by the processes of more primitive parts of our brains. We live in a "remembered present," conditioned by a remembered past and a projected future.[49] The physical state of our body is not only a product of the world of physical survival involving feeding ourselves and fending off competitors, predators and parasites, but also of our internal mind states. Sadness and anxiety can alter regulation of sex hormones and the menstrual cycle. Bereavement can cause depression of the immune system. And, curiously, the reverse is also true: oxytocin released during childbirth and sexual stimulation influences grooming, locomotor, sexual and maternal behaviors. It also facilitates interaction and bonding between mates. The hypothalamus, brain stem and limbic system are involved in body regulation and all neural mind processes. Survival, feeling and thoughts are deeply, evolutionarily intertwined.

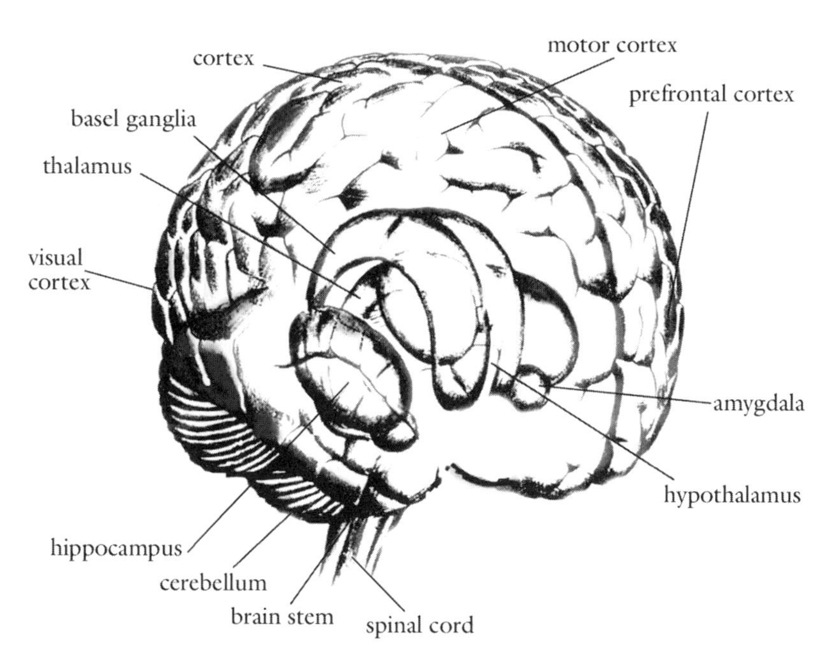

The human brain is composed of a greatly enlarged cortex built around older parts of the brain which in ancestral reptiles processed sensory stimulations and initiated responses. Reptiles' repertoire of behaviors was very limited. In humans these older parts of the brain became more elaborated and inextricably involved in the thought processes which our huge cortex made possible. Dreaming seems to have been a concomitant of the developement of the mammalian brain. As the cortex evolved, dreams were one of the downsides of our ability to think.

Neurophysiology and Meditation

Beginning with sensation and perception, the Buddha describes the sense gates or sense doors. They include eyes, ears, nose, mouth and touch. Human sensory organs such as the eyes sample information from the environment that is needed for survival. Only a handful of the billions of bits of information which hit the eye are used by the brain. Movies are possible because the speed of the passing frames is greater than the rate at which images are relayed from eyes to brain. Eyes also flicker as they see. If flickering muscles malfunction, the image one sees jumps round. The mind responds to a flicker that isn't happening. We do not need to see better than this to survive. Not only are the sensory organs imprecise but, as we have seen, the way the incoming data is handled in the mind has a vagueness about it. It is affected by stored information and primitive reaction responses. One learns very young to pull one's hand away from the fire. It is done without thinking. I get angry with myself when I burn my hand by picking up an iron pan which has been left on the surface of a wood-burning cook stove. I do it over and over again. Having grown up with gas stoves, I don't expect the handle to be hot. I can't seem to learn.

I am impressed with how little we need to know of the complexity of existence in order to thrive. We see only a narrow swatch of the light spectrum. Our hearing is highly selective. Taste is dependent on smell, which is highly individual and unreliable. What we do experience is influenced by unreliable memories. Through survival of the fittest, natural selection has preserved complicated sensory devices just good enough to filter needed information. Humans bumble along.

Not only is our sampling partial and its report in consciousness vague and loaded, but we are unaware of the ordinary. Because of their electro-chemistry, brain synapses become quickly jaded. This means that novel events have fast access to consciousness while routine ones remain in background. Rapid response is crucial to survival in the world of tooth and claw. This sensory adaptation permits instant comparison, calling attention to and therefore arming for survival. We notice the loud, smelly, sharp, sweet, especially when they appear without warning.[50] The Buddha recommended sitting quietly, cross-legged, under a tree, closing one's eyes and meditating by paying attention to what is happening in one's being as defined by the aggregates. Presuming there is no immediate threat and that the meditator has nothing else to attend to, carrying out this recommendation should be a simple task. Needless to say, given the kind of biological entity a meditator is, this is hardly the case. Stimulation continues, but does not require action for survival. The meditator attempts to rest back into the familiar. This is difficult because the mind is built for action and tunes out the familiar.

One meditation teacher, an erstwhile professor of sociology, used the phrase "obstinate familiarity" to characterize things we do automatically.[51] While brushing one's teeth, taking a shower or washing the dishes, the mind is often far away. Meditation begins with bringing the most intimate of familiar things, the breath, into awareness. Some people are so unaware of their breath that they have a difficult time feeling it. After years of practice, I still barely feel the exhalation at my nostrils. Focusing attention is the hardest part of meditation. As one focuses inwardly, one begins to be aware of different qualities. Perceptions become more apparent —how imprecise sensations are and how they are colored. The familiar is tuned in, and how we respond to novel events can be studied. Calmly observing the mind and body in its vagueness and habitual patterns has been called by Suzuki Roshi beginner's mind. It is a fresh way of experiencing life. "In a beginner's mind there are many possibilities, but in an expert's there are few."[52]

My observations on meditating one morning: "When I settle a bit and my mind is not so noisy, a series of feelings become observable. First, my body has many sensations which come in layers, some prominent others in the background. There is a pain in my foot, much stronger than was apparent and not exactly locatable. My breath is slowing down and becoming more noticeable. There is a noise of a bird outside and then the sound of a house motor. Then I feel sensations in the shoulder as if it were a tingling heat flushing. All the time my mind is noting, checking, explaining and querying. Afterwards, I can't remember all that went on as sleepiness comes a bit and the whole process seems to fade. Sitting at the typewriter yawning, now it is too hard to reconstruct what happened. Sleepiness makes my mind lazy, foggy."

With a little meditation practice, it becomes obvious how loaded sensations are. If one observes pain in the knees, it is not just a sensation, but something from which the mind and body can be felt to pull away. This is the second aggregate, feeling. The Buddha recommended that reactions to events at a sense gate be observed as pleasurable, unpleasurable or neutral. Such observations have been the subject of meditation for several thousand years but academic psychologists are only now discovering them. In an article in the science section of the *New York Times*, the psychology reporter, Daniel Goleman, who has both a doctorate in psychology and has meditated for many years, must have had his tongue in his cheek when under the heading, "Brain May Tag a Value to Every Perception," he wrote:

> The palette of sights and sounds that reach the consciousness are not neutral perceptions that people then evaluate: they come with a value already tacked on to them by the brain's processing mechanisms. This is the conclusion of psychologists who have developed a test for measuring the likes and dislikes created in the moment of perceiving.... These tests show that these evaluations are immediate and unconscious, and are applied even to things people have never encountered before.... 'This is all part of preconscious processing the mind's perceptions and organization of information that goes on before it reaches awareness.'... 'The quick and dirty judgments tend to be more predictive of how people actually behave than their conscious reflection on the topic.'... 'In responding to a word the signal would most likely go first to the verbal cortex, then to the amygdala where the affect is added and then back to the cortex.' The circuitry involved could do all that in a matter of a hundred milliseconds or so, long before there is conscious awareness of the word.[53]

The article ends with a researcher saying that one can observe these judgments by watching how differently you react to various things as you walk down the street. For a number of years, I have set my meditation students to carefully observing the aggregate, feeling, as soon as they get up in to morning. Their reports, with three columns labeled "like," "dislike" and "indifferent," get filled before they make it to breakfast, or sometimes even brush their teeth.

If feeling is prior to our thinking about what we sense, are not other aspects of our monitoring the world similarly embedded in consciousness before we have a chance to be rational about them? And is it possible that our rationality is often a rationalization? How we really feel is demonstrated by how we act more than what we say or even think about things. In many hours of meditation, I have often noticed how, at the end of a formal session, I find myself moving without being aware of my intention to move. Slowing your movements down so much that you can watch intentions form and then deciding whether or not to act before you actually act is a standard meditation practice. A meditator is challenged to watch the arising of intention. No matter how slow and concentrated I get, I have noticed that at the point of movement there is always a gap where the sequence of intention-movement precedes my awareness of it. By measuring electrical flow in the brain, researchers found that non-cortical areas responsible for movement are often prepared for and initiate movement before the cortex, where thought operates, is tuned in.[54] Lower levels of mind respond much more quickly than the thinking levels.

This is true in other behaviors as well. According to Robert Ornstein, "The conscious mind sees, hears, or feels the event... just a few milliseconds after it actually occurred. It is as if the brain, below our awareness, spends a half millisecond deciding whether we should be allowed to know about what just happened. If it decides that it is best that we should know, then it also informs us of when the event happened. However, note that, although we become conscious of the sensory stimulus, we cannot use our conscious will to respond to it in less than a half second. This leaves our unconscious mind responsible for initiating any rapid reactions to the world."[55] A similar phenomenon is called backwards masking. In experiments an arousing visual stimulus is flashed on a screen and immediately followed by a neutral stimulus which remains for a while. The second prevents the first from entering working memory so it isn't recorded in consciousness. But this doesn't prevent the first from evoking an emotional response. So one responds to a stimulus without being aware of what you are responding to.[56] With age the time it takes for some stimuli to be consciously imprinted seems to take longer. I constantly find myself doing something which gets interrupted by a distracting thought or event and not be able to remember where I was in the original sequence of actions. "Where did I put my glasses or that tool I had in my hand a minute ago." This happens even if I remind myself to remember what I am doing.

Not-so-rapid reactions are also affected. For years, I have felt that often I don't know what I think or feel about something until the words come out of my mouth. Although spontaneity can be wonderful, I often say things carelessly which unnecessarily offend people. It is not surprising that the brain does not fit together quite as we like to think it does. Our limbic system (the ganglia and midbrain), crucial for survival, keeps more behavior automatic than we imagine. Thought, getting into the action a bit late, takes more credit than is its due.

It is into the gaps that meditation steps. In the example of observing intention, the stronger my concentration and the slower I move in the world, the more likely I am to be able to bring awareness to the formation of intention and its relationship to action. Sometimes I am successful and sometimes not. Although it has limitations we will discuss later, meditation reworks the ill-fitting aspects of brain and mind so as to redirect our animal reactions. The maladaptive characteristics of the human mind are the material of meditation. The Buddha addressed the negative pleiotropies of consciousness although he would never have imagined such a concept.

In Buddhism, the Buddha's ideas are presented in a series of lists. There are the Four Noble Truths of suffering and its cessation. The Eightfold Path includes guides to living so as to avoid discontent. The five aggregates are components of experience. In another list, the Buddha sketches the relationship between intention and behavior. This is called the twelve factors of Dependent Origination. These factors are a causal sequence of the interactions between body and mind, and between acts and their consequences. Engaging in counterproductive behavior lays the groundwork for potential further entrapment. As with the Buddha's other observations, the categories do not quite fit Western categories of thought.

Dependent Origination is a circular sequence. Its factors are body and mind, senses, contact, feeling, craving, clinging, becoming, birth, death, ignorance, karma and consciousness. Dependent Origination is not self-evident and has been subject to many different interpretations. One way of thinking about how the self gets assembled is to picture this process taking place in a living psychophysical organism (mind/body) with six senses which act as agents of contact between consciousness and the objects of consciousness. With each sensory response, likes and dislikes form. These feelings solidify into craving. A meditation teacher recently told a story about coming across a box of tea labeled "Earl Black Tea" at a meditation retreat. Being a lover of "Earl Gray Tea," she became upset by this transformation of the familiar. It was "not a big deal. It just bugged" her. The craving in this case was for the world to be different. "What right has anybody got to change the name of my beloved tea?"[57] As the craving stuck, it became clinging. In her case, the sequence stopped there. She did not complain about the label or brood on the unfairness of life. She was, in fact, amused at how her mind made so much out of something so trivial. But we all know what it is like when we cling to things. It creates more discontent.

After clinging, the twelve factors of Dependent Origination relate more to Buddhist world view. Clinging leads to unbalanced actions. Clinging lays the foundation for "becoming" in the Buddhist sense, which may be thought of as when your state of mind becomes so fixed that it motivates your actions. You "become" a certain kind of person —angry, selfish, or greedy. That person then is reborn with all of the blindness of such state. If you end life angry or greedy, Buddhists believe you carry the blindness or ignorance into your next rebirth. It is lodged in a new mind and body.[58]

The sequence is somewhat confusing. It can be viewed with a different starting point: ignorance lays the conditions for karma formations which condition Consciousness which conditions mind-and-body which conditions the six sense-bases, which conditions contact, and then feeling, craving, clinging, becoming, birth and death. This ordering begins with ignorance of the Four Noble Truths which are the omnipresence of suffering, the possibility of ending suffering, that there is a way to do that and the way is embodied in the Eightfold Path. If a person does not understand the origin of suffering then he or she engages in unclear verbal, mental and bodily action. Talk, thought and action breed a solidifying of self as states of consciousness. These states infect actions later in life, leading to rebirth, and the cycle begins again.

Dependent Origination can be viewed either macroscopically within the general belief in rebirth held by the Buddha or microscopically as a sequence by which experience seems to be welded together so as to appear continuous. In the former view it is thought to unfold over three lifetimes. It is the latter, microscopic sense of Dependent Origination, that I want to focus on here.

In a meditation and study retreat, V. Dhiravamsa, who had been a Theravadan monk for twenty-four years, pointed out that we could observe the twelve factors as we meditated. In each moment of unawareness the twelve factors could be seen. Beginning with ignorance in one moment the other factors arose, leading to the conditioning of the next moment. That is, the "I," or the self, is "reborn" over and over again out of the pile of its immediate past existence. Unexamined, this process makes us appear to ourselves as a unified personality. The sequence has no real starting point. According to a Buddhist scholar: "The beginning of ignorance is not known...but that ignorance is causally conditioned can be known.... The mind follows in the wake of the body....the body follows in the wake of the mind....The psyche is not an unchanging entity but is in a state of dynamic growth and becoming [sic] in close association with the conditioning of the body."[59]

To interrupt discontent, one severs its cause by intervening in Dependent Origination. The Eightfold Path is the method the Buddha recommended for doing this: People are encouraged to cultivate 1) right view, 2) right intention, 3) right speech, 4) actions, 5) livelihood, 6) effort, 7) mindfulness and 8) concentration. Right view and intention at first supply a reason for proceeding. You experience discontent and you want to do something about it. Right speech, action and livelihood are moral dimensions. Trying to meditate when one's life is burdened with the tensions of dishonesty is like trying to undo what one intentionally recreates. For the Buddha, one must address morality without which, he felt, there could be no peace of mind. This can be easily observed. Sit quietly and try to meditate after a day of road rage, manipulating other people, or trying to convince the Internal Revenue Service that you have adhered to the letter of the law rather than what you know in your heart is its spirit. Your insides will be all a-jumble. It will even be hard to pay attention to your breath. In such a situation, the mind is racing, reenacting the day and worrying about its consequences. There is a confusion of emotions. Such actions simply add layers of discontent. It is crucial to tilt one's self in the direction toward which meditation leans. Right effort, mindfulness and concentration are the ingredients of meditation leading to a deeper understanding of view and intention. But like other Buddhist concepts, the list is circular; you can start at any point.

In meditation, one's observation can be aimed at any of the twelve factors of Dependent Origination. One can see whether or not they arise as described, how that arising takes place and the consequences of the factors. From the perspective of behavioral psychology, a deconditioning is taking place: using concentration organized from the cortex, unconscious psychological and bodily processes that are linked in causal chains are laid bare, closely observed, and eventually transformed.[60] These include not only actions but also thought and intention which initiate the process. If meditation is successful from the Buddha's point of view, then one can intervene in the way each step of Dependent Origination conditions the next; the sequence is interrupted. Recall the meditation teachers who instruct their students to focus on sensation so they can observe the feelings of like and dislike as they arise. These teachers want their students to catch feeling in the making so that it will not become fixed as craving and eventually lead to destructive actions. The Buddha claimed that if one can do this in all parts of one's life then behavior will no longer have karma. The infectious spread of actions resulting from unwholesome intentions will stop, and the false illusion of a continuous self will end. The person is no longer an ordinary person, but awakened. And in Buddhism one will no longer be reborn, thereby escaping the cycle of repeated births dependent upon past events.

Soen Sa Nim pictures the trip of penetrating the illusion of self as a circle. (See Illustration 4.) At 0°, there is "i mind." It is the everyday mind which thinks it is in charge and will get its way. At 90°, there is "karma mind," the mind which understands Dependent Origination. At 180°, "no i mind" resides in emptiness, the Zen name of the condition where there is no longer personal discontent, but there is also not a complete understanding of life. (This may encompass parts of what has been called "beginners mind" or "before-thinking-mind" but it lacks a deeper wisdom.) At 270° is "magic mind," the highly concentrated mind with powers like mind-reading, prophecy and teleportation. This is a dangerous place to be

Soen sa Nim's Circle

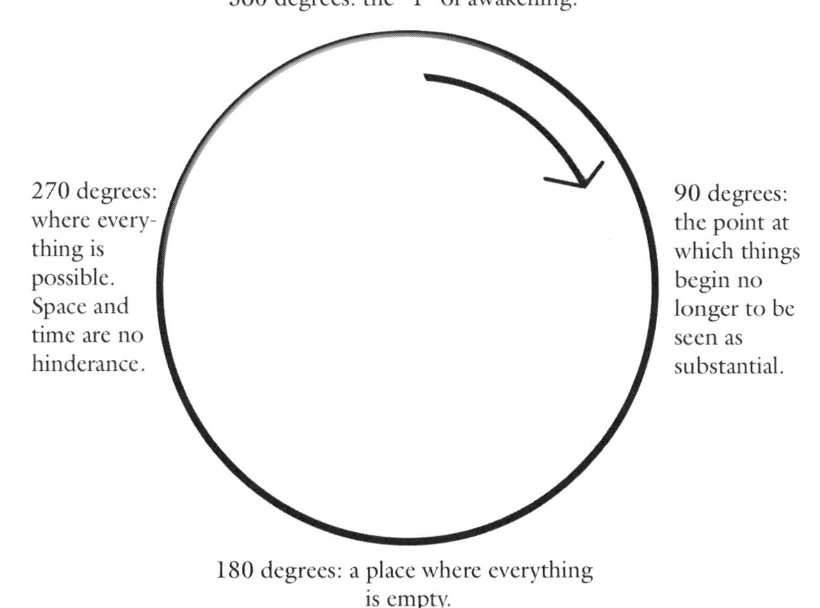

This scheme is based on the *Prajna Paramita,* "The Perfection of Wisdom," a founding *Sutra* of Mahayana Buddhism and later Zen. At 90 degrees "form is emptiness and emptiness form." At 180 degrees there is just silence. At 270 degrees there is complete freedom. At 360 degrees everything is simply what it is, just that, no fuss, no bother. Inspired by Seung Sahn, 1976 *Dropping Ashes on the Buddha.*

without wisdom to balance it. And finally at 360°, there is "I mind", the universal "I" beyond self and not-self. In this state, a person goes about the ordinary business of life in a non-harming way. They do what needs to be done to alleviate the suffering of the world without making anything of it. Zen calls such a person a Bodhisattva, an enlightened being who stays on earth to help others.

Although the idea of enlightenment or *nirvana* is an important concept in Buddhism, it is also a source of confusion and debate. The Buddha's enlightenment inspires faith in many Buddhists that it is possible to transcend the suffering of life on Earth. In some Buddhist traditions, enlightenment is thought to be an extremely rare occurrence while other traditions have many enlightened saints. Historically Buddhists believed the suffering of disease, old age and death is transcended by attaining *nirvana*. I have not addressed that idea here. Instead I explore the ameliorization of the suffering they engender by meditation and emphasize the equally Buddhist idea of non-harming or *ahimsa*. *Ahimsa* literally means non-hurt. Narrowly construed it was taken as not killing or non-cruelty. I use non-harming in the general sense of not doing anything which would harm others or one's self. Whether any specific action is harmful or not harmful could be debated.[61] Without referring to enlightenment it is possible to talk about meditation and how it can be understood as a process of body, brain and mind without entering into an examination of the claims surrounding enlightenment. That there may be no real self is an observation to which even some psychologists lend credence, in spite of the overwhelming commitment to ego psychology of most clinical psychologists. Enlightenment may be a matter of faith, definitional and/or something real.

Earlier we mentioned that physical damage to the brain can impair normal operation of self. Experiments conducted on people whose brain lobes are separated show that, for most humans, speech and language skills are in the left hemisphere. Some people are born without a connection between their lobes, while others have them separated surgically as in treatment of extreme cases of epilepsy. Researchers have devised visual tricks to interact independently with the "person" who inhabits each side of separated brain lobes. If some things are presented to the right hemisphere, such as a pin-up girl to a male he will blush but have no knowledge or recognition of why he blushed. With an unusual person who actually had speech centers on both lobes, experimenters communicated with each lobe separately and neither knew what the other had said. The conclusion that psychologists draw from these experiments is that there is no self-consciousness without language. So, if things are presented to the non-linguistic side, there is behavior but no understanding of where it comes from; that is, no self-consciousness. This leads researchers to postulate that without language there is no thinking consciousness. They see our inner labeling, the "i," in the sense of Soen sa Nim, as mind and the thing that makes speech possible.[62] But these situations also show that there are behaviors to which we have no conscious access.

When the "i" becomes overbearing, it touches pathology. In wilderness travel I have encountered men who fled the snares of society. One was living in a cabin in the Arctic, another in the Everglades. Their grip on sanity was maintained by rehearsing their humanity. They talked out loud to themselves. This is what soldiers who may be captured are instructed to do if placed in isolation for a long time. They are not too different from isolated people in big cities or people in mental hospitals, except that there is really no one to speak with. Some hermits I met either made me feel unwelcome and drove me away, or they turned me into a captive audience. It would be an uncomfortable situation. You are not an other with whom they want to interact but a yes-person who dares not contradict or even interject. While you can have sympathy for these hermits, after a while you just want to escape. And, though not particularly likable, they resemble aspects of our selves.

During my Arctic canoe trip, besides meeting hermits, my brother and I would paddle up to fourteen hours a day. Although it was beautiful, not much was going on. I was unoccupied and so I was very aware of the chattering of my mind. The chattering would focus on things like cookies, something we didn't have. Years earlier working a punch-press in a typewriter factory, I remember a more bored and incessant chattering. To keep the chattering from being painful,

I would do long arithmetic problems in my head, such as calculating how much money Lucille Ball made every second. In both cases I had not yet learned how to observe my chattering mind.

When you sit in meditation and look at your own mind, it behaves like a half crazed hermit. The internal monologue is interminable. Even if it is filled with ambivalence, it brooks little opposition. It recreates social experiences and adds scenarios with no reality. When dispassionately observed in meditation, mind chatter appears like the ranting of an insane person. In everyday life, I get angry at distracting adolescents, loud radios or people talking while walking a trail in the woods. On close examination, the anger feels like the egocentricity of hermits. Other people's noises interfere with my preoccupation with my own mind and I resent the disruption, although when I look at MY noise, it is painful.

The self is a jumble of things. Its perceptions are value-laden. Its way of understanding and communicating is rooted in linguistic and social interactions. It names things and forms intentions. It reports thoughts and actions crudely to others. It is influenced by outside events, subliminal thoughts and feelings. It is filled with memories, dreams, and fantasies.[63] Language mediates the world but more important is constitutive of our self in relationship to people. When we look at our own mind, we see how much our understanding of reality is a function of our changing self. Looking very closely, a meditator can see the process of Dependent Origination in which the self gets born slightly differently over and over again. A new or old constellation of sensations, feelings and thoughts arises, and the mind takes off at a gallop.

From the perspective of brain function, the idea of a unitary self is hard to maintain. Psychologist Robert Ornstein sees minds as "a squadron of simpletons."[64] According to this view the mind creates quick dirty sketches of the world, attends to short-term change and undergoes continual transference. Aspects of our mind are dedicated to simple tasks; they "swing in and out...since minds shift, 'we' are not the same person from moment to moment, not the same 'self' at all." The self is only a small part of mind and may not even be operative. "The self, itself, is just one of the simpletons, with a small job."[65] This is Soen sa Nim's "i." Because of the multiplicity of "i"s and the vagueness of sense signals, a neurobiological "self" is situated in a continuously reincorporated background feeling generated by sensation along with the habitual representations into which new experiences are incorporated. This is what underlies our feeling of personal familiarity and continuity. According to one neurophysiologist, "At each moment the state of self is constructed, from the ground up.... the owner never knows that it is being remade unless something goes wrong with the remaking.... Present continuously becomes past, and by the time we take stock of it we are in another present, consumed with planning the future, the present is never here. We are hopelessly late for consciousness."[66] And that consciousness is not that well organized. The separation of functions between brain lobes and their inefficient communication leave room for a multiplicity of contradictory senses of self. We are flooded with sensations which require memory to make sense of them. The left hemisphere makes sense of the signals even if some don't quite fit. The right hemisphere functions to correct models of reality built by the left, which resists until the anomalies detected by the right are so great that it overwhelms the model builder.[67] With a brain like this it is no wonder a meditator who becomes still is overwhelmed by the onslaught of disorderly thoughts and feelings.

People with strokes that impair the lobes can be persistent in their denial of obvious reality. From the perspective of Buddhist meditation, we live in a situation that psychologists call cognitive dissonance. When first introduced, this concept characterized the behavior of subjects confronted with situations that undermined what the psychologists thought were absurd beliefs. The subjects stuck with their beliefs in the face of "rational contradiction."[68]

Listening to the news, sitting in a university meeting, talking with another human being or observing my inner life, it is rare that the models I have fit the situation. This applies to terrorists, governmental double-speak, street violence, a friend's marital problems, money, automobiles, my sense of who Charlie is and even my attempts to help. Reality is constantly contradicting many of my beliefs but I still cling to them. I am a "good" person but I avert my eyes from beggars. I used to teach epidemiology, yet I eat more saturated fat than is good for me. I feel guilty and vow not to do it again but at the moment of choice some rationalization overrides reason. The left and right hemispheres, along with many other brain and body processes, can't quite get it together.

In meditation, one sees how a normally functioning brain is just good enough to get us by. It is filled with contradictions and cravings which, when we follow them, often lead to trouble. We survive because most of the mistakes the brain makes are troublesome, but not fatal. In this confusion and ineptitude lies the discontent of our lives. The goal of meditation is to step into the midst of this all, observe the dissonance and lack of a coherent self, and rein in one's harmful behavior.

Chapter Six
Emotions, Dreams and Insight

Discontent, the five aggregates, Dependent Origination, no fixed self—where do the emotions fit into this picture? What are emotions anyway? Someone who is driven by his or her emotions seems the opposite of a serene Buddha. Like all other states of being, emotions can be analyzed into the five aggregates. It is interesting that there is no equivalent in Buddhist psychology for our term emotion. When there are sensations and feelings in the Buddha's sense, which then become an object of mind and persist, that might be designated as emotion. In meditation, it is difficult to isolate emotions as entities in themselves. While sitting quietly and attentive, stimuli of one sort or another give rise to an aggregate feeling. It may begin with a sensation or a disturbing thought. Remember that, for the Buddha, the mind is a sense organ. Pain is a common initiator, or a thought I do not like. And there it is in my body: muscles tighten, breath quickens and my mind races uncomfortably. But where is the emotion? No matter how close my attention, I cannot locate a separate entity on which to focus attention.

Psychologists distinguish two kinds of emotions. The first are sometimes called primary emotions: an example is fear in the face of the unexpected. Secondary emotions are what we generally associate with "being emotional," and involve the mind. Both are poorly understood by neurophysiologists. Using PET and fMRI, psychologists are only beginning to locate receptors in the brain that recognize pain. They have been able to trace primary fear in some of its brain components.[1] Primary fear may be genetic, as in the shock-like fear of a rabbit confronted by a predator. Rabbits are born with it. Or fear may be conditioned, as in the case of a dog who cowers when men approach because it has been beaten by men, not women. When such fear stimuli are passed to the brain, they are processed first by the amygdala (a part of the midbrain) where action may be initiated before the cortex is informed. In fact, signals to the amygdala may vanish before the cortex is alerted. You may get a glimpse of something threatening and respond without ever being conscious of what it was.

The pathways between the amygdala and the cortex are asymmetric. Signals from the amygdala to the cortex are more efficient than those in the reverse direction. This keeps thought out of the way when survival is at stake. In the midst of a kill, neither predator nor prey is at leisure to figure out the next move. And because communication from the amygdala to the cortex is better than the reverse, when you try to control things consciously you may not be able to override the fear response. The amygdala is reactive and there may be nothing one can do about it.

When there is fear, signals are sent to the body, and bodily reactions are fed back to the brain. The somatosensory cortex (the part of the cortex which processes sensory inputs) and the brain stem alter brain chemistry. Now there is awareness of what is going on. The cortex has information from the body and from the changes that the body induces in its chemistry. This is what makes up a secondary emotion.[2] By examining brain responses researchers theorize that melancholy and shyness are deeply ingrained physiological variations of fear and hunger. Psychologists also think that the amygdala and cortex can generate signals internally and send them to the somatosensory cortex which act "as if." That is, your thoughts and basic emotional apparatus create or recreate stimuli as if something were happening. This makes rehearsing possible. I imagine a dog coming around the corner and I run the other way. Such "as ifs" may be more or less grounded in reality. As noted, these internally generated stimuli have less detail, definition and variation then stimuli coming from the senses.

There seems to be three reasons why one can't pinpoint emotions: 1) parts of emotions are taking place in systems which aren't easily accessible to conscious observation. As much as one would like to see what is going on, the conditioned response of the amygdala is not observable. You may not know what set it off and you can not watch it do its work. What you can do is to feel its effects in the mind and body. 2) These effects are complex and diverse. They happen in movement and visceral muscular systems and are hormonally mediated. The cortex also gets into the act, bringing in long-term memory and its associations. And 3), the cortex itself sets off reactions. When the viscera responds to the cortex, other emotions may be triggered internally so that the original fear may transmute in anger or sadness. All this presents the meditator with a complex and subtle field, only pieces of which are observable. It may be that thinking is associated with similarly complex processes but without such dramatic expression in the body.

Secondary emotions originating from thought, while less distinct than primary emotions, have real effects. On hearing of a friend's death, the heart pounds, the mouth becomes dry, the skin blanches, the gut contracts, the neck and back tense and the face saddens. There are changes in viscera, muscles and endocrine glands. Endogenous chemicals are released into the bloodstream, the immune system changes and there is contraction and thinning of blood vessels. Signals run back and forth from the amygdala to the autonomic nervous system and back to the cortex.[3] All of this is set off by an association in the mind. It takes a great deal of concentration in meditation to observe such changes.

Several years ago I was at a meditation retreat. I had been there for about three or four days, was calm and focused and rather enjoying myself. I was practicing concentrating on my breath when I was informed that a friend and colleague with whom I had worked for twenty-six years had suddenly died. I remember the scene vividly. The messenger broke the news gently, his face conveyed comfort and concern. On hearing his message, my thought was simply, "Oh, Irv's dead. He knew [The rest of us did not know his heart was bad]!" Beyond that my mind didn't move. There was no emotional reaction. It was as if I labeled the thought as it came into my mind and then went back to feeling my breath. When I realized the bearer of the bad news was waiting for a response, I felt sad and upset. Afterwards I thought, "How interesting: I received the death of this man, with whom I had close but complex relations, with an unmoving mind." I felt that maybe I was "not in touch my feelings." But I know that I have been affected by deaths before and since, and there was no reason why Irv's should have been met with repression. Stilled by days of meditation, my mind did not move and so the chain reaction of thought initiating secondary emotion did not occur.

Such a response may have a neurophysiological basis. The reason why you don't jump at the lion's roar in the zoo is that some part of your frontal lobe overrides the fear generated in the amygdala which is evolutionarily more primitive.[4] Meditation can lead to a conscious undermining of habitual response processes. Although it does not seem possible to beginners, as meditation deepens, mind-body interactions come into clearer focus and begin to change. More precisely, something changes in the mind's response to the brain's chemistry. Fear or sorrow may be initiated, but the thinking part of the mind does not engage them; it bypasses the associated chemical interactions. Then the fear and sorrow recede rather than build.

Let me give a couple of examples from daily life. I suffer from what a friend calls "napitis"—I occasionally nap for an hour or more in the late afternoon. I sleep so soundly I have woken with stiff limbs and drool all over my face. When I wake, I feel terrible. I can't focus. I can't figure out what I need to do. I don't want to do anything. In short I am depressed and often feel that I want to be dead. For years I sought psychological explanations for this experience. The explanations seemed to reinforce the phenomena. The lack of will to live fit nicely into all the things that were wrong with my life.[5]

In college when I was living in books, I used to nap on cold, dark, 1950s, Chicago winter afternoons in the slum-neighborhood where my dingy apartment was located. Upon waking, it felt like the existentialists I had been wallowing in were right: life was horrible, meaningless, and there was no reason to continue. Until the feeling wore off there was nothing that redeemed the suffering. After years of meditation, I began to focus on the process of waking up and other examples of this feeling. First, I noticed some children came out of their naps in a similar cranky mood. I watched their faces and understood exactly how they felt. I wasn't alone. Second, I began to study the general circumstances of my life. I noticed that I was just as depressed on waking from a nap when my life was going well as when it was difficult. Finally, I looked at the events immediately after I woke. I played around with neurochemicals like caffeine and sugar and noticed how input from people affected me. Usually it took an hour or so to wake from the nap. Substances did not make much difference. They just added their effects as a layer over my depression. Once when I woke from a nap, I met a woman to whom I was attracted and told her what waking felt like. She responded with similar anecdotes and while we were talking my mood lifted. I thanked her humorously for having healed me. Since then napitis has recurred. I still have the sinking feelings, but it is easier for me to accept them as merely a function of my body chemistry, not my life.

Although psychological explanations for my napitis may be relevant, meditation takes a different tack. Meditating for weeks and months on retreat teaches one that moods come and go and that your body chemistry is inextricably bound to them. When watching the changes, you can often not quite put your finger on the reason for the changes. The Buddha considered continual change a fundamental characteristic of life. He called it impermanence. When meditating on mood, this often is as good an explanation for changes as any, although some do seem to have specific roots.

In trying to make sense of experience through meditation, we come again to a situation where our rational thoughts and categories are not adequate to describe the kind of knowing that meditation reveals. We actually observe that things are different than our ordinary notions of what they are like. This is psychologically hard to assimilate. According to one meditation teacher, "Once [people] get a glimpse of the possibility of pain and pleasure without hope and fear, they see it as demonic.... There is nothing to latch on to there.... there is no way of relating with it except directly."[6] "Directly" is a trip by the cortex into more automatic parts of the brain or to parts wired to thought indirectly. We mostly know our body and emotions and thoughts in a veiled way. You may have noticed that if your body is not reporting to you, then all is well. But when a part becomes noticeable, you have something to worry about. Most people don't feel their knees except when they ache, and then they want the ache to go away. When you try to bring the parts of your body into focus during meditation, you are asking that normal signals be rerouted. This requires effort, and the images that are presented are not very distinct. It may also set off disturbing secondary emotions.

There is a manual self-published in the 1970s by a teacher who used meditation to change students' relationship to pain.[7] The teacher had studied with a Burmese meditation master who taught a technique based on mindfulness of body. Followers of the Burmese monk were instructed to sweep their attention over their bodies in ever-greater detail, noticing the arising of the second aggregate, feeling. In the manual, meditators are told to study their pain, and to observe, for example, whether it is hard and ball-like or diffuse. They are to examine its boundaries, penetrate its interior, notice its temperature. A person is then directed to imagine an appropriate set of tools. They might include scissors, compacting devices or scoops, depending on the nature of the pain. Using these tools, the person is asked to reduce the pain to manageable pieces which are pushed to the borders of the body and expelled. Such techniques inspired some of the founders of the stress reduction movement in behavioral medicine. These tricks attempt to undo the way the mind freezes painful sensation into first an inkling of the idea that you are in pain, then "knowing" that something isn't right and finally the full blown thought that you are in pain and don't like it. The raw experience of a painful sensation is a primary emotion, which then becomes a perception, the third aggregate, finally setting off thought, the fourth aggregate. "I am in pain" is a secondary emotion which in turn sets off the body.

Emotions may arise in the neural machinery without conscious cause. One can feel depressed without depressing thoughts. In negative mind states as with negative body states, the generation of images is slow, not varied, and reasoning is cloudy. Positive states have rapid diverse images and reasoning may be fast but not necessarily efficient. In my twenties when I first recognized my sunken moods as depression, I also became aware that I could not focus on my study of science except with a great deal of effort. My mind was obsessed with my mood. In later years, the depression subsided and my reasoning became less clouded.

It is no wonder that antidepressants are in demand. They not only ameliorate sunken moods but make it so that moods are less likely to be initiated by obsessive thought. You don't have to be depressed to want relief from the bodily consequences of self-obsession. The neurophysiologist Antonio Damasio underscores just how pervasive the underlying ingredients of discontent are. Sensation and thought are constantly contributing to what he calls "background feeling," a subliminal image of body landscape where emotion isn't prominent.[8] As an understanding of the second aggregate may suggest, while we are alive and the brain is functioning moderately well, there is always the potential of a subliminal classification of life processes into pleasant, unpleasant or neutral. When unpleasant sensations gel into objects of mind which we cannot let go of, we have what the Buddha called suffering. Meditation is aimed at bringing attention to the whole process.

Concentration

So far we have focused on discontent but not how meditation reveals a path out of that suffering. Attention in meditation is used both to concentrate the mind and to understand the nature of being. In some passages of the Sutras, concentration alone is claimed to confer positive benefits.[9] Some of these benefits are akin to mystical experiences. Siddhartha learned how to practice concentration from forest teachers who could enter various states of concentration at will. The Sutras describe him as doing so just before he died: "Then the Lord entered the first *Jhana*. And leaving that, the second, the third, the fourth *Jhana*. Then leaving the fourth *Jhana*, he entered the Sphere of Infinite Space, then the sphere of Infinite Consciousness, then the Sphere of No-Thingness, then the Sphere of Neither-Perception-Nor-Non-Perception, and leaving that he attained, the Cessation of Feeling and Perception."[10]

The *Jhanas* are the Buddhist designations for states of concentration, absorption or serenity. The first *Jhana* is the subsiding of thinking where there is tranquility and oneness of mind. The second is filled with joy. The third is a state of equanimity and mindfulness. With cessation of pleasure, pain, gladness and sadness comes the fourth Jh*ana*. "The *Jhanas* and the mundane types of direct knowledge by themselves do not issue in enlightenment and liberation. [Abiding in the *Jhanas*] only suppress[es] the causes of suffering. To eradicate suffering one must contemplate 'things as they actually are.'"[11] Ascending the *Jhanas* one comes to supra-mundane states ending with "Cessation" which is beyond the states of concentration. It can be thought of as release from the twelve factors of Dependent Origination. Awakened Buddhas no longer make karmic waves. The acts they engage in are harmless.

Although there are disagreements within Buddhism, states of concentration are thought to be one avenue beyond discontent. Among the forest seekers whom Siddhartha met, concentration and other focused practices were used to reach mystical states of unity or transcendence. He could not vanquish his suffering using them and so he settled under the famous tree to inquire further. While neurophysiologists have not explicitly explored the *Jhanas*, they have researched the effects of concentration and prayer on brain. The research is incomplete but it suggests that concentration practices such as trying to exclude thought, performing rituals, repeating prayers or focusing on an image set off a cascade of reactions in the brain which overwhelms neural mechanisms that locate one's boundaries in space and time and differentiate between self and other.[12] This gives rise to a powerful feeling of unity which seems more real than ordinary sensations. The researchers see this as the neurophysiological underpinning of a belief in god or enlightenment. The dramatic effects of a mini-awakening experience such as *Satori* in Zen or abiding in the "Sphere of Neither-Perception-Nor-Non-Perception" is pointed to in mysticism and may be what inspires seekers to continue striving.

In the *Abhidhamma*, part of the Buddhist holy books concerning Buddhist psychology, much is written about the *Jhanas* and supra-mundane states. While concentration may lead to loosening the bonds of attachment, it is not necessarily the case. On the path through the *Jhanas* there are branch points where a meditator strives towards wholesome rather than unwholesome states.[13] Each wholesome state builds a platform for the next, while unwholesome states undermine progress. Wholesome manifestations offers rewards such as joy, whereas the unwholesome ones involve further discontent. Parallel to the *Jhanas* are what are called the seven factors of enlightenment. These are mindfulness, investigation, energy, rapture, tranquility, concentration and equanimity.[14] They intersect Western psychological categories in more apparent ways than the *Jhanas.* In an extensive attempt to understand the neurophysiology of enlightenment, Dr. James Austin of the University of Colorado (and a meditator) has brought together much of the science that might indicate what happens when people meditate. He puts forward a number of hypotheses and proposes many different ways in which the brains of meditators and Zen masters might be monitored to see whether his suggestions are correct.[15] Although a feeling of oneness may result from meditation practice, little current research focuses on the Buddha's main interest of incorruptible insight into discontent, change and the non-substantiality of the self. He claimed it was the successful pursuit of this goal that resulted in liberation from suffering.

Earlier I described the ways attention reinjects the cortex into automatic processes and reconditions or deconditions internal response reactions. Meditators observe this in their meditation. My depression is still sometimes painfully invasive but no longer grips the way it used to. Watching it come over and over again has partially broken the cycle of its effects. On a walk one day with a friend and fellow meditator, he related the story of his paralyzing hypochondria, inherited from his mother and grandmother who were both terrified of disease. It affected his life destructively. He once fainted on hearing an illness described. Finally, he sat for days on a retreat with a pain he was sure would kill him. After living with and investigating the pain and its accompanying terror, the cycle of fear began to abate. Such episodes in meditation can be dramatic.

Sitting with fear, one feels bodily responses. The mind tightens and "you" want to be anywhere other than where this fear is. Your mind runs all kinds of explanations of why there is fear. It constructs strategies for escaping the fear. There is anger towards the causes of the fear, towards yourself for being afraid, towards God and the universe for abusing you. If there is sufficient concentration, the meditative strategy of investigation is like tracking animals in the woods.[16] One backtracks, carefully observing the arising of fear as it is manifest among the five aggregates in the body, mind and consciousness. As with indigenous people on a hunt, the tracking is done silently so as to be completely alert and not to disturb the prey. In the course of doing this, the fear sometimes passes. The passing away of a strong emotion is like a moment of grace. As one pays attention, the feedback loops between thought, the amygdala, and the somatosensory cortex dissolve and one's chemistry changes. You can feel it in your breath. The importance of observing the breath is underlined by the fact that an entire Sutra entitled "Mindfulness of Breathing" is dedicated to it. At the point an emotion no longer has any kick, the breath slows as in a sigh of relief.

The cortex not only monitors information about the external world and the condition of our bodies, but also controls heart rate which may be why breath is so crucial to meditation. When the breath slows, it signals changes in one's being. This has been noticed by meditators for thousands of years but it eluded science until meditators became interested in neurophysiology.[17] Meditators notice that thoughts usually rise on the in-breath and subside on the out. Sometimes you can delay breathing in and thereby suspend your thoughts. Now it has been demonstrated scientifically that breathing out quiets the brain and breathing in activates it.[18] The amygdala that is so important to emotion is similarly affected. If you emphasize breathing, it increases the swings. When meditating, experienced Zen monks spend less time breathing in and more out and their rate of breathing becomes slower than when they are at rest. Experienced Transcendental Meditators who repeat a mantra over and over as the core of their meditation practice, have periods when thoughts stop. These overlap with pauses in breathing and their brain signs are not at all physiologically like the drowsy prelude to sleep. They are quite alert. In all of this are the beginnings of a scientific understanding of meditation and what happens when it delivers the Buddha's promised release from suffering.

The breath measures our being in a manner similar to how my dog, Sasha, mirrored me. I hand-fed her as a pup and raised her using a method borrowed from the socialization of wolves. She bonded deeply to me and would spend days lying near the entrance to the Insight Meditation Society while I was meditating inside. As she lay there, her eyes met those of the meditators coming out the door. She became so attuned to their feelings, they would ask me if she was enlightened but I knew she had plenty of wolf-like habits. Her killing and eating a feral cat in the woods would have shocked her admirers.

It took me a number of years to realize that she often knew what I was feeling before I was conscious of my own state. She would look sad for days. I would ask her what was up, only to realize she was mirroring my mood. I had been sad and had not been aware of it. In her last years, I would turn to her as a reflection. It is similar with the breath. You may have all kinds of ideas about what is going on inside, but turning to your breath is a way of anchoring the attention in reality. You may not be able to name the state, but its expression will be evident. When I sit with my undergraduate students at the beginning of our meditation class, I ask them to examine what they have carried into the classroom. A collective sigh of settling is often felt.

As a Tibetan teacher put it, "Meditation provides some gap in the movement... That gap creates a sort of chaos in the psychological process...that...helps us see what is underneath all these thought patterns, both explicitly and unconscious.... What is underneath may not be particularly appealing.... This layer is like the cloudy mind.... Any kind of thing you wanted to ignore...is put in this bank of confusion.... The probability is that the beginning meditator will have to go through all sorts of emotional and aggressive thoughts."[19] While addressing known difficulties, meditators open themselves up to hidden ones. For both, the remedy is the same. The Buddha and other teachers claim that clear seeing or real investigation will end forever the dominance of a troubling state. This has not happened among meditators I know. What does happen is that the context changes within which the emotion recurs and the emotion creates fewer destructive consequences. Still, troublesome behaviors can occur out of the blue, and old dormant habits recur after long intervals. When this happens, they tend to run to completion.

The Limits of Meditation

Uprooting of fear may not mean that fear is gone forever, but that it is held differently in one's being. Pain and pleasure are crucial ingredients in biological evolution. The mechanisms in which they are embedded were laid down long before thought and its fellow traveler, discontent. Fear and arousal are instinctual, so it is unlikely they will disappear due to changes in the thinking cortex. An advanced meditator is still subject to the operation of systems that create emotions. It may simply be that the power of the practice transforms how emotions ramify. Studies of fear and the amygdala indicate that it may be impossible to eradicate some conditioned responses. Reactions get built in subliminally and, while they sometimes subside, they may also be resistant to extinction or recur, particularly when a person is under stress.[20] Meditators are familiar with difficulties in their lives they simply cannot overcome, and meditation centers have tales of practitioners escorted from retreats to mental hospitals. There are limits to how much cortical reconditioning is achievable.

The tree that fell on my van

Little things like laughter are out of our control. Some highly regarded meditation masters are famous for their irrepressible sense of humor. Laughter as a secondary emotion initiated simply by a zest for life may be contagious. But responding to being tickled is at a primary level. When you are tickled by another, the somatosensory cortex is active. When you do it to yourself, the cerebellum signals that you are the source of the tickle, and the somatosensory cortex dampens its response. Response to tickling is almost biologically determined. Cortical reconditioning has a hard time overcoming this set up. When I was young, my sisters used to tickle me until it was painful. In order to get them to stop, I was determined not to respond so they wouldn't get a charge out of torturing me. Since then I have not responded to tickling, except when it came to one of my sister's children who would repeat the story of their mother's tickling their uncle and then jump on me. I couldn't resist. They conquered their unticklish uncle. With discipline I could almost control responding to tickling, but there is no way I can overcome the cerebellum's signal to the somatosensory cortex and its response.

When it comes to negative reactions, dysfunctional conditioned responses can be so inaccessible that psychologists and doctors look to drugs to treat them. Traumatic stress disorder may be ameliorated but its complete eradication is often impossible.

Moreover, the presentation of fear by meditation teachers is often out of accord with our knowledge of physiology. Fear is characterized as a state of anticipation. Meditators are instructed to carefully examine fear. By getting to know it, it will no longer control them. This paradigm of the fear mechanism is illustrated by a teacher who had lived as a monk in the forests of Thailand. He was terrified of snakes, which Thailand has in abundance. His fear conjured snakes at every turn. And when a snake was actually present, his fear shifted to being bitten, and when bitten, it became fear of dying from the bite. But anticipation is only one aspect of fear. As a secondary emotion, that is, as emotion in which the cortex stimulates the limbic system, his anticipation is quite capable of initiating fear and then increasing it as thoughts become obsessive and the cortex continues to bombard the amygdala. But fear is chemical in the body and can occur at levels below thought. Subliminal fear can undermine survival as in rabbits who die of fright when confronted by a predator from whom they might be able to escape. I do not know whether the neurochemistry of rabbits indicates that they are in a state of perpetual anxiety. If they are, then few would claim that the source of their anxiety is conscious thought. For humans there are many non-anticipatory instances when there is just fear—inchoate, gripping fear.

In December of 1995 I was sleeping in my van during a powerful California rainstorm. Early in the morning a blast of wind blew a large hunk of a tree some forty feet, landing on my van. (See Illustration 5.) I heard the crash and saw branches surrounding the windows. I looked at the ceiling of the van expecting it to come folding in. When it didn't, I exited through the only unblocked door into the driving rain. I saw my van almost completely surrounded by the branches of the tree. Then the fear came. I went into a house nearby where I meditated for a half-hour. I was completely safe. I wanted to really feel what the fear was like and not dissipate the energy through nervous action or talking to my friends about it. While meditating, I observed it as just fear coursing through my mind and body. There was no story line.

This fits with known neurophysiology. Visceral responses to primary emotions take time which may be why there is often clarity in the midst of emergencies. Afterwards one gets rattled. I have calmly saved myself from raging torrents, chosen to drive off a cliff to avoid an automobile accident, and gotten a drunk away from an oncoming train. Only later I felt it. Fear tightens the gut. Terror loosens it. As secondary emotion, you can create fear by anticipation. When I finished revising this section and the subject "being afraid" was very much on my mind, I was in my northern California kitchen cooking dinner when the San Andreas fault, a few miles away, let go a 5.0 quake. It was like an explosion. I ducked with heart pounding, very aware how quickly my mind imagined the hill above me tumbling down. I ran out of the house and looked up at the quiet, undisturbed hillside. I didn't know it was an earthquake. I was scared. Nothing made sense. This fits more with the teacher's description of fear. Before these events, I often wondered whether the hill behind my house was safe.

Another example of inchoate fear comes from a beginning meditation student who had an undiagnosed dizziness which upset her. She was also struggling with a series of other traceable disorders. In her meditation journal she reports:

> I felt more "motivated" tonight, whatever that means. I got back to gentleness, concentrated on my breathing for a long time, saw that it is not such a big deal when I get carried away on thought and worries... I can just return and say "thinking" and not try to analyze the whole thing. I spent a while focusing on the motion, really following it. It was/is so intensely chaotic, that

> it got really difficult and brought up so much anger and anxiety: raw, in-the-moment anxiety, not the kind that comes from thinking about the past or future. It was like the kind of adrenaline-based fear that comes when your body knows you're in danger or something's not right with it. It seems so much more difficult to deal with because it seems more valid or reasonable or real. I can't just call it thinking and move on because it's my body's natural reaction to the very real and powerful and disturbing dizziness/motion.

Her observations were more accurate than her meditation instructor's description of all fear as anticipation. That only applies to fear when it occurs as secondary emotion.

Powerful emotional events can happen at any time in life. There are famous meditation teachers, high lamas and Zen masters who seemed to have overcome anger, lust or greed in their lives, but when confronted with circumstances never before encountered, committed transgressions and showed a shocking loss of honesty and even viciousness.[21] Even one of the Buddha's disciples whose awakening he attested to treated other followers unkindly after the Buddha's death.[22] So, while meditation can lead to reconditioning in extraordinary ways, that reconditioning happens within a biological system subject to change. While the system is functioning, any of its features, even meditation achievement, is subject to breakdown.

Although the Buddha presented his techniques as having universal application, teachers recognize that meditation does not work for everyone. As if with a wave of the hand, Buddhist rhetoric would have it that the people for whom it doesn't work do not try hard enough or that they give in to destructive tendencies. That can be an accurate assessment, but it also risks blaming the victim. A more thoughtful explanation comes from the idea of karma. Ajaan Cha tells a story about one of his monks. In Thailand monks go walking to nearby villages on alms rounds where they are offered food for the day. On one alms round, this monk accidentally looked at a tribal woman who was making an offering. He was smitten and could not escape his longing no matter what practices he undertook. Ajaan Cha sent him to a distant monastery to forget about the woman but he was obsessed. Later he abandoned monkhood, found the woman and married her. Ajaan Cha's comment was that they had been married in a past life and there was no resisting the force of karma.

This was seen as an unavoidable failure. In Buddhism, a person has many rebirths on the path to awakening. One may be able to progress only so far in a certain lifetime. Wherever people end up in their struggles, that is their karma which is acceptable because there will be rebirth and new opportunities. This reasoning assumes that it isn't meditation that is limited, it is just the person's karma which limited their application of it. So the failings of a teacher of high meditative achievements are not laid at the feet of meditation itself, but the teacher's flawed karma.

What about natural changes rather than bad choices? I am unaware of anyone who has seriously investigated how meditation adepts handle the breakdown of the brain and mind as a result of disease and old age. When my mother became senile in her last years, I began to ask meditation teachers whether meditation masters suffered similarly. I wanted to know whether senility was a matter of psychological withdrawal which had a remedy in the cultivation of meditative awareness, or whether it might be an organic disorder over which even a great meditation master might have no control. I got very few responses. One celebrated Tibetan Lama brushed the question aside as if demeaning to the dignity of Tibetan Buddhism. When pressed, he hinted that there were senile old Lamas in the monasteries but offered no descriptions. From his answer, I inferred that there are indeed organic limitations to meditative powers but that it wasn't a topic for public discussion.

In his early sixties, Ajaan Cha began to suffer from water on the brain. He declined into a coma where he remained for ten years until his death. Among the various stories about him during this period was that of a former monk who went to see him while he was still alert. The student jokingly said in effect, "Well old man, it is just the body," implying that great meditation accomplishments would protect one from suffering arising from disease. Ajaan Cha called him up short with a commanding, "Never, never underestimate how difficult it can be!" While Ajaan Cha was in a coma and in a hospital in Bangkok, his monks attended to him. One day when nobody else was around, one of his Western monks who possessed healing powers tried to revive him with some new age technique such as breathing along with him softly and, saying "ah" on the exhale or rubbing his chest. While the monk was doing this, Ajaan Cha reportedly sat bolt upright, roared, and then fell back into the bed. This terrified the monk. He got a clear message not to mess with what was way beyond his understanding. There are a number of stories about how Ajaan Cha expressed disapproval, particularly when worldly concerns were brought up in his presence. At another time when two monks were arguing about politics in his presence, he spat a great wad of spittle, which stuck on the opposite side of the room.

I asked one of Ajaan Cha's caretakers about the quality of mind Ajaan Cha had after his stroke. Although Ajaan Cha could only communicate by wiggling his toes, the caretaker could sense from Ajaan Cha's eyes that he seemed to understand and was in a very clear space. The monks felt very bonded with him. It appears that no one thought about setting a yes-no form of communication with him through the wiggling toes. After a number of years, he seemed to recede; i.e., not respond as much. It was as if he had withdrawn so inwardly that he could no longer project his presence, even through his eyes. Before the withdrawal, the caretakers thought he enjoyed it when monks would come and chant to him. It was as though he became peaceful. Also during medical episodes, he seemed to withdraw inward. The doctors would take this as his failing and would apply more effort. The monks were finally able to convince the doctors that he was just relaxing or releasing in the face of the medical incident. It was not a sign he was dying.

How are we to understand the relationship between Ajaan Cha's meditation abilities and what happened to him? His monks, who were devoted to him and cared for him for many years, felt he had a very special presence of mind in the midst of his ailment. Because Ajaan Cha was so well known at the time he became ill, the secular authorities took control over his care. Although Ajaan Cha had expressed the wish not to be kept alive, he had become too famous to be allowed to die even though an equally renowned forest monk of his generation felt that this was wrong. Was he at peace, especially as he degenerated, or was there no person there to meditate at all?

Another meditation teacher, Lama Yeshe, was a revered Tibetan Buddhist monk. He described what happened to him after an almost fatal heart attack. He was so traumatized that it was impossible for him to meditate or maintain his presence of mind.

> Due to this medicine my mind was powerlessly overcome with pain every two hours and my memory degenerated.... Some days I could not do my commitments.... My ability to recite prayers of ordinary words degenerated and, after considering what to do, I did stabilizing meditation with strong mindfulness and introspection. By the power of this there arose clarity of mind. Within this state I continued stabilizing meditation with great effort, and this was of much benefit, though the enemy of lethargy often overcame my meditation.[23]

While Lama Yeshe recovered enough to overcome the disorientation of medicine, pain and disease, he describes his condition as having become, "the lord of a cemetery, my mind like that of an anti-god, and my speech like the barking of a mad dog."[24] He also asked the monks to whom he wrote this description not to reveal it to anyone lest others not understand.

Finally, there is the story of a Chinese Zen monk who was tortured by bandits for his robes. Although most Zen stories praise the fearlessness of monks in the face of death, this monk's screams were supposed to have been so loud and awful that they brought another monk hearing them to enlightenment. We don't know if the murdered monk was able to maintain equanimity while being tortured.

In considering the limits of meditation, we come to a dualism between science and meditation. From the point of view of natural history, meditation is a function of the mind, and the mind is a biological manifestation of the brain. So it would be no surprise that meditation may be limited by built-in characteristics of the brain or by malfunctions in the brain's operation. From this perspective as science learns more about the brain, it may better describe how meditation works and may even substitute pharmaceuticals for meditation practice. Of course there is the other view implicit in Ajaan Cha's subliminal communications with his monks. There may be more than meets the natural historical eye in meditation. I have encountered accomplished meditation teachers whose illnesses ought to have killed them. Yet, they communicated clearly with an energy that seemed to come from some place other than their body. There is no simple answer.

Dreaming

Although it is not immediately obvious, an examination of the processes of dreaming can reveal much about how the mind works and meditation's place in it. Dreams occupy an honored role in the human drama. The Hebrews were granted freedom because Moses interpreted Pharaoh's dream. Australian aborigines dream time and space. Hunter-gatherers dream of animals and demons and regard their dreams as omens. Dreams are part of the Native American "spirit-that-moves-in-all-things." Since Freud, dreams are thought to have a special place in the unconscious.

In contrast, Buddhist sacred texts do not emphasize dreams. A detailed index of subjects covered in the Buddha's middle-length discourses does not list dreams. In the only textual reference I have come across, the Buddha said dreams are the "fire which smokes by night and flares up during the day."[25] An early commentary says dreams are due to bodily disturbance, mental indulgence, wish fulfillment and the interference of devas; or that they are prophetic.

In their stories and biographies, the monks of Southeast Asia speak about dreams in this way. Monastic dreams indicate directions for meditation or alert the dreamer of the risks of violations of precepts. Dreams are thought to be a guiding spirit in practice. In Buddhism, dreams are an uncomplicated, special kind of knowing, as real as anything else. They are states of mind, part of the fourth aggregate, but different from thought. The *Visuddhimagga*, the great fifth-century codification of Theravadan practice, cautions monks who choose to do walking meditation in the charnel grounds. A monk or nun is instructed to walk up and down with eyes half-open noting what is seen so that the images will not haunt him or her at night. Charnel ground meditators are not supposed to eat certain things which non-humans like, and "even if such beings 'wander about screeching, [a monk] must not hit them with anything.'"[26]

Given that dreams are ranked nearly as important as thought, it is an odd omission that so little attention in meditation is paid to how they occur and what kind of phenomena they are. Although psychotherapeutically oriented Western meditators give credit to dreams, this has not been true of some meditation teachers. J. Krishnamurti, an Indian teacher, whose teachings were close to the meditation part of Theravada Buddhism, said something to the effect that the unconscious mind is thoroughly as trivial as the conscious mind. Like other meditation masters, he saw thought at the root of fear, anger and sadness. Thought for him almost equaled discontent, and dreams were just more of the same.

A poet's dream which was put to music is a classic example of Krishnamurti's assertion. The Chancellor in the Gilbert and Sullivan operetta, "Iolanthe," sings,

> ...your slumbering teems with such terrible dreams that you'd much better be waking;
>
> For you dream you are crossing the Channel and tossing about in a steamer from Harwich,
>
> Which is something between a large bathing machine and a very small second-class carriage,
>
> ...
>
> But this you can't stand so you throw up your hand, and you find you're cold as an icicle,
>
> In your shirt and your socks (the black silk with gold clocks),
>
> crossing Salisbury Plain on a bicycle:
>
> ...
>
> You get a good spadesman to plant a small tradesman (first take off his boots with a boot tree),
>
> And his legs will take root and his fingers will shoot, and they'll blossom and bud like a fruit tree,
>
> ...
>
> But the darkness has passed, and it's daylight at last....[27]

When I look closely at my dreams, they are no more coherent as in the following:

> I am being taught to make doughnuts, rolling them out properly by a big cook whom I don't know. I am thinking about how there is sugar in the dough and listening to the animal living in the walls which sounds like the sighs of people making love and thinking about the Gilbert and Sullivan song about the nightmare of crossing Salisbury plain on a bicycle, when I am woken by the cat scratching next to me. Then I see the bright moon heading toward moonset. A bit later, lying in bed while trying to meditate myself back to sleep, I have a strong image

> in the tiredness and attempt to be clearly aware of mixing a pot with oatmeal on one side and spaghetti sauce on the other, with a large beef bone sticking out of it. My thought is that this is quite a yucky mixture. Then I think how accurately Gilbert and Sullivan portray dreams. If they are closely attended to, they are like a mish-mash of daily thoughts.

If we shift the focus of our discussion from the content of dreams to the process of dreaming, using neurophysiology and meditation, we can see that dreaming exposes ways the brain and mind work. Ajaan Cha spoke about the strange and wonderful animals which appear at the still pool of awareness revealed when one becomes concentrated and silent in meditation. Sleep and dreaming contain some of these. One meditation adept gives his version of sleep and dreaming.

> In deep sleep consciousness retires into a state of repose, as it were. When consciousness is absent there is no sense of one's existence or presence, let alone the existence of the world and its inhabitants, or any idea of bondage and liberation. This is so because the very concept of "I" is absent. In the dream state a speck of consciousness begins to stir...one was not yet fully awake... and then in a split second in that speck of consciousness is created an entire world of mountains, rivers and lakes, cities and villages with buildings and people of various ages, *including the dreamer himself.* And what is more important the dreamer has no control over what the dreamed figures are doing.[28]

Studies of meditators' brain waves show they fall in and out of sleep. Nineteen percent of the time they are drowsy and forty percent of the time they are in various stages of sleep. There is no evidence that while meditating they are in the active dreaming states of sleep, but this may be because the studies were done in the afternoon when that rarely occurs.[29] Anyone who meditates knows that sleep is a big problem. Besides the fact that meditators fall asleep a lot in the first few days of retreats, meditators go in and out of sleep states even when they have settled into a continuous schedule of meditating. It was shown in studies of Transcendental Meditators that they become very still and alert to even the comings and goings of sleep. So, by using meditation, it is possible to observe how we drift in and out of sleep.

What is going on in the brain when the exotic imagery of dreams fills the mind? The brain has three main states: awake, deep sleep and what is called REM (Rapid Eye Movement) sleep. There is a neurochemical switching system in the base of the brain which controls the upper brain where feelings and thoughts occur, and the lower brain where sensations and motor control are lodged. When awake, one of the systems predominates; when asleep, the other. During sleep, an active signaling may take place which is similar to waking input from the senses, except that the signals originate in the mind. The mind may also respond by sending out motor signals to the muscles but these are interrupted in their passage. Different stages of sleep have been identified by sleep scientists. In the deepest sleep, people do not dream or have simple unemotional dreams. Dramatic dreaming in humans is indicated by Rapid Eye Movement, REM, or occurs as one falls asleep in the late evening and just before waking. REM sleep can be found in other animals. It is usually, but not always, associated with certain neurochemical changes and a burst-pause discharge pattern in the brainstem that gives rise to the rapid eye movements and twitching. While sleep may be necessary to repair oxidation damage to the brain, REM sleep's function is unclear. REM sleep may be an evolutionary concomitant of the development of a mammalian cortex crucial to mammalian success.[30] Because human minds are so interwoven with brain functions it gives rise to some extraordinary effects. Studies of dreaming beginning in the 1950s involved subjects' brain waves. Subjects were awakened during REM and asked to record their dreams.

There are several schools of thought as to what REM sleep is and what kind of mental-physical activity it represents. One sees dreams as arbitrary and irrational, another as not too far fetched and sufficiently consistent to be psychologically meaningful.[31] My own experience and understanding of dreaming, as we will see, lies somewhere between the two. We'll begin with J. Allan Hobson, proponent of the first view.[32]

During sleep, most motor control is shut down. While dreaming, small twitches of the face, head, and limbs may be associated with dreams of running, flying, and swimming. (Sleeping cats assume postures as if they were being threaten or startled.)[33] Rapid heart beat and sharply increased breathing indicates panic and anxiety. During REM sleep, the neurochemicals adrenaline and serotonin are absent and the brain/mind comes under complete dominance of the sleep system. "[A]nything goes in the image, idea and feeling factory. Her brain-mind was like a pinball machine with all its lights flashing."[34] The motor pattern generators of brain stem and upper brain are pumping out, but the signals are not relayed to muscles. Attention fails because there is no way to focus on overactive image production. The brain has switched from a top-down to a bottom-up control, "where volition is swept away....Thinking is like a motor act in that it requires top-down control. But with those [electrical] waves rolling over her brain and those showers of [neurochemicals there is a dampening of] any spark that persists of the 'I' that watches, weighs and wills her actions when she is awake."[35] She knows who she is in a dream but cannot control anything.

This disorientation in dreams implies how crucial to survival and procreation are the features of orientation in our waking life. Incongruity, discontinuity and uncertainty predominate in dreams. In dreams, we are drowned in a vague sort of "i" and a dizzying array of signals. We lose the ability to orient because there are no space or time cues. In a waking state, the sense organs are always feeding novel inputs and one doesn't get habituated to novelty.

Dreaming may be like the decline of the brain as we age. In my search to understand my mother's senility from the point of view of meditation, I concluded that with age the five external senses—seeing, hearing, smelling, touching and tasting—become increasingly unreliable. In advanced senescence people lose most of their taste buds. Food becomes tasteless. This leaves only the sixth Buddhist sense, mind, as a source of input. And now it is speculated that long-term memory storage fills with age, so in old age there is little room left for short-term memory items to be stored. Thus some senile dementia, such as living in the past, is a response to the facts both that the other five senses become unreliable and that recent events are not retained. Persons retreat into their minds where there is little current reality. They live as if in a dream. And dreams are disorienting places in which to live.

It's hard to convince ordinary people that their dreams are not significant. And mythical stories which are at the heart of hunter-gatherers *Weltanschauung* read just like Hobson's dreams. Examples from the Haida include girls going berry picking and coming across a one-sided man whose heart and lungs jumped out of his missing side when he breathed or a father dancing until he broke in two; then feathers and serpents emerged from his body. Or people arriving in canoes landing on the upper tier of a house and when they depart they proceed to the upper tier where they lash themselves in their canoes, lose track of themselves and awake to find themselves in open water.[36]

Hobson's version of dreaming is contested on two grounds. The neurochemistry has been questioned and research into dream content shows much less of the bizarre than Hobson's studies indicate.[37] Hobson discounts the latter because he feels that dream memories are not very accurate. The amines of the waking state chemical system which stimulate memory cells are not present so the cells are dulled in sleep.[38] If you observe closely when you wake up from dreaming, you will notice that memory of the dream rapidly fades unless you rehearse it in your mind or quickly write it down. It has been hypothesized that dreaming functions to reorganize and reinforce memories from the day. They may be part of the process which turns short-term into long-term memory. As anyone knows who has tried to figure out their dreams, we want our dreams to mean something, yet Hobson comes up with a contrary conclusion.[39] He feels that the plot imposed on dreams may be a way of fixing memories and relating them to action scenarios, and thereby classifying them for storage in long-term memory. REM sleep may also be an intense exercise of brain-mind internal operation necessary to prevent memory decay. It may also rehearse potential actions. In contrast to waking fantasies, dreams are two-and-a-half times more crowded, impersonal and less detailed. They are more distant from present events and more exotic. Characters' identities are less distinct but both known and unknown persons in dreams sometimes have peculiar characteristics, like the Haida half-man with his heart sticking out of his chest. Hobson claims changelings never occur in dreams. The distinction between problem solving and fantasy we easily make while awake is blurred in dreams.[40]

Then there is the phenomenon called dream splicing, established in numerous studies. Any sequencing of dream subplots makes as much sense as any other because there is no overriding plot. Researchers conclude, based on these studies, that the extraordinary meaningfulness we assign to dreams is misplaced because the random images of the dreams can be made sense of in so many ways.[41] Hobson's critics sees dreams as closer to ordinary thought, with less of the bizarre. The differences of opinion about whether or not dreams have meaning may be unnecessary. However confusing and arbitrary the content of dreams may be, they are made up of perceptual, emotional and mental materials from a person's life. Although in dreams it may not be possible to locate a well-defined ego in the sense of Western psychology, and while dreams can be taken as support of the Buddhist idea of no-self, we do not dream each other's dreams. So dreams may be used as Rorschach tests on which to drape meaning even if they are a more or less disorderly process of brain and mind.

One of the interesting things about dreams is their tendency to contain more anxiety, joy or anger than sadness, shame or remorse. Studies indicate that thirty percent of dream content is anxious and twenty-six percent joyful, but a total of sixty-eight percent is unpleasant.[42] Hobson thinks dreams are like the waking state of someone with an organic mental disorder. It's a delirium psychosis associated with disorientation, distractibility, spotty memory, confabulations, vasomotor hallucinations and delusions. The religious thinker Emmanuel Swedenborg heightened dream awareness by keeping a dream journal, sleeping irregular hours and depriving himself of sleep. While dream content is individual, the form in which dreams occur is universal. From the perspective of evolutionary adaptation, dreams may be rehearsals of learned behaviors in many possible circumstances without the risks of real life consequences.[43] The rehearsal would have to have subliminal effect because we don't remember most of our dreams and the control parts of the brain are inactive. If this is the case, then the anxiety component of dreams, although unpleasant, has survival value. Like other antagonistic pleiotropies, the brain may overdo it.

It is not difficult, from watching the mind in meditation, to conclude that the waking mind is equally insane as the dreaming mind. Spending a few weeks sitting quietly in a meditation hall watching your thoughts come and go, watching your moods swing from pillar to post, trying not to add to the bustle by engaging in more thinking about what is going on, gives some perspective on the nature of mind and thought. Without attacking clinical psychologies that place the subconscious and its communication via dreams in a pivotal position in the psyche, I know many meditators who've gone through years of therapy without finding real relief until their meditation practice deepened. It is not that one doesn't learn from therapy, and it is not that meditators who have not had therapy do not need it, it's just that placing thoughts and dreams on a pedestal may reify them even though some troubling content may be alleviated. Meditation, on the other hand, aims to dismantle them.[44]

In Tibetan Buddhist meditation there is something called dream yoga.[45] It involves visualization practices. As one can use concentration to bring attention to the five aggregates, attention can be brought to dreaming. Some of Swedenborg's practice was to engage in lucid dreaming. This involves techniques to alert yourself to the fact that you are dreaming while you are in a dream, and then exert some influence over the direction of the dream. A trick Don Juan points out in the Carlos Castaneda books is to decide before you go to sleep to look at your hands in your dreams. He says that when you can do this you can control your dreams. Other tricks include looking for the incongruities in the dream, trying to look around in the dream or making dream life more active. Dream yoga utilizes techniques similar to lucid dreaming. Although the Tibetan teachers' ultimate reason for doing dream yoga is to transcend dreams as elements of the fourth aggregate, they sometimes become fascinated with dream content as Buddhist metaphors.[46]

In a meditation hall after lunch on a warm afternoon, energy plummets and there are plenty of opportunities to observe falling asleep. Qualities of awareness begin to get hazy and, without noticing it, focus becomes dull and attention does not really wander, but cannot hold an object. Many times you think that you are perfectly attentive to what is going on only to have your body jerk sharply as it catches your head falling toward the floor. When so startled, you realize that you have been asleep and did not know it. At night you can set yourself to observing "falling asleep." In meditation interviews, some precise teachers ask students whether they fell asleep while inhaling or while exhaling. I do not know whether I have ever caught the exact breath, but I have observed the way in which a woozy sort of fantasy overwhelms attention and thoughts go from being recognizable as thoughts to being a chain of dreamlike associations. In the middle of a dream you can become aware that you are dreaming, even labeling what is going on as dreaming and you can exercise control. There are two kinds of control. One is the kind aimed at in lucid dreaming wherein the content of the dream is given shape or plot. The other is not really control but attention in the spirit of the Buddha. One sees the dream for what it is, a movement of mind. It arises. It is pleasant or unpleasant. One becomes attached or pushes away.

I had a lucid dream one summer while sleeping in my open van on the side of Lama Mountain. I dreamt I fell out of an airplane. As I was falling toward the ground, I was terrified. Then I realize that it was just a dream and that I would wake up before I hit the ground. I thought, "What nonsense the mind is filled with, why not just live out the absurdity of the dream?" So instead of waking up in fear before crashing, I purposely crashed myself into the ground. Nothing happened. I woke up laughing.

A final place where one can see dreams as mind objects is on waking from a dream state. Sometimes when you wake in the morning and then fall back asleep, you dream but not from a place of REM sleep. On a recent morning, I was a reporter in Vietnam and then all of a sudden as I transited to wakefulness, I was reporting on hotney (a nonsense word) in prison. My first thought was that I was thinking nonsense but then my sleepiness let the original line of thought continue as if the dream were going on. As you transit from sleep to waking, your sleep-awake chemical systems are switching gears. It is easier to be aware of the dreaming process here because your waking mind with its more pointed attention and better memory is coming into play. In REM or deep sleep, dreams are more remote. As you emerge into wakefulness, you can observe that your dream is just you talking to yourself. And as we saw earlier, self-talk is nothing other than one thought following another, and we credit each as a different voice. Dreams then may be regarded as thoughts generated by, and washing around in, a sea of disorienting neurochemicals. Even Darwin thought dreams were a continuation of thought, although he was puzzled by their confusing images.[47]

Many poems compare waking life to a dream. Although referring to the characters in a play, Shakespeare could have been alluding to life itself when he wrote,

> We are such stuff
>
> As dreams are made on, and our little life
>
> Is rounded with a sleep.[48]

In Buddhist literature, both daily life and sleep are often compared to dreaming. Some of the psychologists I have cited feel there are similarities between dreams and wakefulness.[49] Gerald Edleman calls what we live, "the remembered present," and Robert Ornstein thinks because of the time it takes to process stimulations and because we are always flitting between memory and planning, "We are hopelessly late for consciousness."[50] The ability to lucid dream is facilitated by questioning the reality of one's experiences during the day. This fits well with these views and dream yoga.[51] In a study of highly trained Transcendental meditators, a majority could lucid dream, witness their dreams or witness their sleep. Over years of meditating on dreams first the actor was dominant but it still felt like the dream was out there and the self in here. Then the meditator became aware that out there is really in here and they could engage dreams. In the third stage dreams were more like thoughts and the meditator could note them and the dreams were not gripping. In a fourth stage, called witnessing, dreams had no sensory aspects, no mental images, no emotion. "I am aware of the relationships between entities with the entities being there.....Just expansiveness....like light in an ocean...."[52]

It is not clear what to make of this neurophysiogically since the frontal cortex is so inactive during dreaming. Studies have yet to be done to show whether during dreaming these meditators' brains are behaving differently than non-meditators and, if it is so or not so, what that means. Taking the testimony of meditators at face value, it appears that one can be aware of one's dreams and observe how the various layers of experience fit together. We begin to see how different we are from the way we usually conceive of ourselves.

It may be that our waking memories and thoughts provide no secure place to stand other than as tools for survival. They do have dreamlike qualities. Nonetheless, the neurochemistry of dreaming is quite consistent with meditators' awareness of it. The chemical norepinephrine is essential to attention and it is diminished in dreams and boredom. The thalamus, using norepinephrine, controls attention by tuning out extras and tuning in the point of attention.[53] No wonder it is so hard to meditate while dreaming: the neurochemical involved in concentration is scarce. A dose of mild amphetamine helps focusing, but creates other problems such as frenetic thinking. There is a tradeoff between attention made possible by the waking system and the richness of association connected with the dream system. These are analogous to clear manipulation of sense data and thought, versus background speculating and caressing of plans and memories. The former ties us to survival; the latter is important in planning edifices and dreaming our dreams. It can lay the foundation for great accomplishments but can also be a world lost in fantasy.

Insight

Our lives begin with a series of physical and mental developments and unfold in moods, emotional reactions, thoughts and actions. These all occur in bodies over which we have limited control and minds in which a sense of self resides. The biological imperatives under which we operate include the maintenance of our bodies with food, protection from predators and the elements, and the reproduction of our species. As a social animal, we meet these challenges connected to other humans. That we are obsessed with food, money, weather, sex and relationships is natural. To deny our animal roots is to miss how we are constructed. Fear and greed are necessary for survival but have destructive downsides. Over the last several thousand years, many people have observed that completely giving in to our animal nature is to surrender the possibility of transcending suffering. Although it might strike some as repressive, one forest monk kept the food at his monastery very simple. He asked his students, "What have you ever learned from a good meal?" He felt that observing the desire for food more tasty than rice is a better teacher than mindless gluttony.

From the point of view of meditation, learning comes from an intimate ongoing examination of how our lives are put together. Pleasure, pain, wakefulness and sleep modulate all of our states of being. Because of rhythmic neurochemical changes, attention is sharpest in the morning. For most people this is a good time to get work done. In the evenings, we tend to be more fanciful, sensuous, sliding into dreams. It is a time for eating, story telling, intimacy and sex.

Carlos Castaneda in *The Teachings of Don Juan* talks about the time of day when one is most vulnerable. For me that time is just after sunset when the quality of light changes in a particular way. I remember distinctly when the light hit that particular point while backpacking supplies in the mountains of northern British Columbia. I would begin to feel a bit stoned, unable to focus as I was trekking under a heavy load. It would be frightening and I would look about me for expected danger. A similar thing happens when I do walking meditation out of doors all night long. Around two in the morning, I can feel the hair on my neck rise. It reminds me of the admonitions to monks doing charnel grounds meditations. They are warned to be alert to the power of their imagination at night. The quality of light and the time of day affect body chemistry.

Research offers interesting suggestions about other ways neurochemicals influence behavior. The genetically determined physiology of dopamine receptors appears to be connected to novelty-seeking behavior. Psychologists test for temperament with categories such as the tendency for harm avoidance, dependence on reward and persistence. They feel that heredity and random experience affect how much of each quality a person has.[54] For the meditator all this is open to observation. Each of us has a different portion of fear, desire for approval, patience, and specific biorhythms.

Few meditators I know are interested in clinically recording their observations. There have bigger fish to fry. Meditators, like the Buddha, seek the origin of their discontent and how to undo it. Nevertheless, aiming a PET machine at an experienced meditator practicing the technique of carefully labeling the coming and going of mind moments might produce interesting results.[55] Boredom is chemical. Experienced meditators have spent incarnations with boredom and know it intimately. They are familiar with the feel of its energy and have watched the mind hold onto boredom and push it away. A friend who was a monk in Sri Lanka and Thailand for a number of years describes endless days of stifling boredom while sitting in the woods. In the course of his practice, boredom became interesting.

Similarly, the first few days of a meditation retreat are often filled with sleepiness. Then sometime during the second or third day of the retreat something striking occurs: the sleepiness recedes. One can observe the breath clearly and alertly. A noticeable transition in the chemistry of one's mind and body has just occurred. Is it due to a change in the neurochemical norepinephrine? Perhaps. When one is sick, it is difficult to meditate. With the flu, higher body temperatures suppress REM sleep, "leaving our minds in the thrall of the persistent obsessive thinking of non-REM sleep and the microdreams of repeated and unsuccessful sleep onsets."[56] For those who suffer insomnia or sleep shallowly, this description feels familiar. The sick and dying meditation masters mentioned earlier represent the limits of meditation's ability to offset imbalances in the brain.

And dreams happen. We want to hold onto those which are entertaining, pleasurable or support our sense of self, and avoid those which are unpleasant. I once fell into conversation with a street person I met in a liquor store on a hot summer day in the California Central Valley. It was one of those occasions when you communicate from the heart. He told me it was only the pleasure of his drunken dreams that made his life bearable. I was touched. As we departed, he went off to find a cool bridge under which to get drunk. In contrast, I often wake from discordant dreams with my muscles tense and mind all a-jumble. I know it is just a dream, but wonder why my mind so disturbs me when dream events bear little relationship to how I could possibly lead my life. As meditators know who have sat long retreats, dreams in the first days of a retreat can be extraordinarily disturbing. In the early years of doing intense retreats, I hacked my bother into pieces and fed the pieces into a fire. On a recent retreat, I wrote a detailed treatise on the petroleum paleo-chemistry of organisms I had never heard of, concluding with a poetic statement about chance in evolution and history: had evolution been different John D. Rockefeller would not have been able to make a fortune on the oil that was so necessary to grease the bearings of the later parts of the industrial revolution. At the end of the dream, the gates of hell were blown open by a terrible wind which swept past me. I swear I saw curtains, which were not there, billow. I woke with an incredible cramp of fear in one foot and heard a couple of raccoons fighting just outside my open window.[57] I later checked to see whether or not my dream writings on geochemistry were grounded in reality. I could not find any reference to match the names my dream gave to organisms. As I woke from the dream, it felt as though someone else was inhabiting my mind. I didn't know the things I was writing. When I look closely at my waking thoughts during a meditation retreat, they possess a lot of dreamlike qualities.

As clairvoyant as our dreams may be, we would not want to rely on them to make our way through the world, because our survival depends upon acute observation of our natural surroundings and the quickness with which we physically react to events. But waking thoughts also have many setbacks. Daydreams, obsessions, wanting and negativity are a piece of each day. I suggested that, in mimetic mind, there may have been a quality of Soen sa Nim's before-thinking-mind. I don't think that anyone would argue our ancestors, who both hunted and had social life using only mimetic communication, did not engage in violence and cruelty. As I cite in various parts of this book, archeological remains attest these were present before our verbal, thinking cortex reached its current dominance. What meditation contributes to easing the burdens of human nature is to use our cortical dominance to address the ways in which our thoughts and our underlying animal responses create discontent and harm. It allows us to tap into the mimetic aspect of before-thinking-mind to create a space around the excesses of the thinking mind. It reveals a gap of silence. In that gap, we have some access to our automatic processes. While access may be incomplete and the processes sometimes too deeply conditioned in buried parts of the brain to be affected, meditation nevertheless creates the potential for knowing and changing what is occurring. The change can come from a spaciousness which undermines the tendency of unobserved behavioral sequences to roll to completion, or it can come from a purposive intervention. In either case, the sequence of Dependent Origination is interrupted, and the discontented self, itself a byproduct, does not solidify.

In the silence are revealed some of the wondrous beasts who come to drink at Ajaan Cha's still forest pool. There are many more than we can imagine. In meditation, stillness is the secret of awareness and insight and thence, hopefully, wisdom. Since the being experiencing stillness is a human animal, the stillness is a product of its animal operation. As much as we might like to detach our spirit from our flesh we are not, as the Buddha emphasized, made that way.

Our flesh can perform like no other animal, especially when it comes to using brains. But as we have seen, evolution has left us with traits that are excessive if not destructive. It may be that meditation is the best we can do to cope with this antagonistic pleiotropy.

Observation

In the chapter on Darwin, I tried to show that explanations using natural selection often fell short of observed phenomena. The idea that adaptation neatly fit species into environmental niches is frequently belied by the fact that species were often adapted just well enough to survive in relatively unchanging conditions. Also, species sometime seems constructed on an *ad hoc* basis, not the way a rational engineer would have made them. In the last three chapters, we have looked at human traits which are arbitrary or downright troublesome. When natural historians look at life and try to make sense of it, they constantly have to go back into the field to look with more care. Life is incredibly complex and awe-inspiring, but it is neither as efficient nor as reasonable as we sometimes imagine it to be.

The cultivation of awareness in meditation reveals analogies to the way natural history observes. Sitting in a meditation hall year after year, one becomes aware of properties of being which are embodied in the five aggregates. Sometimes the awareness is subtle and comes gradually and sometimes there are moments of blistering insight. Time and again one notices how the mind draws back from unpleasant sensation or follows a pleasant thought into fantasy. It becomes obvious that discontent happens because we hold on too long to some experiences and push others away. This is canonized in Buddhism: attachment and aversion lead to suffering. Presented as a part of a system of beliefs, whether true or not, it is nothing more than another thought or mental object. It is not what is observed. The metaphor often used in Zen is that the finger pointing at the moon is not the moon, or as Rajneesh once said, "The word love is not love. The word God is not God." If meditators rest back in the generalization that attachment and aversion create suffering, they miss what is really taking place. In other words, the way in which our minds freeze experience into dictums is never quite adequate to the demands of life.

One good neurophysiological reason for this is that our thoughts are only fuzzy approximations of experience. But more important, as in evolution, explanations invoking natural selection never really account for enough of the variables or the particular historical circumstances to fully explain outcomes. And life, as the Buddha characterized it, is ever-changing. Nonetheless, the pursuit of temporary explanations cues natural historians where to look next. In a similar manner, meditative insights are only good enough for the moment at hand. If one holds on to them, one misses the deeper looking which the next moment requires. Insights are good pointers, but bad conclusions. An ingrained habit of the mind is to grab hold of cognition and make it into a solid object of mind. One often tricks oneself by treating thoughts about what has just occurred in meditation as if they were somehow superior or meta-thoughts and not just other thoughts, which is what they are. This is a variation on the illusion of inner dialogue. "Should I keep writing?" "No I have done enough." The pervasive habit of hierarchalizing thought may not be so evident if you have never seen it in your own mind. People who make their living or their identity out of the content of their minds have a hard time grasping this and often give up meditation. It is also very important in the internal dialogue relating to addictions, self-esteem and ego. Let me give two examples of its role in observation.

If my mind becomes quiet, I can observe the coming and going of sensations, thoughts and feelings. If I feel a pain somewhere, I can watch my mind rebel against it. If I am calm and concentrated enough, I can label the whole process "aversion," and be aware of the discontent the aversion brings. This unit of awareness may create a kind of space around what is occurring in my mind and body and give some relief. It also may not. If I freeze the labeling in my mind into the Buddhist truth that aversion creates suffering, then I have stopped observing what is going on in me, and I am back in my thoughts. If I let the characterization be an insight for the moment and stay with what is present, then I have the chance of seeing deeper. And since it is the mind's habit to claim experiences and turn them into thoughts, the skill of meditation practice is to let go of all mental credits like, "what a deep insight," and return to observing the process.

The second example has a Zen flair to it. On a meditation retreat I had an insight while sitting on the toilet. For many years, I used the mantra, "Who is lonely?" as an inquiry and a way of developing concentration. During the retreat my attention had become very concentrated and the concentration was accompanied by unusual physical sensations as described by the *Jhanas*. Since concentration sometimes gives rise to odd experiences, I was more entertained than disturbed. Then, one morning, while I was paying close attention to defecating, the answer came to me with intense clarity. "'Charlie' is lonely!" Each time that I made "Charlie" happen, I also built in "loneliness." It was such a familiar habit that I did not see it happening. For the moment, I realized that I did not need to do this any more. This is an example of the Buddhist idea of "no self." For a moment, my habitual identity was suspended. The insight made a space in my life. I was quite happy for a while. But the insight was too simple and immediately became a mind object. No self, no more loneliness! That would be like asking life to stop. So, the next step is even more careful attention.

When the mind quiets down in meditation it is possible to bring attention right up against what is being observed, such as breath, at which point there may be no room for the mind to run away with thoughts. In this case thinking may either go into abeyance or be like chattering in the next room, noticed but of little importance. As a meditation teacher said in answer to a question about whether the Buddhist notions of impermanence and no-self were concrete descriptions of reality, "Certainly the experience [of them] is essential to freedom, but the words about it remain simply concepts... in the deepest places of practice that I've experienced, the mind is silent. So those places are not announcing themselves as being one thing or another. The announcements always come after when the mind is talking again."[58] What science can understand of this state of silence—the gap in which meditation really happens—has yet to be seen.

At this juncture in our book Buddha and Darwin can be seen to have a similar understanding of the relationship between observation and knowledge. While the conclusions of both may have the appearance of permanent truths, they are, I believe, best interpreted as instructions for ever closer attention to changing life. In the next two chapters we take our natural historical and meditation descriptions of mind and look at how the changing circumstances of human life in the last ten-thousand years have affected the way in which discontent occurs; then we will be in a better position to address what to do about that discontent.

Chapter Seven
Attention and Wisdom

Hunter-gatherers

Hunter-gatherers present a challenge to civilized humans. They link us to a mostly forgotten past. The remnants of hunter-gatherer groups still inhabiting the world remind us how much civilization has altered the planet. For the Buddha the goal of meditation was to live with less discontent and rein in harmful actions. Since meditation makes contact with qualities hominids possessed before speech and because meditation uses the mind which was co-created with speech, it is illuminating to look at people who stood much closer to that divide than we. We turn now to hunter-gatherer attention and wisdom. We will subsequently look at what happened when humans invented agriculture and built civilizations.

Hunter-gatherer Attention

Included in the 1960s counter-culture rebellion was a back-to-the-land movement. Some drop-outs adopted rural lifestyles and became organic gardeners. A few fled to the wilderness to see if they could sustain themselves living intimately with nature. In the early

1970s, my brother and his family tried homesteading in the coastal mountains of northern British Columbia. Taking a leave of absence from the university, I picked up my friend Janet at the commune where she was living in Oregon and, loaded with sacks of grain, we drove north to spend the summer helping my brother and his wife build a log cabin. The homestead was just off the road to Telegraph Creek, a ghost town inhabited by a few whites and several hundred Tahltans, who belong to what is referred to in Canada as a First Nation.

Telegraph Creek is nine-hundred miles north of Vancouver, at fifteen hundred feet above sea-level on the banks of the Stikine River. Nearby are the glacier-covered mountains of the Coastal Range rising to fourteen-thousand feet surrounded by miles of emptiness. The area is a veritable Himalayan wilderness. Despite its remoteness the road proved too busy for my brother so, two years later, we built a new cabin, this time a seven-mile hike into the mountains. While working on the second house, Janet and I went walking in the bush. Wanting to dine on the sockeye salmon that had begun to run, I tied a gaff hook to a branch to use as a fishing spear, stripped down and stepped into the icy waters. With the image of an Indian standing still and attentive, spear poised, I waited to catch our dinner. (See Illustration 6.)

The task was not simple. The water was so cold I could not stand in it bare-footed. We remedied this with an old pair of gumboots some traveler had used to cross the stream. Again I waded into the creek. Then, because the sun was low, it sparkled in the creek and I could not see the fish until they were too close. So Janet perched on a rock above the creek for a better view. She shouted over the rush of the stream when a salmon approached. I tried my best to stand quiet and alert in the freezing water. I would yank the poised hook when something seemed to pass. After much frustrating effort, all we had to eat that evening was the dried food we brought.

Spearfishing by Janet Essley

I learned from this adventure that there's more to the Native Americans' foraging skills than pictures in children's books convey. In fact the literature on hunter-gatherers contains an awed respect for the skill they bring to sustaining themselves. Very few of the first Europeans to observe Native American fishing could spear a salmon, heavy enough to pull a person in, as dexterously as the Indians. As some who try foraging discover, you can starve out there. A man from a heavily wooded suburb of Boston and an Indian friend descended from the area's original inhabitants were unable to find enough to eat when they tried to do so for several days. And a young man tried subsisting a few miles off of a road in Alaska one summer relying on a book about native uses of plants he had swiped from a library. He died from eating a part of an edible plant which accumulated toxins at the end of its season, a fact known to the natives but not recorded in the book.[1] Two ingredients are missing from such modern attempts. One is the knowledge accumulated over generations of what is there and where to find it. Although my brother had learned from the local Indians how to net returning king salmon, I knew little about when and how to spear the sockeye which had just begun their run. The second missing ingredient is the proper kind of attention which would have enabled us to apply that knowledge.

In previous chapters, we saw how attention is a brain function. One psychologist sees it as an adaptation to the need to act effectively. That humans cannot do more than a couple of things at once with awareness may be a limitation of the brain's evolutionary construction. We compensate for this by relegating repeated behaviors to lower brain levels. By so doing we're able to drive cars safely while engaged in chatter, but interestingly not while talking on a cell phone.[2] Routinized behaviors are called automatic; those needing attention, deliberate. Animal behavior studies reveal that some mammals engage in deliberate actions. Chimps may even make rather complex plans.[3] The more highly developed minds of humans enable them to plan to pay attention. This trait allows us to use experience to direct further action. We sometimes undermine this capability because we spend so much time in our heads thinking, planning, day dreaming and obsessing that, without our noticing it, much behavior which might profit from being deliberate becomes automatic. Here we have a balance between the adaptive power of our cortical evolution and its antagonistic pleiotropy. Even the act of deciding what to do can become automatic, molded by the unconscious or memory. So many of our actions become automatic that, although we may like to direct our attention to where it seems most important, our habits channel our attention elsewhere.[4]

Reacting to the contemporary world some people look to indigenous cultures as models of how to live in balance with nature. Anthropologists and writers report the impressive abilities of native people to sustain themselves without harming their environment.

Their foraging skills seem out of the reach of more civilized citizens. As Barry Lopez describes it, "[W]hen a Nunamiut hunter goes out, he leaves his personal problems behind as though they were a coat he had left on a hook. He slips instead into a state of concentrated, relentless attention to details, the bend of the grass along trails in a certain valley, the movement of raven in the distance. It is customary for most biologists, on the other hand, not only to bring their mental preoccupations into the field but to talk about them while they are walking along. Eskimos rarely speak when they are on the move."[5] The Achaur, hunter-gatherers of the Amazon rainforest, can identify more than 262 wild plants, not all useful. They have 600 animal names of which 42 are for ants. Of the 240 kinds of animals they know are edible, they regularly eat 60. They can identify 33 species of butterfly. Western science has identified only half the cat species they distinguish. When presented with pictures of a member of a species, they can describe the depicted organism's behaviors.[6]

While hunting caribou, the Nakapi Indians of northeastern Canada don't talk. Half-running, every man takes into account the wind, weather and features of the terrain and relates such features to the position of their game. They hunt without spoken strategy. Each man reacts to the others' movements. Nakapi hunting has similarities to that of wolves with whom they feel a special kinship. Hunting is a sacred activity. Among the Yanoama, neighbors of the Achaur, the hunt is also done quietly.[7]

Through mime, games, stories, imitation and apprenticeship, young hunters are taught skills viewed as crucial to survival and central to their way of living. Some contemporary hunter-gatherers survive solely on foraging, while others have gardens. Prior to agriculture all humans foraged. We might learn more about human nature if we could enter the mind of our hunter-gatherer ancestors and experience how they saw the world. From analogy with modern foragers, it may be guessed that sometime after the rise of *Homo sapiens* one-hundred to three-hundred-thousand years ago, humans acquired an increasing ability to hunt and gather using attention, discrimination, and strategic skill while not necessarily relying on language. From the reports of modern hunter-gatherers, it seems that mimetic memory and communications are more than adequate for foraging and hunting. The preverbal human-like being in my mirror vision dates back to this stage of preliterate wisdom in the wild.

A student of Eskimos sees their hunting abilities as founded on careful, but not fully articulated, observation and knowledge of the environment.[8] They predict weather, note the handedness of polar bears and imitate how bears hunt seal. Some know the history and behavior of individual animals and species so well that they seem to read animals' minds. Their lives have much in common with those of their prey and, like them, have experienced severe privation. Their skills appear mysterious because they are alert to aspects of nature necessary for survival which elude their anthropological observers. Hunter-gatherers do not see themselves as separate from the natural world. In fact, many do not even have a separate and separating word for nature as we do.

Hunting, observing, telling tales and speculating about sources of food are specialized knowledge analogous to that of scientists. Such knowledge is a centerpiece of hunter-gatherers' daily life rather than part of an occupation separate from home and friends. The almost magical nonverbal way in which foragers hunt is based on mimetic and accumulated knowledge not easily accessible to the anthropologists observing them. Hunter-gatherers call upon the mind's ability to remember, categorize, abstract and plan actions before engaging in them.

Animals too possess large amounts of information. It may be instinctual, taught or learned through trial and error. Birds learn to identify prey from among tens of thousands of insects. It might take a lifetime for an entomologist or forest dweller to come close to making such refined distinctions. Different finches of the same species on an isolated Pacific island forage different foods with distinct techniques. They learn the techniques from their parents and other species. Their capacity to learn in this latter way is limited so they often hunt clumsily and frequently miss their prey.[9] While learning among birds is little better than stringing together a few behaviors, the Achauers' classifications extend into the realm of narrative memory and include things not necessary for survival. The Achauers' ability is not different from a sports fan who can remember thousands of baseball records or a taxonomist's knowledge of tens of thousands of plants. The mind of the hunter-gatherer and the scientist are the same mind. They are simply applied to different domains.

What is special about hunter-gatherers' attention is how it is applied.

According to Richard Nelson:

> While [the Koyukan Indians of Alaska] highly value the teachings of elders and hunters, they believe the only way to make knowledge whole and meaningful is through direct experience with the world. The brown creepers' call is twisted together with the forest and the cove, the breeze, the mist, the moss under foot...the unfolding of this day and the experiences of many days before.... I remember the Koyukan people's keen awareness of the changes in the terrain around them, based on what they had seen during their lifetimes and what the old-timers had seen before them....people could remember when their cabins stood where the middle of the Koyukan River runs today...they had watched lakes become ponds, ponds become bogs, bogs become forests. The land came alive....[10]

The attentiveness of Indians standing quietly, spear in hand, comes from this mimesis, from the way they learn their skills and the lack of separation between themselves and the surrounding world. Eskimos see themselves as much a part of the world as the animals with whom they live. Like us, hunter-gatherers are capable of having busy minds. But they seem to be able to suspend that busyness more easily. It does not arise when they engage in activities of survival which demand great attention and to which they feel deeply connected. They rest easily within these activities because the activities define their relationships within the world. As we will see there is more to their world than survival.

Attention and wisdom are fundamental elements of the Buddha's path. If hunter-gatherers have special skills of applying the former, do they use those to develop the latter? What can we understand about the wisdom possessed by our ancestors using simpler technologies and living closer to nature? In what way was discontent part of their existence and how did they deal with it? Is the Buddha's claim for the universality of the problem of suffering relevant to hunter-gatherers?

Historiography

In addressing these questions we run into the same limitations that apply to the issue of attention. All we have to reconstruct the inner meaning of prehistoric foragers' lives is crude archeology and analogies from recent hunter-gatherers. The extent to which the lives of recent hunter-gatherers are like our common ancestors is not known. Regarded as "people without history," their lack of written history was taken by their Western conquerors as a sign of cultural and moral inferiority. More sympathetic recent observers see hunter-gatherers as embodying eternal verities of nature. They trumpet tales of tribal elders as examples of wisdom which claim authority from ancient lineal transmission. But this too may be another way of treating a tribe as a "people without history." As early anthropologists turned a blind eye to changes their "primitive" subjects experienced as a result of contact with Europeans, so modern enthusiasts do not give sufficient credit to the historical changes natives' ancestors underwent which may not be known to the generation now living.

A classic example is Darwin's mention, in *The Origin of Species,* of the succession of vegetation on the mounds left by older Indian societies in the southeastern U.S. These forest-covered mounds were the remains of thriving city cultures whose inhabitants were probably killed by smallpox brought to the New World by the Spanish when they first landed in Florida in the early 16th century. The greatest of the mounds is in Illinois, the remains of a 13th century community inhabited by forty-thousand people. The explorer DeSoto, on his rampage across the south between 1539 and 1542, saw people in highly stratified societies living in areas of dense population, with many villages and miles of cultivated fields. The population of Florida may have been close to three hundred and fifty thousand when the Spanish landed.[11] One hundred and fifty years later, the English and Spanish found a wilderness populated by scattered groups of Indians who hunted, gathered, grew some crops and traded. They knew nothing of the people who preceded them. Their myths and stories may have been less than several generations old when the 18th century French and English settlers were told their traditions were ancient teachings. Moreover one expert on indigenous story tellers feels that myths retold by story tellers are highly individual.[12] They may draw on characters and events which are widely known and even used by different peoples, but the tales are not preserved by story tellers as core cultural values. They are the individuals' take on life and can vary a great deal both in the sequences in which things happen and also emphasis on what the moral is, if there is a moral. It would be difficult to assess the continuity of myths among Tahltan of northern Canada because all the recognized history keepers died in the plagues of the mid-19th century.

Besides of the uncertainties of history, our understanding of indigenous traditions may be affected by current political agendas and the fact that contact may have altered traditions in ways we cannot map. An ethnobotanist friend, while on a harrowing trip to study tribals in the upper Amazon before the recent invasions of Mestizo colonists, was shocked to find T-shirts and other consumer items in the possession of people who still obtained their evening meal of parrots using blow-guns. Despite the complexities of their circumstances, we need to rely on evidence from recent hunter-gatherers.

Hunter-gatherer Emotions

Since wisdom is inextricably bound to human emotions, we will look at the role emotions play among hunter-gatherers. To do this, we first review some of the ways humans fit into nature. Human bodies and minds evolved in a series of adaptations to the environment in which proto-humans and humans lived. Changes were partially in response to the environment but more likely because mutations made for more successful survival, or fitness in the narrow sense of evolutionary biology. The body became more generalized, thereby surviving in a greater variety of settings, and added special adaptations such as the efficient metabolism of Pima Indians who only had access to nutrient-poor desert foods, or the super-circulation of Eskimos, excellent for life at freezing temperatures.[13]

As we saw in the chapter on the natural history of disease, old age and death, the setting into which the evolving human mind fits has some terrible aspects. Gaia, the great natural selector, dictates that anything goes as long as it contributes to the successful survival of the species. Gaia is completely unsentimental. Adaptation includes cannibalism, infanticide, mutilation and gruesome parasitism. There is also nurturance, cooperation and altruism. If humans fit into nature along with other creatures, then they share some of these traits. As the human mind developed, it encompassed and facilitated those behaviors. *Australopithecus* may have had much in common with chimps. Although meat is not necessary for the survival of chimps, some groups hunt colobus monkeys and exchange meat for sex.[14] Other groups of chimps are mostly vegetarian. The differences are cultural. *Australopithecus* foraged plants but may also have scavenged meat and killed small animals. *Australopithecus* faced predators more dangerous than do modern day chimps. Its life was very risky.

Hunting and gathering are two different kinds of activities. For most of hominid history, gathering was the major way we sustained ourselves. In recent hunter-gatherer societies, gathering accounts for two-thirds of the caloric intake of groups, save for Eskimos and some Indians who survive mostly on game. For most hunter-gatherers, gathering is an irreplaceable source of livelihood. Gathering is also mostly an activity of women. Reports of gathering are replete with descriptions of noisy infants, playing children and talking adults. Gatherers use impressive memory. The Kalam of New Guinea identify 1,400 plants and animals. In an ornithologically contentious warbler species where the morphology is indistinct, their neighbors the Fore' distinguish species by behavior and song. Solomon, a Haida elder on the Queen Charlotte Islands of Canada, remembered details of birds he saw forty years earlier or places where he had gathered particular coastal edibles long ago.[15] It seems that most attention used in gathering is automatic attention. Since the object of attention is relatively static, gatherers can use their eyes to cast about for the particular plants which are remembered growing in a particular place while carrying on conversation. Although we have no way of knowing the kind of attention *Australopithecus* possessed while gathering, it was likely to have been more focused because there were fewer distractions created by thought and speech. Like us, contemporary gatherers have a much greater mental capacity and can gather while distracted, just as teenagers do their chores with music blasting in their ears.

Hunting is different. While gathering was improved by digging sticks and grinding stones, hunting was revolutionized by weapons. Tools, handedness, thought and speech were intertwined. There are debates about which came first and how it might have influenced the others.[16] By the time *Homo erectus* came along less than two-million years ago, stone tools were used for processing dead animals. It may be that major changes in human behavior occurred when the technology of hunting improved. Traps, spears, fire and the spear thrower made hunting more efficient and much safer. Without them, killing a large animal was extremely dangerous. Early in human evolution, hunters used stealth to get close enough to the prey to gain an advantage. This required extreme quiet, caution and alertness.

Hominids lacking discursive thought may have held themselves like recent, native hunters. Tools, weapons, strategic planning and thought developed hand in hand. Hunting could now be done at greater distance, more effectively and with greater safety. As the brain developed, hunters undoubtedly used thought, but they continued, I argue, to rely on the earlier quality of meditative-mimetic attention as a crucial ingredient of the hunt. Along with improvements in hunting technology, humans developed mimetic memory. There is mimicry in the animal world, but nothing like human masks or the stalking and imitating of animals using their hides and bones.[17] Using Darwinian reasoning, we can take vestigial traits tentatively to be evidence of ancestry. Mimetic memory is a vestige but also a foundation for the narrative mind which was built on top of it. No matter how skilled the hunter, hunting is dangerous. Although hunter-gatherers feel at home in the woods, they encounter many threats. The gatherer and scavenger, *Australopithecus* was cat food for lions and tigers; the skull of an adolescent in New York's Museum of Natural History bearing the indentations of a leopard's saber tooth provides dramatic evidence. Hunting is a risky business. Accidents happen during a chase. Prey strike back. Even though you know the woods, you can get lost. Wolves, ants, bees and squirrels do. And starvation is something with which even modern hunter-gatherers are familiar.[18]

The Fore' do not express tender feelings toward wild animals, but are sometimes afraid of them. Fore' are appropriately cautious of dangerous snakes but have no generalized fear of either snakes or spiders. In one instance, a reticulated python killed and ate a fourteen-year-old boy. For them, indiscriminate fear belongs to stupid

Europeans who cannot tell the difference between snakes. Among the Indians and Eskimo of the Yukon, one group had little fear of moose while another, who lived where moose were more aggressive, were wary of them. Bears are regarded as dangerous, particularly grizzlies who are feared. Some contemporary Indians hunt black bears in their dens, shooting them through holes dug in the top or as the irritated bears emerge. They don't hunt grizzlies in dens because grizzlies, who seem to know when someone is nearby, may come charging out. The Indians feel grizzlies have an eerie prescience and don't really sleep while hibernating.[19] Imagine how much more difficult bear hunting was when the only weapons were spears or bows and arrows. In the spring and in years with no berries or acorns, bears can be particularly dangerous. Today Native American descendants carry rifles as protection against chance encounters.

Although William Long, whom we encountered earlier, incorrectly thought that animals were immune from fear, he put the case poetically for humans. "The only real fears you will ever find in the woods are those which you carry in your own heart, as the price of being a man."[20] As we saw, the neurochemistry of fear is present in mammals. The squirrel my dog, Sasha, scared out of a tree fatally blundered from fear. Predators have fear while killing. What humans add is fear as a secondary emotion, something magnified by thought.

The older parts of the brain, including the amygdala, control unconscious flight-or-fight behavior. The more modern neocortex creates the imagination into which the fear is incorporated. As humans lived in nature in more mediated ways because tools, thought and speech allowed them to survive more easily, the mind began having a more independent life. Fed by sensation and the responses of the older part of the brain, it created a reality of its own where its images and the accompanying emotions could be out of proportion to the stimulations which initiate them. As the immediacy of reality recedes through technology and social organization, the life of the mind becomes ever more dominant. Here, in human evolution, a bigger sense of self develops. While traditional hunters use before-thinking-mind, they also possess thinking minds which are just as troublesome as our own.

Many ethnologists would argue that contemporary hunter-gatherers are emotionally and intellectually the same as civilized peoples.[21] This makes sense given the uniformity of our neurological makeup over the last thirty thousand years. Observing humans as they would animals, some ethologists claim that universal human behaviors are exhibited in early childhood. Before children can speak, they learn rules of give-and-take with particular individuals. The bonds thus created are defended like possessions but also compromised and contested. In differing cultures investigators found children similarly competing with siblings and displaying ambivalent feelings when they were accepted or rejected.[22]

Modern ethologists provide supporting evidence for Darwin's observations of the commonality of important gestures and expressions. Many hunter-gatherer societies are touch-rich. The Yanoama touch and lick the genitals of their children of the same and opposite sex without any apparent psychological damage. Grooming is a frequent activity among chimps and apes. Touch produces more calming neurochemicals than speech. It is in early childhood that strong bonds are formed through touch and expression. Some scholars feel that as the mind and social life developed, gossip and intimate speech became a weak surrogate for touch.[23] Most talk in tribal society concerns social problems. It may be that speech is more emotionally detached than acting out and thus diminishes conflict.[24]

With speech and thinking, emotions such as fear, anger, desire and joy became interwoven with thought. Hunter-gatherers sublimate, repress and nurture emotionally charged memories just like inhabitants of industrial societies. Speech and thought do two things. In comparison with non-verbal predecessors, speech and thought add a slight amount of distance between feeling and response, allowing for cortical control. But speech and thought also add a new dimension of stimulation. They become a factor which sets off behavior. A Zen saying illustrates before-thinking-mind: "When hungry, eat. When tired, sleep." In the mind these easily can become food obsession or avoidance of life.

Hunter-gatherers live in nature, which is their world. New Guineans studied by Jared Diamond talk mostly about relationships or subsistence. They have no regard for the pain they inflict when they capture animals live for convenient transport to their villages. They cut out decorative designs from bat wings without bothering to kill the animal. They torture live bats for amusement and treat their captured human enemies similarly. The wanton killing of enemies, the taking of captives as slaves and wives seem as ancient as the human race. While archeological evidence of cannibalism is controversial, it seems that it occurred as early as three quarters of a million years ago and wars may have been fought thirty four thousand years BP.[25]

The Achaur of the Amazonian rainforest live in spread-out family groups. They have no clan structure. During disputes between family groups, men recruit relatives and gather into forts. Conflicts may last several years. Anthropologists describe them constantly at war, with no word for peace. One out of two male and one out of five female deaths are attributed to feuding.[26] Some hunter-gatherer cultures seem to magnify aggression while others, like the !Kung, who have been called "the gentle people," appear more peaceful. But even among !Kung, homicide and violence rates are similar to big cities, though they do not engage in warfare as do the Achaur.[27]

I noted before that many of the nuns in the early Buddhist poems of the Elder Monks and Nuns came to meditation practice for release from suffering resulting from devastating family tragedies. In fact, the earliest record we have of human feelings is that of mourning the death of a child more than 20,000 years ago. (See Illustration 7.) !Kung women confront similar life situations. Disease, old age and death are sources of hunter-gatherer discontent. Marjory Shostak who lived among the !Kung relates the experiences of her informant.

[Nisa] told me about her sister-in-law whose "heart had been ruined by children"; she had given birth to ten but only four were still alive. After the youngest died, she drank the medicine that makes a woman sterile; their deaths had caused her too much pain and she didn't want to have to go through it again...

> Sometimes your heart is filled with rage. Other times your heart is miserable. You think, "A baby isn't itself a painful thing. Why should a painless thing hurt so much [in childbirth]?" ...but once it is lying on the sand, a baby is a wonderful thing.
>
> [Nisa's husband] was sick for only a few days, not even a month.... [The sickness] touched him in the night entering his body.... The next morning he awoke and was in pain. [After three days] it was hard for him to breathe and he was throwing up blood. The next night we slept beside each other, but soon after the rooster of the early dawn called for the first time...he died. I couldn't hear him breathing. I touched him again and again and again but there was no breath.... Why did this happen? The two of us gave so much to each other and lived together so happily. Now I am alone without a husband.... Why did god trick me and take my husband. ...god's heart is truly far from people.
>
> When you give birth to a number of children, you know one of them will probably die. [One daughter dies in a few days from a broken neck inflicted by her husband and a teen-aged son after three months of sickness from chest pain.] When your child dies you think, "How come this little thing I held besides me...has died and left me? ...This god...his ways are foul!... You cry out like that and cry and cry and cry. ...You look at other people who are surrounded by their families and you ask yourself why your whole family had to die. Your heart pains and you cry

> and you can't stop....I'm afraid. Dying is a horrid thing."...What about getting old? "That's bad too. When you are old what can you do?"[28]

The !Kung are not totally isolated from modern society. At the time of this report, they still hunted and gathered, but had occasional access to medicine and engaged in trade with their agricultural neighbors. Their life expectancies and resources were greater than prehistoric hunter-gatherers. Twenty percent of the !Kung reach sixty. One-half of those are dependent on the care of others. They have difficulties walking, respiratory disease, worn down teeth and reduced vision. At sixty, they have a ten-year life expectancy. Forty percent are widows.

The kind of suffering the !Kung experience has an ancient legacy. In East Indian cave paintings from ten to four thousand BP, there is a stick figure scene, presumably of the burial of a child. Two crouched adults are portrayed and there is no mistaking that they are mourning. Besides a vivid picture of a pregnant woman, this is the only picture whose meaning is unmistakable.[29] Prehistoric hunter-gatherers probably had as much or more discontent as moderns.

Barry Lopez relates an incident illustrating how hunter-gatherer minds create the trivial irritants that plague us. "Owl Friend set out on his own one morning to reach the southern band. He wore a red robe and deer skin leggings with a great deal of porcupine quill and beadwork in them...but he expected clear weather. In the afternoon a thunderstorm came up which changed to sleet and, later, to snow. By dusk Owl Friend suspected that he was lost. He was also distraught over his ruined clothes."[30] Even though he was a Native American, living in nature, Owl Friend's ability to attend to his surroundings did not completely overcome his ordinary concerns.

Sometimes what modern interpreters take to be indigenous values grounded in the old ways steeped in nature can be alternatively seen as familiar discontent. Gary Snyder tells of an old Indian who

Cave paintings from Bimbetka India. They are upper-Paleolithic, dating from 40,000 to 20,000 BP. One is the earliest known depiction of human mourning, presumably because of the death of a child. The other indicates that while Paleolithic hunters armed with bows and arrows were indeed prodigious hunters, they also risked injury from their prey and other animals. Wild water buffalo, gaur, cattle and boar are all extremely dangerous.

refused to meet with the only other living person who spoke his native tongue. "She wouldn't want to come over here. I don't think I should see her. Besides her family and mine never did get along." Snyder explained his response as, "Louie and his fellow Nisenan had more important business with each other than conversation. I think he saw it as a matter of keeping their dignity, their pride, their own ways...until the end."[31] Because he was a Native American, his people's ways were seen as analogous to the relationship between predator and prey: fixed in nature. Louie too would live out that relationship to preserve the old ways. On the other hand, being human, maybe he was just a stubborn old man for whom family grudges outweighed any interest in his native language. His nostalgia for the old ways was less important than family conflicts in the confines of tribe and village.

Prehistoric Wisdom

Once they possessed tools, mimesis, narrative and modern minds, our *Homo sapiens* ancestors behaved as we do. Since hunting required a special presence of mind, it may be that that attention engendered a kind of wisdom. Hunters know there is something special about the moment of a kill. Tom Brown, the self-styled tracker of the New Jersey Pine Barrens, describes his first kill done as a hunter-gatherer might. "Deer because of their keen hearing, eyesight and sense of smell are the true test of the predator. ...I went into the woods to hunt my deer with only a knife. I stalked him for a week." Brown dropped down on the deer's back from a tree over one of the deer's runs. In the struggle, he stabbed it. "All the way in, I could feel the terrible force of predation driving my hand and I understood how it was to be part of the dance of the eater and eaten that made up so much of the flow of the woods.... The agonies of the kill stayed with me, and I believed that the forest had made me a present of death so that I would avoid inflicting it in the future."[32] Another time he was threatened by a pack of feral dogs. "I hated them at that moment in a way that I did not understand. I have come now to realize that hate is just fear worn inside out, and it was my fear, and not my hatred, that drove me."[33] (See Illustration 8.)

Thousands of miles from the Pine Barrens, while kayaking in stormy weather, a student of mine and her boyfriend were stranded on an island of the Queen Charlottes. They were unable to go to sea to fish, a deprivation the traditional Haida Indians, who once lived there, often experienced. Like the Haida, they were threatened with starvation until they managed to trap a deer in an incoming tidal flow and stab it to death with their knives tied to sticks as spears. As she told the story, the listener felt the fear, desperation and rage of both hunter and hunted which came together in the act of killing.

The first time I captured an octopus using traditional Haida methods, I held the live squirming animal. It wrapped itself around my arms. As I held its arms open, my companion cut out its beak so that I could empty its innards and then beat it against a rock. It was not until after several minutes of beating that the octopus stopped squirming. Even though it was dead, its suckers still held on to me. Although I did not think that the octopus could hurt me, the experience filled me with fear and excitement. I was wrapped in a life and death struggle, the struggle for existence.

The underlying feelings in these situations are instinctually driven, primary emotions. As noted above, humans possessing mind and language often hold these in abeyance while hunting. Nonetheless, the secondary emotions evoked by hunting leave deep traces in the mind. Much conversation of hunter-gatherers concerns hunting. There are stories about real and imagined hunts and the comparative abilities of hunters. There are tales of danger and conquest. Thinking creates a space where the hunt can be rehearsed both as a technical or strategic activity and as an act in mimetic memory. Hunters discuss the efficacy of weapons and the habits of game. In ritual preparations for the hunt, we find vestiges of earlier human experiences of the aggression and fear. Hunting rituals are mimetic ways of steeling courage, assuaging fear and relating to a context of violence.

Tribal shamans often play a role in these preparations. Shamans are specialists. Their skills come from inheritance and training. Their powers often reside in their ability to engage in mimetic acts recognized by others as entering into the intense energies evoked in the hunt. We do not know what shamanism may have been like in hominids who communicated only mimetically. The shaman activities which have been studied also involve language. Somewhere in the transition to speech, more sophisticated tools and thought, humans began to feel that language had power over the world. It was as if language could create material changes the same ways tools and cooperative efforts did. Shamans became specialists in applying mime and language to affect the world.

The Achaur impose kin structures on nature, placing plants and animals in a continuum of relationship. The most difficult animals with which to communicate are loners. Jaguars and anacondas are believed to reject this communication and wage war on humans. Only shamans can understand them.[34] An ordinary person requires a shaman to negotiate relationships with them. Hunter-gatherers feel they need to learn the language of animals. It gives them power in the relationships. Dreams and visions are ways of engaging with these others.[35]

Although the meanings of dreams and visions may be available to everyone, shamans have special understanding. Shamans use rituals, austerities, physical prowess, ecstatic states and psychotropic substances to penetrate the recesses of the human mind. Acting for the social group as a whole, they explore extremes of fear, greed, lust, desire, joy and the uncertainties of nature. As Wade Davis puts it, the training of a shaman takes "years of hard work. First one had to master the basic visions.... You had to learn to bend the visions with the songs. Then there was the terror... Only after many such terrible journeys did the initiate meet God. He stood before a solitary tree and a door that opened into nothingness. The initiate had to walk into that emptiness. Only when he realizes what lies beyond the door can he...be the protector of his people."[36] In the parts of the brain thus accessed, strong feelings may be lodged in mimetic memory or more ancient structures. As experiences with psychedelic drugs illustrate, the human mind contains much that is not easily conveyed in words.[37] Bringing this back into ritual and ordinary life, the shamans create both real psychological and imagined material protection for the tribe.

What sense can we get of the role of attention and wisdom in the activities of shamans? The experiences of meditators and some users of psychedelic drugs share traits with hunter-gatherer ritual and shamanic work. Some meditators trace the origins of their interest in meditation to their drug experiences. In the early 1960s, Richard Alpert and Timothy Leary gave lysergic acid (LSD) to Harvard Divinity School students who then had what they considered to be true religious experiences. Some people taking LSD experience, for the first time, their thinking coming to a standstill. Reality is unmediated by thought. As Alpert, who became known as Ram Dass, often tells it, while in India searching for explanations of the meaning of the effects of LSD, he gave an eccentric old guru a handful of LSD tablets, enough to blow most people's brain circuits. The guru swallowed the LSD and calmly continued their conversation for hours. Startled, Ram Dass realized that there existed an entirely different relationship to mind than he could imagine. The guru seemed able to be steadily present with the extreme parts of the mind which are let loose by LSD.

While sitting in the meditation hall on a long retreat became a way for some to get high without turning on, as Ram Dass and others quickly learned, that was only a fraction of meditation experience. Some people who have taken LSD again after years of meditating find its effects less interesting. The displays of colors and distortions of reality are not worth the hangover and other side effects of the drug. What can occur in meditation are the energies which drugs or shamans evoke: confusion, desire, fear, bliss and demons lurking in the meditator's mind and body. Both shamans and meditators probe similar dimensions of the human mind.

The training of shamans in one Amazonian tribe has analogs to meditation practice. In a talk the anthropologist Wade Davis described young neophytes are taken from their families and kept for years in the dark. During their training they had to overcome deprivation and come to terms with their minds in a disciplined way. After years of apprenticeship, they are taken out into the light and see the world in a wholly new way. This prepares them to be guides for their people. Before Wade Davis gave his talk, I was outside eating my lunch and overheard a young woman describe to her friends a highly disciplined Zen meditation retreat she had attended. It lasted for several months. The attendees rose every morning at four AM, meditated and chanted until daylight. Then they worked in the garden. This was all done in silence. The evening was again spent in meditation until bedtime at ten or later. The last several weeks of the retreat were spent doing intense meditation. Everyone meditated, had their meals on their meditation cushions and slept in the meditation hall. Allowed very little sleep, they alternated between meditating and walking rapidly around the meditation hall paying close attention to their steps and breath. Her experience was a less intense version the Amazonian shaman's training. On emerging the world looked completely new to her, as if she had not really seen it before.

In both meditation and some shamanic training the protected setting allows the trainee to witness how sensations set off mind objects and how the mind objects erratically drive human behavior. In states of deep concentration, as in the Buddhist *Jhanas*, the mind no longer has to respond to the sensation; sensation itself becomes a kind of shadow. As they rest in the quiet of concentration and repeatedly observe the uncontrolled nature of self, meditators' training begins to reprogram their minds so as to undermine the unmonitored intention and conventional unaware behavior that so often accompany the interior life. Some of the respect accorded shamans comes from the fact that they are not driven by their human weaknesses as ordinary people often are; they have been trained not to.

How shamans use their understanding is a result of many different influences. One would have to look at the particular circumstances of a hunter-gatherer group and the individual shaman to see whether the actions of a shaman in probing fear or hatred were put to alleviating suffering or used for power, personal gain or vengeance. There clearly are wise people among tribal shamans, but there are also wicked ones. As described by someone who knew them well, one Amazonian tribe's shamans use of psychotropics is "a business.... Most of them just sell colors."[38] There is little to their shamanistic activities other than the psychedelic high. One difference between shamanism in general and the Buddha's goals in particular is that Buddha strove to alleviate discontent. Meditation looks at anything that interferes with living a life of clarity and harmlessness.

What do meditators and shamans do when they probe what has been called the antipodes of the mind? We have evidence from meditators but we can only guess at shamans' internal processes. For meditators, emotions which forge the links of Dependent Origination (especially when sensation leads to craving) are to be examined as aggregates which solidify the self. When form, sensation, perception, mind states and consciousness are seen for what they are, when they are seen in their arising and passing away, then the fact that there is no permanent self becomes more evident. When the impermanence of life is understood, there is space for meditators to be less attached to their mind states. With this comes an internal peace and the possibility of freedom from creating more harm. For meditators, the dreams and visions, so important to shamanistic ritual, are simply mind states which may sometime be prophetic but are fundamentally transient and need to be seen as such. Neurophysiologically, visions share characteristics with dreams. They are mostly colorless (except under the influence of psychotropics), lacking in detail and abbreviated.

An anecdote is relevant here. A friend accompanied her Tibetan Buddhist teacher to a meeting with a group of Native Americans. Contemporary Native Americans are not hunter-gatherers, although many claim continuity with those traditions of their forbearers. During the meeting, the Native Americans proclaimed kinship with Tibetan Buddhists. There are uncanny similarities between Tibetans and American Indians, and a few Native American practitioners have even been initiated into Tibetan practices. During the meeting, the Lama nodded politely, saying little. When he stepped into the car afterwards, my friend asked him what he thought of the encounter. His two-word answer spoke volumes: "Bon Po." Bon Po was an indigenous religion in Tibet before Buddhism arrived in the 7th century. Lamistic Buddhism and Bon Po adopted many things from each other and the two continue to be rivals. Bon Po often exists in the form of village shamans treating ills and empowering talismans. The Lama felt the Native Americans represented something similar: they merely addressed the transient in life, while his Buddhism looked into the nature of life in order to heal its fundamental suffering.

Another way of characterizing the difference between Buddhist practice and shamanism is to refer back to Soen sa Nim's circle. The circle starts with little "i mind," the mind of me talking to myself. But 270° is when concentration is so strong that it can be used to affect the world. It is "magic mind" or power mind. Shamans probably touch this domain. For the Soen sa Nim, this is a dangerous place because the powers can so easily be abused.[39]

Non-harming and Nature

While shamans are the ritual specialists, all tribal members have access to experiences through which they understand nature to be kindred. For the Koyukan Athebascans of the Yukon territory, their woods are not empty wilderness but a community of beings filled with different spirits. Hunter-gatherer artifacts often contain talisman images. Eskimo animals are intelligent, and will take revenge if you denigrate them.[40]

Hunting stories help to honor these relationships. They tie language as power to manipulation of real things. After I killed the octopus, skinned it and shared it for dinner with my friends, I embroidered an octopus on my work shirt. I recreated the violent changing colors of the animal's death throes. I still feel bonded to the octopus in ways I can't explain. It seemed appropriate to honor the octopus as part of my totem, my being. The embroidered shirt represents an experience few people I know understand. It represents what I imagine to be my place in nature. Life and death in hunting reinforce the feeling that hunters are part of nature. Success in the face of death demonstrates prowess. The death of hunters, sad or frightening as death might be, is part of the natural order.

The Conversation of Death may be just as applicable to hunter-gatherers as to predator and prey. It is a death that is appropriate given each creature's role in nature. There is a difference between fear in nature expressed in brains that do not possess mimetic and narrative memory, and fear in minds where mimesis and thought do play an increasingly dominant role. When considering hunter-gatherers we need to distinguish between times when their embeddedness in nature gives them qualities more akin to before-thinking-mind and times when their minds become more complexly involved.

We saw in the last chapter that our guts have anciently rooted minds of their own. Fear in war and hunting leads to constipation which helps in the chase. Terror, a product of mimetic and narrative mind, has the opposite effect. It triggers diarrhea, a hindrance when survival is at stake. This is another antagonistic pleiotropy of mental representation. Since hunter-gatherers have fully developed minds, even their at-homeness in nature does not completely protect them from the mind going awry. When an anthropologist who had crossed the academic line from observer to participant left his Yanoama wife alone in the jungle to return to her mother's people, she was terrified and did not know whether she could find them because women never traveled alone. Her fear was magnified because unprotected women were the objects of sexual predation.[41] And Peter Freuchen describes "kayak sickness of the Greenland Eskimos on still days with the sun low on the horizon getting hypnotized... [The paddler begins] feeling like he is sinking. 'Horror stricken, he tries to stir, to cry out, but he cannot...he just falls and falls. Some hunters are especially cursed with this panic and bring ruin and sometimes starvation to their families.'"[42]

Using the perspective of meditation, we can understand some of hunter-gatherers' sense of at-homeness in nature and their feeling that words have power. We have seen that both hunting and the activities of shamans rely on parts of the mind that developed before speech. We have also seen that with the development of tools and thought, dangers of the world could be held more at bay. In addition, when the mind became capable of thought and speech, humans found it both useful and easy to live in their thoughts. Because the useful is adaptive, the increased efficacy of consciousness was accompanied by an increased distance between thought and raw nature.

As the world became more an object of contemplation, life began to be lived more in the mind. Although mimetic fear represented by expressions and gesture without words is as terrifying as fear in the narrative mind, it has far less of an independent existence, especially when the causes of fear are more likely to end your existence quickly than give you time to dwell on it. You simply don't have the verbal-mental apparatus to make fear into a repetitive story that can be rehearsed over and over again. Thus, at-homeness in nature is part of the hunter-gatherer's existence which harks back to an earlier embeddedness. Responding to the world in this way involves less solidification of the five aggregates. The first two, form and feeling, occur because hunter-gatherers' bodies respond to stimulation. But the third and forth, perception and mind, which make the self into something that holds on, stay undeveloped. So when it comes to discontent created by the mind, hunter-gatherers probably experienced less suffering in the Buddhist sense. On the other side, in the parts of hunter-gatherer life which leaned toward narrative experience, there was a larger role for the mind and a more defined self. This gave rise to the sort of discontent easily recognizable to us today.

The mind of hunter-gatherers that developed tools, improved hunting skills and observed nature had other liabilities. The Nanamiut Eskimos, profoundly tied to their landscape, saw themselves just as observant as the scientists who came to visit them. From their observations of nature, they figured out what was important to their world. They felt that they too were scientists. Their skills do lead to deep understandings.[43] But this wisdom derived from nature has limits. Circumstances arise which overwhelm it.

We encountered epizootics and zoonose in the chapter on the natural history of death. The former is an epidemic among wildlife, while the latter are infectious and parasitic diseases shared by humans and other vertebrates. Some animal diseases in the United States may have existed in the wild from prehistoric times; others were brought by European colonists. An outbreak of disease among caribou which caused them to die of bloody vomiting, terrified the Ottawa River tribe. Then, at the time of the 1781 smallpox epidemic in New England, when the Indians were decimated, the wolves and dogs that devoured the bodies were found dead with mange-like symptoms. In the Rockies, as smallpox swept through the Indians, bison and deer disappeared, as did moose, ducks and gulls. In the Canadian boreal zone, there were massive wildlife die-offs but no one knows why.

In 1797, Indians reported distemper among the beaver to explain why there were so few pelts for traders. They reported great numbers of animals floating dead on the rivers. An 1801 epizootic was described by an adopted Caucasian living among the Ojibwa. Dead and dying beaver were found everywhere. Beaver lodges were empty. Some even died at the feet or trees they were attempting to fell. The Indians also got the disease when harvesting wild rice in the fall of 1803. This resulted in bronchial infections and a forty percent mortality rate. Their coughing scared off game and made hunting much harder. They had never seen this illness before.[44]

Whether these outbreaks were from European or indigenous sources, the Indians believed that animals brought the diseases. This belief came from having witnessed animal diseases before the Europeans arrived. Although the Indians applied rituals and other traditional means to remedy the onslaught of disease, their efforts were of no avail. So they felt that either animals were rebelling against them or that they were at fault. A group of Indians who tore into the tents of their enemy during a sneak attack fled in terror when they found their intended victims lying mutilated by smallpox. Indian rituals aimed at maintaining a balance of nature no longer worked. Their understanding of their world broke down. Their traditional relationship with nature that relied heavily on mimesis became inadequate to the circumstances. That left the Native Americans overwhelmed by the confusion and fear of their narrative minds. They had little beyond their understanding of nature to address these emotions and that understanding had failed. This contributed to their suffering and the breakdown of their cultures.

Impact on the Environment

It is claimed that because hunter-gatherers are so observant and live in rapport with nature, they are natural ecologists.[45] A number of anthropologists record rituals of reciprocity in which hunters carefully ask permission of their prey. When hunting they only take what is needed and do not deprecate the prey to whom they feel in debt. Although this was the behavior of Native American hunters of recent memory, it may not have held true for their predecessors. Before and during the fur trade, periods of hunger plagued the Cree of northeast America, yet they did not manage their resources.[46] We do not know if pre-contact scarcity was caused by overhunting. The Cree did not preserve food and usually exercised minimal effort in obtaining and keeping food. They simply moved on when stocks of game diminished.

To obtain more furs, early European traders raised the prices they were willing to pay. The Indians did not always respond. They stopped trapping when they secured the trade goods they wanted. Higher prices did not induce them to continue trapping. European concepts of money-based supply and demand, commodity trading and private property meant little because surrounding nature was so supportive. Despite this, within a generation or so the Indians became dependent on trade goods, which included guns, knives, axes and pots. By the late 18th century, some colonists felt the Indians would starve without guns.[47] The new trade goods gave the Indians an easier life and leisure time. Although the fur animals they trapped were part of their diet, they killed mostly for the market, eating only the delicacies of beaver and caribou. In the 1700s they killed more beaver than they could transport to market. Both the Cree that settled around the trading posts and other tribes in the interior killed more buffalo and caribou than they consumed. They also ate only choice parts. Some anthropologists think the Cree regarded moose, caribou and beaver as inexhaustible. Others feel they did conserve beaver.[48] The Greenland Eskimos were also known to waste resources. They took advantage of the fact that Greenland sharks move sluggishly. Snagging the sharks, they hauled them to their kayaks and slit them open for their livers, the only part the Eskimos consume. They then released them alive.

The first anthropologist who proposed that the American Indians were "aboriginal ecologists" studied the Algonquins in 1915. He repeated the Indians' story that the conservation measures they were currently using came from pre-Columbian traditions. His conclusions are another example of treating hunter-gatherers as a "people without a history." That the Algonquins did not know their own preliterate history is one thing, but that a scholar simply assumed their actual history was identical with their myths was naive. The idea of reciprocity between Indian and animal comes only from contemporary Indian sources. In the 19th century, Indians did institute conservation measures once animals were depleted, but there is no evidence for reciprocity in 17th and 18th centuries. They told traders "the more they killed, the more they would kill in the future."[49] Since animals had souls, one needed only to propitiate them by ritual. They did not even need to do it for individual animals but could ritually propitiate a whole catch. Also, in some cases animals were seen as adversaries. They had to be killed or they would tell others who would then not present themselves. In this instance, killing what one came upon asserted proper human dominance.

The first game shortages resulting from overhunting were seen by Indians as periodic shortages. But when animal populations collapsed, leading to starvation, the Indians changed their strategies. After the 1820s, with the help and encouragement of the Hudson Bay Company, they set up trapping territories and began conserving. This is apparently the source of today's reciprocal use beliefs that are recounted as if they had great antiquity. There is abundant evidence that, outside of North America, hunter-gatherers and early agriculturalists of the past thousands of years have severely altered the composition of animal species in the places they inhabited. Landing for the first time in islands of the South Pacific, humans hunted out species after species. While there are scholarly arguments about whether the first human inhabitants of the New World, more than fourteen thousand years ago, hunted mastodons and other megafauna to extinction, in New Guinea there is evidence that as each new hunting technology was adopted some species were eliminated.[50]

Conclusion

We have seen that hunter-gatherers are people like us. They have emotions and discontent and unless they or their ritual specialists purposely direct their behavior towards the source of their discontent and towards non-harming, they are as likely as we to experience suffering and hurt each other. What is more interesting in trying to understand our own lives is how their connectedness to nature affects these feelings. Hunter-gatherers possess extraordinary powers of observation and a deep presence in the natural world. These do not protect them from fear of changes not understood. Nor are they insurance that the hunter-gatherers won't affect nature in ways they can not imagine. In this latter, hunter-gatherers are no different from other species. Some individual cactus finches on the Galapagos Islands eat the tops of the female cactus flowers before they are pollinated, obtaining a bit more food before other cactus finches do. This makes it impossible for the cactuses on which these finches are dependent to reproduce. Were finches that did this to thrive in sufficient numbers, the species would go extinct. Luckily for them (and the cactuses), only a few appear afflicted with this predilection in each generation.[51]

Even though there are limitations on hunter-gatherer attention and wisdom, it is important to emphasize that hunter-gatherers related intimately to nature in ways that are difficult for us to access. They understood their own membership in the natural world in ways we no longer comprehend. This is exemplified in a return visit to the !Kung of the Kalahari Desert by the anthropologist Marjory Shostak, who had studied them some years earlier. She was sitting with her informant whom she knew well when:

> My attention was caught suddenly by a flying ant with long worm-like body and almost transparent wings. [It flew near the fire and Nisa removed it twice. The second time] she picked it up and pierced it through half the length of its body with a thin twig leaving the upper half...free. She planted the stick...upright

in the ground and tapped it gently with her fingers. The insect's wings burst into motion, as if in flight propelling the free parts of its body around and around the stick...I watched horrified. On my first field trip I had come to terms, though not easily, with the killing of wild animals for meat. I had also seen this species of termite being roasted....But what Nisa was doing was different. It seemed like an inexcusable torture.

As Nisa sat there, she mimicked the insect's painful contortions. "My revulsion did not disappear... The incident reminded me of the cultural gulf between Nisa and me."[52]

This gap is the gap between hunter-gatherers' connection to nature and civilized citizens' disconnection from nature. Nisa's dance illustrated her identification with the insect whose life became hers when she saved it. She understood with her gestures what its experiences were. This mimesis so crucial to the !Kung's survival may have made disease, old age and death easier for them in a world where nature was more immediate. For Nisa there was no category "nature" as something different from her existence. Life and death seem to bother us more. Even for Annie Dillard, so attuned to the processes of nature, "nature" is an entity whose meanings are to be plumbed. It was more distant than it was for Nisa. "I used to kill insects with carbon tetrachloride...and pin them in cigar boxes....I quit one day when I opened up a cigar box lid and saw a carrion beetle, staked down high between its wing covers, trying to crawl, swimming on its pin."[53] It is to this difference we turn next and look at what happens to attention and wisdom as humans become civilized.

Chapter Eight
Civilization's Discontent

Transition to Civilization

In 1978, I was headed to the Boston airport to fly to Chicago for a high school reunion. When I got to Route 2, I unaccountably turned west, the wrong direction. Instead of the airport, I drove to the Insight Meditation Society, where Ajaan Cha, on his only trip to America, was leading a meditation retreat. I made one of those odd choices which deeply affected my life. Ajaan Cha spoke to the group from a life we could hardly imagine. For years he had been a forest monk practicing ascetic meditation in the jungles of Thailand and Laos. He spoke with such simplicity, wisdom and humor as to be an irresistible teacher. He had the folksy style of a wise old man sitting next to a pot-bellied stove in a Vermont general store. At one point he looked around the meditation hall with a twinkle in his eyes. He said he found America to be a curious place. He remembered that in Thailand before the Vietnamese war there were few electric lights in the countryside. It was dark out of doors, but the people were bright inside. With the changes that Vietnam wrought and the subsequent years of development, electricity and the modern world had come to rural Thailand. As the countryside was increasingly illuminated, Thai villagers became darker inside. Now in America, a country filled with lights, he found the people very dark inside. From the feel in the meditation hall, one could tell that he touched a sensitive spot. The lightness of his mind contrasted remarkably to the heavy minds the American meditators had brought with them.

Ajaan Cha's perspective was by no means that of a hunter-gatherer. He came from a village of subsistence farmers and his outlook on the world was deeply Buddhist. As a result of years wandering in the jungles studying his own mind, Ajaan Cha came to see some of the changes in our mental maps that civilization produced. In trying to understand these changes, we face a gap in the historical record similar to the one that plagued our search for ancient hunter-gatherers' outlook on life: there simply is little evidence about the experiences of humans during the transition from hunter-gathering to early agriculture and then to highly developed civilizations. We know that various stages of this transition coexisted for thousands of years. We also know that, as civilizations failed, some of their inhabitants went back to earlier modes of subsistence. We know that many changes which underlay civilization were irreversible and affected even those people who continued to live as hunter-gatherers.

We can get some insight into the transition from hunter-gatherer existence to early civilization from the Europeans who first contacted the more primitive of the New World Indians. In the early 16th century a Spanish expedition was shipwrecked on the gulf coast of Florida. Of the four men who survived the wreck and Indian attacks, two wandered for seven years until they got back to civilization. They traveled from one Indian tribe to another, sometimes captive, sometime free, until they arrived in Mexico. One, Cabeza de Vaca, returned to Spain where he wrote a journal of his ordeal. Bearing in mind the prejudices of a Spaniard of his day, his journal gives an interesting portrait of primitive societies which ranged from hunter-gatherers to agriculturists living on the outskirts of the Aztec Empire. Unlike the image of noble savages, Cabeza de Vaca found tribes who struggled for existence and could be extraordinarily cruel to each other and the Christians who became their slaves. He describes fighting among different tribes, their mutual fears, cruelty towards women and the slashing of a child who cried in the captives' presence.

At times he lived with tribes who endured long periods of cold and starvation. "The people suffer so much from hunger that they cannot do without roots and will wander two or three leagues in search of them. Occasionally they kill deer and at times catch a fish or two; but this is so little and their hunger so great that they eat spiders and ants' eggs and worms and lizards and salamanders and snakes and vipers such as kill the men that they bite; and they eat earth and wood and everything they can lay their hands on, and dung of deer and other things I will not mention; and I firmly believe that if there were stones in that land they would eat them....the women and the old carry heavy things, for they are the folk that these Indians consider the least."[1] One survival tactic was to gorge when food like the prickly pear fruit of cactus was abundant. For the Spaniards, the Indians' diet was hard to stomach and their daily routine oppressive. Cabeza de Vaca probably inferred more hardship than the Indians actually experienced.

The captured Spaniards walked around naked like the Indians. And even though they lived the Indians' life for six years, they never acclimated to its harshness: "the sun and wind developed great sores on our chest and backs...." The loads they carried cut their arms and the forest thickets broke their skin. Cabeza de Vaca respected the natives' endurance and disposition. "Often their bodies are pierced through by arrows, and they do not die of their wounds...rather they heal very quickly. They see and hear more and have the sharpest senses...of any people in the world. They are very patient sufferers from hunger and thirst and cold, like folk who are more used and inured to it than others.... They can run from morning until night and run down a deer.... They are cheerful people no matter how hungry they may be...."[2] Like other pre-agriculturists, they transformed their environment for sustenance and comfort. They burned the woods to catch edible lizards and drive deer to slaughter. When mosquitoes tormented the Indians, they burned the woods in order to destroy them.

The tribal lifestyles Cabeza de Vaca observed changed as he traveled the Gulf Coast to east Texas and northward toward Austin. The coastal peoples were agricultural while those further inland were migratory hunter-gatherers. The latter were the most primitive Paleo-Indian. They engaged in "cyclical harvests and communal hunts for insects, reptiles, and occasional larger animals...they knew little of agriculture..."[3] Later, among the semi-nomadic people living on the edges of the "Pueblo" Indians, Cabeza de Vaca hunted buffalo and possibly feral cattle.[4] As he headed farther south into Mexico, he encountered increasingly prosperous agriculturists and finally the conquering Spanish who were brutally destroying Aztec civilization. Of the last two non-civilized groups he met, one was agricultural but still hunted and the other was village agricultural speaking a variant of Aztec.[5] Just as Roman allies and trading partners living on the northern borders of their Empire were harassed by more remote barbarians, these relatively civilized villagers were raided by the Apaches. The Apaches themselves had previously subsisted by foraging on foot until they caught and broke the horses that had escaped from the Spanish and run wild in great herds. The domesticated horse utterly transformed the Apaches' lives.[6] They became terrorizing horse bandits.

The complexities taking place at this late date suggest what must have happened five to seven thousand years earlier, as civilizations emerged out of sedentary villages in the Near East, India and China. A number of factors may have contributed to the change, including new technologies, changing sources of foraged food, migration, population growth and disease. These factors were present long before the most profound change of all, the domestication of plants.

We described some of the effects of this process on the human body in Chapter 4.

Domestication of plants took thousands of years. It developed from accidental domestication connected to humans' toilet habits, the disposition of garbage and the disturbances of vegetation from repeated human occupation of foraging camps.[7] Like Australian Aborigines, a people still without agriculture in the 18th century, humans who had not domesticated plants assisted friendly foraged plants and hindered less friendly plants by burning large areas. Starting about ten thousand years ago, domestication changed human life as more stable sources of subsistence allowed humans to settle. As we saw, this settling required different uses of the body and had significant effects on disease, old age and death. To examine the effect of this transformation on the human mind, we again draw on what we know of recent hunter-gatherers.

As was noted, hunter-gatherers have a kind of attention that binds them mimetically to the natural world. They also have fully developed minds which they use to manipulate the world. People often did not understand that the eventual effects of these manipulations could be destructive. We have seen that hunter-gatherers experienced discontent as modern peoples do. There was greed, hatred and delusion. What may have eased these for hunter-gatherers is that their manner of sustaining themselves, their daily life cycle and their sense of being located in nature all contributed to keeping the life of the mind somewhat within bounds. In Cabeza de Vaca's journal we have a tour of peoples who stood on the cusp of change. Their lives were being transformed by the agricultural civilization to the south and the Spanish whose power grew from their writing, gunpowder and navigational science. What peace of mind they might have derived from their relationship to nature was being torn asunder. Without reading, science-based technology or psychology, without media or complex social organizations, hunter-gatherer minds rested in an accepted natural setting. Their reign of the mind was less pervasive than in the societies which displaced and eventually destroyed them. Although they used their minds for survival and daily life, they relied more on their mimetic talents. Complex abstract concepts on which action might be predicated were less important for them. Because they lived less in their heads, they may have experienced less of what the Buddha understood as suffering.

We now turn to the different role the mind plays in civilization and how that role affects discontent. In so doing, we should remember that whatever social systems succeeded hunter-gatherers, those systems emerged from hunter-gatherer society. Thus, no matter how integrated and natural was hunter-gatherer life, hunter-gatherers were thrust into new ways of living. We'll look at the parts of human nature which led them to make these changes.

We noted in the chapter on the body that craving for salt, sweets and fat are evolutionary adaptations to the scarcity of these foods. Consistent with this is the fact that no known ethnic group rejected sugar when it was introduced.[8] Before the arrival of the colonists, the Indians of coastal Massachusetts did not even use maple sap for sweetness, and they thought the English craving for sugar strange.[9] Despite this early attitude, sugar quickly became an Indian trade good, as did rum which was made from it. Similarly, although the modern world was often forced on hunter-gatherers, they also adopted it on their own. The quality of attention in hunting, an important ingredient in how wisdom was cultivated among hunter-gatherers, was one of the first things to go in the transformation to civilization. Reviewing this loss is depressing to me. Although some of the tribes Cabeza de Vaca encountered still subsisted primarily as hunter-gatherers, agriculture and trade goods had been making their way north from Mexico for hundreds, perhaps thousands of years.

The Indians of the hinterlands of North America lived in the penumbra of the civilizations of Mexico. Corn originated in tropical Mexico and by the end of the first millennium it had spread to New

England where it displaced local plants that were in earlier stages of domestication.[10] Corn was more fully domesticated and yielded greater food value. Some tribes both adopted and adapted these new seeds. Where adaptation did not succeed, as among the Abnaki of Maine, they traded for the corn with agriculturists to the south. For cultural reasons we do not know, some Indian groups chose not to participate in the spreading changes. Nonetheless, resistance to more productive means of livelihood was less prevalent than adoption.

We cannot begin to guess how the Apaches' shift to horse culture, like that of the Mongols a thousand years earlier, affected their relationship to discontent, but we may be able to understand what happens to the mind of contemporary hunter-gatherers when they encounter modern life. Let's begin with foraging. As we noted, hunter-gatherers have voluminous, detailed and subtle knowledge of wild edibles and wildlife.[11] In addition, we saw that the mind is designed to respond to novel inputs and becomes jaded easily, relegating the habitual to automatic attention. Gathering becomes routinized while hunting retains its excitement, largely because it requires continual attention to changing circumstances and is risky. Technologies that introduce novel stimulation without risk are seductive to the mind.

In the 1920s, Robert Flaherty made "Nanook of the North," a famous documentary of Eskimo life. When he showed his Eskimo subjects the film, they sat around talking because they did not see anything on the screen. They did not know how to watch a movie. Such perceptual incapacity does not last long. The Biami of Papua who purportedly had never seen their reflections were first scared and then intrigued by mirrors. They had to be taught how to look at Polaroid pictures of themselves. When they learned how to see Polariods, they were frightened at first but quickly got into taking pictures which they proudly showed to others. They also loved to play tape recordings of their own voices.[12]

Once exposed to the new technology's greater stimulation, nature is less interesting. "My son...at an early age...became an avid naturalist. But when he was fourteen, I bought a personal computer.... His exploratory instincts found an even more attractive outlet in...the computer than they had in observing the natural world."[13] Yacqui and Pima children of the Arizona-Mexico border regions can no longer identify plants. As a biologist who worked with them for years observed, "Not only does the exotic, unreal world presented on the television screen threaten to distract children from less spectacular local wildlife, but it fails to provide the personal stimulus awakened when you yourself choose the images; sounds, smells and ideas that direct your experience of nature."[14] My nephew grew up on a homestead where moose, bears and foxes wandered through the yard. When was a teenager living in town, I invited him to go on a walk in the hills. His response was, "Walking is boring." Years later his interest in nature returned. Once they attend school, the children of stoneage, New Guinea parents show little interest in the forest. They move to town to work.

Many social factors contribute to these changes. Economic and environmental forces crowd out old ways. The habitat which supported hunting and gathering is altered and natives are made to adopt cash economies. Although these influences are important, indigenous people are also attracted to the new stimulation. It is analogous to sugar craving. When people no longer have to endure hardship to sustain themselves, the mind can indulge its built-in attractions. I recall sitting around a cabin in the Coastal Mountains of northern British Columbia listening to a conversation between a white man who had married into an Indian tribe, and his Indian father-in-law. At one point, the father-in-law casually mentioned a walk he had taken when he was twelve. On a whim, he and a friend decided to walk to the next village. Without supplies, they sauntered for four days across fifty miles of mountains inhabited by wolves, grizzly bears caribou and moose. All on a whim! Around a campfire, on another occasion, an Indian elder talked about his life in the interior of the Yukon just before the Alaska highway was built during WWII. There were no roads. He traveled by foot or dog sled. He recalled being caught in a storm when the temperature went down to minus seventy degrees Fahrenheit. He told of his surprise on seeing the parkas and boots worn by the men flown in to build the highway. The only footwear his people had worn was moccasins in the summer and mukluks in the winter. These comments went unnoticed by the others, but astounded me. I was struck by how much life can change in one generation.

Much has been written about the gap between native elders and their progeny. Elders do not understand how the young can be so attracted to the new stimulations, and the young are bored by the old ways. The following was written by a naturalist who grew up in a family that foraged the woods of Germany after the devastation of World War II.

> Much of nature is subtle, and it is difficult to appreciate it if one is used to the grandiose. I doubt I would have stopped to watch a mere beetle, a bird or an ant if I had a toy train when I was young, a train that rumbled and tooted and sped on fast tracks at the touch of a button. I became attuned to spending hours watching a bird just to see what it brought back to the nest, getting pleasure from discovering the subtleties. It is the subtlety of a bird or a carabid or ant multiplied a few million times over, that makes the whole. If one is not attuned to the first subtlety, then all the rest can pass unnoticed, also just as one sees only the train with the loud whistle rumbling past.[15]

Although attention may be as riveted on mechanical stimulation as on natural events, as we will see, the context of machines is more uniform and their stimulations are more well-defined than nature's. I remember an adventurer talking about the Alaskan wilderness, emphasizing that life in the northern wilds was uneventful. Unless you had the patience for little happening, you did not belong there: you would be bored. While the mind which tackles a mathematics problem needs to be as attentive as a hunter's, a mathematician's attention engages only the mind. Similarly, while the computer gives enormous feedback, it engages the senses less and entails less risk than following an animal's trail.

Before we explore further how civilization locates experience in the mind, we need to look at what happens to risk and discomfort in the transition from hunting and gathering. In preceding chapters, we saw that the riskiest point in the life history of animals centered on birth. This has been true for humans throughout much of our evolutionary history. It is only in the last hundred years, in advanced industrial societies, that infant mortality has radically declined. So it is to be expected that if technology offered safer childbirth, humans would chose it.

Two examples support this conjecture. In the 1960s and 1970s, the Canadian government decided to end what little remained of the migratory life of Indians and Eskimos. The government offered health services and housing at specific locales in exchange for settling down. The natives accepted the offer. A major consideration was childbirth. With a clinic or hospital nearby, birth was less dangerous. In a different part of the world, an anthropologist fled back to America from the Yanoama of the Amazon in the company of a Yanoama wife he had been given while studying them. For her, the transition from hunter-gatherer to 20th century America was as abrupt as it was, at first, incomprehensible. As was the custom of her people, she had once cared for herself alone in the jungle while miscarrying. Now in America, because it was safer, she agreed to have her first live birth in a hospital where she was hooked up to machines. Her children by the anthropologist grew up in New Jersey and, like other American children, they were plugged into television.[16] Although the transition for her was wrenching, she was able to do it. Her children, despite their trips to the Amazon, had little understanding of the world from which she came. For most humans a life that becomes safer and more abundant far outweighs any importance a connection to the natural world might have had.

This story has a second ending which indicates that while cultures rarely return to their idyllic past except when agriculture fails, individuals sometimes do. Disquieted by the isolation of suburban life, the anthropologist's wife chose to leave her husband and remain in the Amazon when they were there working on a National Geographic film. The anthropologist's father died and rather than going to the funeral, he stayed with the film crew. To the Yanoama woman, such behavior was unthinkable.[17]

Choosing safety also applies to hunting, the symbolic heart of hunter-gatherer life. Improvements in weapons and tactics have been developed over hundreds of thousands of years. As was mentioned, stone spear points, the spear thrower and traps made hunting more effective and less dangerous. In Neolithic agricultural societies, copper and then iron were used for knives and plowshares. The hunter-gatherers who survived on the edges of farming cultures readily adopted the new implements when they became available. Until recently, dangerous tribes in the Amazon were lured into contact with gifts of steel knives and axes. Firearms are the foremost hunting technology. They so multiplied the efficiency of the hunt that they contributed to overhunting. They also significantly reduced risk.

The Alaskan Kutchin still make their living by hunting and trapping, have had guns since the 1840s and have completely integrated the gun into their subsistence. They live on the eastern edge of the great Yukon flats. The landscape is stunted northern pine and juniper woods bordering permafrost tundra to the north. The Kutchin hunt moose using boats, snowmobiles or snowshoes. They can pick out moose at a distance. They make moose sounds. They can smell them and track them. Attracting a moose by rubbing an old shoulder bone against a tree is something left over from bow-and-arrow days. The old men are the ones skilled at this. They trap and snare other animals, an activity requiring considerable knowledge of their habits. Before the gun, muskrat were rousted out of their dens and then clubbed to death. It is much easier to hunt with a rifle. Modern technology has tilted the balance away from patience, attention and fear of survival. Even though a Kutchin "spends a lifetime learning more about the landscape.... The Kutchin see hunting and trapping as a hard way to make a living and generally look forward to the day when they can give it up for good.... 'Oh, job way is better.'"[18]

When I was hunting the octopus, I tried to use a traditional Haida barbed, pointed stick to irritate the octopus so that it would emerge from its lair that was exposed at low tide. As soon as an octopus emerged, it could no longer use its tentacles to hold onto the sides of the lair and resist to the death being pulled out. When exposed, the stick was lunged into the octopus, and it could then be hauled up on the barbs for killing as described earlier. The only difficulty with this technique was that I could not get the octopus to come out by gently jabbing at it. Losing patience after a while, my friend and I resorted to modern technology, pouring a little bleach through a piece of surgical tubing, the end of which was pushed into the lair. The octopus emerged quickly. With ease comes comparative safety and less need for attention, so the hunter retreats from mimetic memory. Less mimetic skill and mimetic protection from the dangers of the hunt are necessary. While hunting in foraging societies, even with modern weapons, takes skill, hunters consistently choose technologies which make their work easier and safer. These require greater use of their narrative minds. A modern sports hunter armed with a satellite geo-positioner, high-powered rifle and telescopic sights risks neither starvation, the counterattack of prey, nor even getting lost. They have little need for ritual mimetic protection.[19]

As hunter-gatherers became gardeners and then farmers, they continued to practice rituals to locate themselves in the natural world and to placate natural forces. As agriculturists, humans no longer saw themselves as just one kind of animal among all others. Since wild nature no longer sustained them, humans turned their feeling that words and rituals had power to the new things that mattered, the earth and the sky. These nurtured or destroyed their domesticated plants. While animism still survived, the sky and earth gods emerged as important objects of supplication. It was easier to relate on equal terms to animals whose habits you understood than to the sun, wind, cold, rain, earth and water whose unleashed forces drove you into starvation. Migratory hunter-gatherers retreated into the hills when great floods washed into the valleys. In contrast, the livelihoods and homes of sedentary alluvial farmers were destroyed. For the Haida of the Queen Charlotte Islands, the starvation which came when storms kept them on shore was not something to be feared, but rather to be suffered through. In the *Rg-Veda*, the earliest of the Brahmin texts, and in the Biblical story of Job, both products of agro-pastoral societies, the gods of the terrifying elements had to be appeased.

Priests related charismatically to angry sky and earth gods just as shamans had used their special skills of mimetic and narrative communication to deal with anti-social jaguars. In this transition, the agency of power becomes words or thoughts. The animals to which hunter-gatherers felt related were anthropomorphized in myth and story, but they were still real. Hunter-gatherers interacted with them in daily life. Raven, the creator of the Haida's world, talked to the Haida as they paddled their canoes over the ocean. While the Haida projected all kinds of human qualities on Raven, Raven talked back to the Haida. It is difficult for civilized people to comprehend such conversations. I have been present when people conversed with ravens. It was eerily real. With the sky gods, projection is more pervasive. Sky gods are often made in the image of humans and the conversation is cerebral as in prayer. Words in the conversation are tokens of power. Though the dialogue may have mimetic dimensions, it is mostly in the mind.

The relationship between animistic and heavenly, human-like gods has been much discussed. Even Darwin felt that belief in god was a personification resulting from collective feelings of terror and the sublime.[20] What is significant here is that, with respect to human's relationship to nature and the elemental feelings which arise there, as we have gone from hunting and gathering to civilization, we experience life more in the mind. In our current world, this has magnified. Although the change was rooted in the minds of hunter-gatherers, the tree which grew out of that soil has produced fruits which make us even better candidates for the insights of the Buddha than were the hunter-gatherers. We will look at how the way we now live in our minds creates a potential for greater discontent.

The Modern World

In the midst of unparalleled prosperity, modern life has brought alienation, unhappiness, stress and social breakdown. The raw physical power that industrial societies can bring to bear on the planet is difficult to comprehend. Abundant possessions, travel and convenience are taken for granted. The wealthier classes which consume an outsized share of the output of industrial society and sap the third world of resources and labor, are rarely aware that an evening's dinner in a moderately-priced restaurant represents a substantial part of a year's income for the poor. This situation deserves analysis from the perspective of the Buddha's teachings. Other thoughtful writers have assumed that task. Wealth and the inequities of the world make for a haunting background of the examination of what has happened to attention and wisdom in modern life. What happens to our minds as industry more completely dominates our existence?

In nature, hunter-gatherer attention reached a profound depth. Volumes chronicle how improved technology led to withdrawal from nature. The process started slowly with the domestication of plants and animals, then picked up steam during the industrial revolution and took off after World War II. Farming, agricultural labor, specialized skills working cloth, clay, wood and metal all helped lay the foundation for future factories, offices, schools and agribusiness. It was not the first time. Early Neolithic goat herds had transformed the countryside. Solomon's temple, cities and navies had stripped accessible forests of trees. Some writers suggest that when civilizations weren't waging war on each other, they engaged in genocide of hunter-gatherers. From China through India and Europe to the Andes, the land was cleared and the mind was used to create the material foundation of empires. For five thousand years and through the medium of millions of people, intense attention was focused on making nature bear fruit. Until the industrial revolution, most humans were farmers of one sort or another.

I spent time with a Canadian immigrant who had grown up on a farm in eastern Germany. Until they were expelled after World War II, her family had made it yield bounty for four hundred years. That was longer than ancient Sumer which destroyed its environment by stripping the hillsides of trees, diking the Euphrates and hauling water which led to salination of the soils and terrible floods.[21] No matter how wretched their lives, the people at the bottom, tilling the soil, subjected to taxes and priests, lived more in their minds than did hunter-gatherers.

This discussion easily glides over the immense gap between hunter-gatherers and post-industrial societies. Images of hunter-gatherers are much more romantic than those of dreary farm laborers before machines lightened their burdens. For six thousand years people stooped to plant and pick weeds or walked through rows of crops swinging the scythe. The weary laborers gathering dropped grain in rural France in the mid-19th century, as painted by Millet in "The Gleaners," could have been working the land twenty-five-hundred years earlier at the time of the Buddha. (See Illustration 9.) Farm laborers have mostly been "peoples without history," so it is difficult to lay bare how they used their minds or the character of their

Illustration 9

The Gleaners by Millet

discontent. The Buddha came from the gentry, if not nobility. Although his father was a king, his people, the Sakyas, were thought to be a more egalitarian society than a highly stratified one. His patrons were merchants, gentry and large farmers. Slaves and soldiers had to be released from their obligations before they could become monastics. The discontent he addressed was that of an agricultural civilization. Many of the metaphors he used in his teachings were drawn from agricultural life.

> [W]hen a cowherd…fails to pick out flies' eggs, fails to dress wounds, fails to smoke out the sheds, he does not know the watering place…he is unskilled in pastures, he milks dry, and he shows no extra veneration to those bulls who are fathers and leaders of the herd…. [H]e is incapable of keeping and rearing a herd of cattle. [Correspondingly a monk milks dry when he] does not know moderation in accepting [etc.].[22]

The issues troubling people were those of early agricultural civilization, famine, ravaging diseases, old age and death. The realm of hungry ghosts may have been the heritage of both the scarcity of hunter-gatherer life and early agriculture's times of feast and famine, like the seven good years and seven bad years of the Bible. Early funereal ritual often included burying food and implements with the dead to assure them sustenance in the afterlife. The emotional impact of these distresses coming after shorter, more vigorous and possibly easier hunter-gatherer life must have been enormous. Because there was neither the technology nor medical knowledge to affect the material causes, the Buddha turned to the mind where he could change how these tragedies were received.

How successful the Buddha's followers were in heeding his admonitions is hard to assess. The subsequent history of Buddhist countries makes it difficult to support the claim that the Buddha's message inspired people to live with greater wisdom than those who believe in, for example, shamans. As among shamans, there have been Buddhist meditators who seemed to have transcended ordinary human unhappiness and lived lives of non-harming but Buddhist societies have also engaged in violence, and, as we saw, there are limits to meditation. The central characteristic in the Buddha's message is its aim toward non-harming as a goal, so meditation bends in that direction. It might seem that an awakened Buddha becomes otherworldly by not engaging in acts with harmful karma. A Buddha's effect on the world is much discussed in Buddhism. In spite of the fact that Zen and other Buddhist schools emphasized being in the world, the Buddha of the original *Sutras* was otherworldly. Rather than being an impediment to our examination of the mind in modern civilization, the Buddha's otherworldly stance allows us to see the mind's contemporary character more clearly.

As technology improved, the mind became more prominent. Writing and abstract design underlay technological change, and humans brought into being whole realms of existence which were dependent on mind. As the mind claimed more territory, nature faded further away. Not only did humans need to spend less time in nature, but the products of technology were used to shield them from its impact, and humans liked it that way. Rather than the place where everyday life occurred, the out-of-doors became an impediment. It is both humorous and frightening to see the extent of this today.

Once, at a party in Cambridge, Massachusetts, a friend who was living in the back of his pickup truck in a swampy, blackfly-infested area of southern New Hampshire, chatted with a an accountant from the Lotus Corporation. (Remember, there are good reasons from evolutionary survival for us to be interested in weather. It may be the most important selective mechanism.) The accountant told my friend that she avoided going outside whenever possible. She was proud of her expertise on the climate inside windowless, air-conditioned office buildings. She knew the breezes and the warm places. My friend listened with disbelief. He could not bring himself to tell her how he lived. A few years ago, it was estimated that the average American spent 95 percent of their time in their cars or indoors. My guess is that this number has increased. Most Americans live and work indoors and are uncomfortable outside. An announcer on public radio on November 15, 1993 said, “It is almost 76°, the record temperature for today. I am going to actually enjoy going out to feed the parking meter where my car is parked.” This is his contact with nature. When the weather is other than perfect, he doesn’t like it.

While I was waiting to board an airplane to Florida during a cold New England winter, a man angrily declared that he was going to Florida to escape the minus-40° weather, a great exaggeration. And to a nurse in a nursing home in Florida, I said I liked most kinds of weather; I found that if I didn’t resist and dressed properly, I did not feel imposed upon. She apologized for her aversion to cold. “Don’t think that I am a coward.” So, being out of doors in the cold is an act of bravery. In the nursing home, temperature, humidity and light were regulated to reduce discomfort. Life becomes adjusting the interior climate, as in Lotus Corporation’s office. Consistent with this aversion to uncomfortable weather, reporting temperatures on television inflates the bitterness of cold by citing wind chill factors. Commentators who introduce weather forecasters make them responsible for the weather. As we withdraw from the elements, our dislike of changing weather has increased. The growing migration to Florida would not take place without air-conditioning. One can do little about cold and snow except stay indoors. People who move to Florida dip into the pleasant parts of the year, but escape the heat and humidity with air conditioning. On fresh spring days, I've noticed a high proportion of closed car windows indicating the use of air-conditioning.

When people travel, they no longer see much of nature. Unlike older roads, modern expressways are dug through hillsides or sheltered, removing the surrounding countryside from view. On commercial airplanes it is worse. Once, flying over the Sierras in mid-winter, the mountains surrounding Lake Tahoe were glistening below, covered with snow around a pure blue lake. In the plane's cabin, a video began and people wearing earphones sitting next to shaded windows watched a picture of clouds and sky, and then tropical vegetation. They ignored the real thing, if a view from an airplane window can be called that. The airline had placed computer screens to the back of each seat so that a passenger jammed in had unavoidable stimulation. To keep from being drawn into the computer, I put my sweater over it.

Civilized people also have trouble with the other parts of nature. As Woody Allen purportedly said, "I don't like the outdoors— there's live things there." And it is those live things which make possible our lives on earth. Without bacteria, protists, fungi and plants, planetary thermoregulation would go awry and garbage would pile up. Sometimes slimy, unpleasant and noxious beings fill niches which are crucial to maintaining life's balance. We could not live without bugs and weeds. A seemingly minor distortion of the upper atmosphere, ozone holes let in a flood of UV light, thereby leading to increased rates of cancer and threatening our existence. The Japanese poet,

Nanao Sakaki, who has lived in mountains and on agricultural communes, points to our alienation.

Future Knows

Thus I heard:
Oakland, California—
To teacher's question
An eleven-year-old girl answered
"The ocean is
A huge swimming pool with cement walls."
On a starry summer night
At a camping ground in Japan
A nine-year-old boy from Tokyo
complained
"Ugly, too many stars."

At department store in Kyoto
One of my friends bought a beetle
For his son, seven years old.
A few hours later
The boy brought his dead bug
to a hardware store, asking
"Change batteries, please."[23]

Many people no longer like nature in the raw. They don't know how to operate in it, and they regard it as is dangerous. In a radio interview with an expert from the New England Wildflower Society, the interviewer talked about preserving native species, many of which he grew in his yard. His friends referred to them as the "uglies." In one study, researchers found that most people would like to get rid of insects and other unpleasant animals because of fear and anxiety. They expressed disdain and loathing toward them. One mushroom expert feels that most people are fungophobes, regarding mushrooms as "the vermin of the vegetable world."[24]

A German youth leader noticed that children on an outing in the woods stumbled as they walked along the trails. She felt that their clumsiness was because they had never walked on uneven surfaces. They only knew how to walk in sneakers on asphalt. They were frightened of wild animals at night and were afraid of the silence, "which seemed the most terrifying of all."[25] Even when children relish the outdoors, their parents may be anxious. "At first the children liked the snow but the thrill wore off.... 'You can't take the children to the park because the swings and climbing things are all covered with ice, so they can't run around without killing themselves.'" In Connecticut, a mother reluctantly sent her three sons outside in bicycle helmets to survive their increasingly dangerous luge runs down a neighbor's icy driveway. "On a snow-day, part of me wants to send them out for the entire day.... But then I think about all the things that can happen in ice and snow, and I am suggesting they play another computer game."[26]

The Mind as Tool

Attention in our technological society is used in two interrelated ways. The first kind of attention involves conceptual thinking which brings technology into being and makes it run. The second has to do with entertainment and the evocation of emotions. As an example of the first, consider mathematics, the most abstract kind of efficacious thinking. Mathematics originated in referents to the natural world such as lines, circles and numbers of objects (particularly in merchants' inventories). In the hands of Greeks, Arabs and mediaeval Europeans, mathematics became a symbolic activity bound by rules of logic. From measurements and diagrams on pieces of paper, mathematics came to be pictured entirely in the mind. As the structures of mathematics became more abstract and complicated, visual pictures were no longer adequate. Until the arrival of computers, mathematicians did their work in their heads, making calculations and keeping track of the trail of logical conclusions on paper.

In the mid-1960s, an U.S. Congressman attacked a famous mathematician (mostly because he was an outspoken critic of the Vietnam War). The charge was misappropriating grant monies because the mathematician's activities didn't seem mathematical. In defense, his department chair responded, "...the duration and intensity of [mathematical research] simply cannot be measured by examining overt behavior patterns [such] as sitting at a desk....The heart of the work...goes on in the mind...and it is a common experience that this actually intrudes into periods when the individual can be observed walking, eating, and sometimes even sleeping...." The mathematician was exonerated. His most famous work had been done while on a Brazilian beach.[27]

Henri Poincaré, a famous late nineteenth, early twentieth century French mathematician, rode Paris street cars pondering mathematical problems. And Hermann Weyl, an abstract mathematician who contributed to the foundations of quantum mechanics, was able to see mathematical objects as if he were walking through a landscape. His written works were difficult to follow because he gave only a glimpse here and a suggestion there. He could see how everything fit together, but a reader with less power of thought was overwhelmed. When I was working on my doctorate in mathematics in the early 1960s, I would walk the streets of Berkeley, immersed in thought as if it were the center of reality. At the end of the longest period of that obsession, I had the "Eureka" experience: after much thought I suddenly found a solution. The problem and its solution seemed quite real.

Attention in mathematics is focused directly on the objects of mind. It is sometimes accompanied by silent speech or, as with Poincaré and Weyl, images and symbols. In my experience, mathematics is like other mind objects. It stimulates the senses and emotions. In the Buddha's language, it involves the other four aggregates. That this kind of attention manifests in the physical world is simply an extension of the efficacy of hunter-gatherer thought. Now reasoning is done in the context of an incredible accumulation of already successful thoughts that have led to successful manipulations. Efficacious thought is reinforced by other people working in the same context. The systems of feedback among humans, and between humans and nature, continually refine thoughts which have application. Thoughts which fail to work, either in logical discourse or in application, are put aside and those which do work are added to the corpus of thoughts labeled "knowledge." Thinking feels more real as it has further effect in a world of preexistent workable things which have also been organized by thought. This has been true of efficacious knowledge since thought became part of human life. It may have been true for mimesis too: expression and gesture came to have the feel of real things in place of what they represented.

It is unnecessary here to argue about theory and practice. If theory is thought and practice is manipulation, then they have gone hand-in-hand since the time of stone tools. Remember that the evolution of mind is interwoven with the use of tools and speech. The uncountable accumulations of actions in the material world and the understandings which go with them that, say, are buried in my computer or implicit in nuclear physics, would take more volumes to recount than anyone might like to read. In any case, the details that moved each improvement forward are simply too many to trace. I remember reading a history of grinding wheels in the late 19th century. The number of refinements of materials and techniques would only interest an antiquarian or historically inclined machinist. At each step of the way, part of this knowledge is lodged in narrative memory and part in mimetic. To patent a device in the United States an inventor only has to supply enough detail on how the device works for a skilled technician to make one. Much of what such a person knows has been learned mimetically. I think the same is true for mathematicians, but this arena needs exploration.

Attention in the hunt felt like the most real activity that hunter-gatherers undertook. So in the modern world reality is accorded to thought which brings about efficacious manipulations and the devices that do those manipulations. As these developments became more the locus of reality, the natural world of hunter-gatherers further recedes. In a laboratory or on a five-thousand acre farm, the minutiae of molecular biology or the versatility of a four-hundred-horsepower harvester use the mechanisms of nature to convert her material into food. This is dealing with nature intimately. But the scientist only peers into an electron microscope or manipulates specimens with a machine, and the farmer sits in his air-conditioned cab focused on his instruments. Their contact with nature is profoundly mediated by technology and is comparatively safe. In this complex of changes, the mind and thought become more disembodied.

Even natural history has experienced some of these changes. Paul Elton, who coined the word ecology, felt that despite Darwin's love of observation out of doors, many of his followers became bench scientists. "But the open air feels very cold [and lab work feels so normal that the biologist] finds it a rather disconcerting and disturbing experience to go out of doors and study animals in their natural condition."[28] For German biologists of the 1870s lab work was true science and Darwin an aging romantic. Much current research on bird migration is done using radar, satellite transmitters and computers. "'You can sit at your gawdamn desk and tap on your gawdamn keyboard without ever going outside,' grumbles a lean, grizzled ornithologist...weathered from years in the sun and wind. 'The ultimate in fat-man biology, that's what it is.'"[29]

In Chapter 4 we saw the way the body is now used does not fit with some of the traits with which evolution endowed it. This is a problem for modern thinkers and doers because their minds are located in their bodies and they have no choice but to deal with bodily malfunction. What is important here is that thought is experienced as if separate from the body. The hunt required the two to be coordinated in ways that did not draw so much upon the mind. In the world of technology, the mind now seems to have power of its own. Ideas leap into manipulations of the material world via machines and control mechanisms. If I could think directly into my computer, I could overcome my slow typing and dyslexia. I could withdraw into my mind and the book would be published without my having to deal with the messy physical world, without risking carpal tunnel syndrome, myopia, "theater knee," headaches, poor sleep and back spasms.

There are many other examples of this disembodiment. In the early 1980s, a high official of the revolutionary Sandinista government in Nicaragua came to the United States to persuade the Reagan administration to stop its war against his country. While in the United States, he saw a movie in which the hero, a pubescent boy, defeated a Russian sneak attack on the U.S. The boy hacked his way into the Pentagon's computers and destroyed the Russian planes. Each time the boy hit his target, the audience cheered. The Sandinista official was horrified. He felt that he had no chance of affecting opinion in a country where the life and death of his people would be decided by childish minds in a disembodied way. He felt that the Yanquis, the Americans, no longer understood the meaning their actions-at-a-distance had on real people.

Another example goes further. During the Gulf War, U.S. soldiers were astonished at how devastating their modern weapons were against the Iraqis who were using outdated U.S. and Russian equipment. Coming over a rise toward Iraqi tanks and firing on them in the one real battle of the war, the Americans wiped out the Iraqis without suffering casualties. The computerized technology of U.S. firepower was so devastating that each shot annihilated its target. The U.S. soldiers were astounded how effective their weapons were and how little risk they incurred. Even more devastating was the destruction from U.S. planes. Pilots sat at their instrument consoles and played video games with the lives of thousands of people. The war halted when the U.S. military commander realized that it had become a slaughter. The inequality of technological might was evident. A battle between technological equals might resemble the trenches of World War I, where machine guns manned by soldiers at a distance mowed down anyone who was exposed. We have come a long way from trying to kill an animal with bare hands, stone blades or spears.

One final example: when I wanted to learn about computers in the early 1990s, I joined the Boston Computer Society, the largest such organization in the world. I didn't remain a member for long. I had difficulty communicating with people in BCS. What I found in their offices disturbed me. There were rows of computers, a library of manuals and software and members working at machines or conversing intensely. When I asked for help or information, I did not feel that I was recognized as a person. I did not speak the right language and was treated with impatience and sometimes irritation. Then I noticed an element of acknowledging another person's person-ness was missing. The place had the atmosphere of minds communicating with minds about computers. Members were riveted to computer screens and talked about what was happening there. The body language of the members had a disturbing quality to it. Many members looked uncomfortable in their bodies. They carried themselves as if they had bodies only by accident. They did not need to care for their bodies with exercise or proper eating. All that seemed to matter was what took place on the computer screen or went on in their heads. They felt like disembodied minds.

Because of my sociological interest in the presentation of self in everyday life, I may have been sensitive to what is now speculated to be a kind of low-level autism present among highly talented people who work in centers of computer innovation. The offspring of programmers in Silicon Valley are exhibiting increasingly high rates of autism. It is thought that many of the parents have what is known as Asperger's syndrome which seems to have a genetic basis. Some forms of this condition include an extraordinary ability to do repetitive abstract thought such as being able to maintain a visual image of thousands of lines of computer code. In so doing, the programmer discerns something out of order and can correct errors in the code. Along with this ability goes an indifference to their bodies and a low level of social skills such as the inability to maintain the kinds of eye contact and nonverbal rapport so crucial to emotional communication.[30] Because the constellation of skills, machines and people is now accorded status and financial rewards and cited as a model of productivity and progress in a post-industrial world, there may be a selection for it with the unanticipated side effects of social disconnection and tragic autism.

Stimulation and Fantasy

Thinking in order to manipulate material reality is one way attention is linked to disembodied mind. Technology has also led to the disembodiment of mind in more general realms of narrative and mimetic experience. The two are interrelated. One Christmas Eve I sat at dinner with two psychotherapists, their boyfriends, a grandmother and three men in their twenties. Two of the young men were computer programmers and the third a computer maniac. The two programmers were independent contractors hired by an advertising firm to develop a website on the Internet for a soft drink company. They were working fourteen to twenty hours a day, trying to do their best job. If their clients were happy with the results, then their careers would be made. They loved what they were doing. Only occasionally were they distressed by the number of hours that they sat before computer screens. They expressed no doubts about the worth of the task at hand: entertainment in order to sell soft drinks to people who would spend hours at the computer. Money was to be made by the company, the advertising firm and them. The consumer would be satisfied. This is the wave of the future.

The third young man was probably clinically disturbed. I lived in a room next to his for many months. He was athletic, a bicycle rider. He had an eating disorder. He had dropped out of graduate school and worked in a video store. Every hour he was not biking or working, he spent in his room, playing computer games, communicating on the Internet and watching movies. He sometimes did two or three of these at once. He lived in the world of fantasy created by these media. I have no idea of the origin of his mental disorder. There was no evidence that his use of computers or television caused it. His symptoms manifested in asocial behavior, inappropriate catharsis and personal disorderliness. There was some concern in the household that he could be dangerous to himself or others.

This young man may have been delusional independent of the technological level of society. That is not the point. Technology had made possible for him a real, imaginary world with which he could interact in a disembodied way. While he was playing computer games, I often thought a second person was in his room. His shrieks of laughter, the mumbles of discontent mixed with the voices of the machines came through the closed doors into the small hours of the morning. He went to parties where people dressed up as various characters in the computer games and acted them out. His absentee landlady, also a therapist, once went with him. Her comment was that the others at the party were pretending, but for him it was real; in fact, his only form of live interaction.

This young man is an extreme case of the way in which technology offers a progressively encompassing life in one's head. Listening to people at work talk about television programs, it is hard to tell whether they were participants in what they are talking about or were only an audience to its presentation. They relate the events with great affect as if they were real. They care about the imaginary dramas. This phenomenon is not isolated. There is a recent psychological study subtitled, "How People Treat Computers, TVs and the New Media as Real."[31] The stories of elder hunter-gatherers around a fire were the beginning of the elevation of vicarious living. Written works, from the Bible to novels, enlarged its realm, while television and movies make it even more encompassing.

At one time a friend came to me worried about her five-year-old daughter who had grown up without television. When they were visiting friends, her daughter was drawn to television. She observed her daughter become riveted to the screen, face dropping into the vacant look the mother had seen on other television-watching children. She was worried that her daughter had become hooked. She wanted to know what was going on that she could lose her child in such an unreal world. When watching soothing shows designed for them, the metabolic rate of preadolescent girls drops to a rate as if they were only semiconscious.[32] It is a bit like hibernating animals which is why overweight is correlated with television watching. Toys employing manual dexterity are diminishing in popularity. The Lego Manufacturing Co. in Denmark has fallen on hard times, losing market share to computer toys and video games because children are more intrigued by toys that appeal to their minds.[33]

Adults also choose mental-only stimulation. They sit in movie theaters, body anesthetized, attention locked on a larger-than-life, vivid, over-colored screen with amplified, high-tech surround sound. The fantasies of computer animation stimulate sense receptors of sight and sound. These excite our built-in responses to the world in a way that challenges anything nature can come up with. People identify with these experiences. A computer expert on the radio defined virtual reality as when a person is enmeshed in the data. "People talk about the new virtual reality games as if they had just returned from their vacation." And fantasy is what many choose for their vacations. Las Vegas is the second most popular destination in the world after Mecca. It is set up as an adult playground with themes for adults. The lights are so bright that they can be seen from a Mars orbiter. Since casino gambling, from the point of view of mathematical probability, is a loser's game, people are going for the thrill of illusions. For the Buddha, daily unawakened life is an illusion. He recommended that people use meditation to see life as it really is. The examples I offer here show how the modern world leads into greater and greater illusion and therefore, from the Buddha's perspective, greater discontent.

Recall that our brain dulls to routine, repeated signals. It is good at catching novelty. It had to be in order to survive. Our bodily responses to danger are instantaneous, often bypassing the cerebral cortex, only informing it later. Animated space travel, with explosions and frightening creatures suddenly appearing, keeps audiences on the edge of their seats. As in dreams, secondary emotions are those ignited or reinforced by stored images which are filled with narrative meaning. They excite receptors of fear and desire lower down in the brain. Movies or soap operas with multiple stories weaving in and out of each other so that there is always emotional action means that attention is always gripped and the amygdala and other primary response regions are stimulated.

The video one evening at Lama Foundation, eight thousand six hundred feet high on the side of a mountain next to the Carson National Forest, began with an advertisement for MGM films flashing rapidly from one scene to another. The audience excitedly identified the different movies from which the few seconds of film came. It was over-stimulating for me, too much sensation too quickly. I lasted ten minutes. Going outside, the stars above the mountainside were quiet and luminescent. Nothing was happening. It seems nature is boring without the attention hunter-gatherers needed to survive, and even the life that a modern audience leads pales in contrast to what is on the screen.

Consumer businesses are very aware that to sell things they need to engage natural human connectors. The food industry synthesizes compounds which specifically evoke human taste responses. "Go-away" is the technical term for non-persistence of taste or texture. Add oil and sugar to peanut butter and improve its "go-away." In contrast, eat organic peanut butter when the natural oil has settled out and you will experience non-"go-away." It sticks to your palate so strongly that it takes effort to open your mouth. Likewise the entertainment industry grabs on to the ways the mind creates internal stimulation. Its executives intentionally try to hook people on their products. Children have no problems liking commercial peanut butter or sugar products. They become addicted to them with a response in the body which reacts anew each time until it becomes jaded. This may also occur *in utero* according to a study which showed that the infants of women who watched soap operas while pregnant were attracted to the same shows. The unborn infant hears the music and may be conditioned by the mother's response. The same is also true for people in front of television screens. The mental receptors respond until they become over-stimulated. Give them a rest and they are ready to go again. Experts are raising the alarm that young children raised on television have troublingly short attention spans.[34]

This is clearly not all that's going on in the modern world. Many things balance life and some people live engaged lives. There is love and compassion. Still, what I have described is a significant part of "progress" in advanced industrial society. The economy depends upon consumption of illusions. What would the GDP (or gross domestic product, a measure of prosperity) be without entertainment, the Internet and tourism which vies with oil as the world's largest industry. Also, given the kinds of social breakdown, distraction and isolation, these elements play a larger role in people's lives. Children are deposited in front of the television set at a very young age. What is important in this analysis is that in the realm of leisure, as in the realm of material manipulation, civilization is heading in the direction of reality experienced as disembodied mind objects, as fantasy.

In the chapter on mind, I mentioned Michael Cohen's nature education. He spent thirty years taking groups of young adults around the country to natural places so they could experience living together in nature. In doing this, he slept out-of-doors most of the time and learned how to bond his charges together. He developed exercises to reconnect people to the natural settings which gave birth to their ways of experiencing and responding to the world.[35] In the university, I taught courses investigating humans' place in, and our effects on, nature. Most of my students had grown up in the suburbs or large Eastern cities. They had little experience with nature. What follows comes from my notes after leading a college class through one of Michael Cohen's exercises and then discussing its implications with them.

I and the ten students of my class each crawled, with eyes closed, along his or her own hundred-foot length of string we had laid out in the woods. It was the end of the fall, and we were near some office buildings at the edge of the campus. The dark brown oak leaves on the ground crunched under foot. Crawling along the string, the ground felt very different than it appeared when laying down the string. The instructions for the exercise had been to pick a safe place without thorns. That was what it had looked like, but on hands and knees it turned to have abundant prickers. The eyes of civilization did not see what the hands and knees felt in intimate contact. The ground was much wetter than it had appeared to me. Comparing notes later, some students found it drier or softer. My spot appeared garbage-free but turned out to have a hunk of Styrofoam and a bottle hidden under the leaves. The sensations they set off were strikingly different from natural objects and the ground. The Styrofoam felt so strange that I couldn't resist opening my eyes to look at it. The Coke bottle was instantly recognizable.

In the discussion afterwards, the students had a hard time talking about the subtleties of the variations they experienced. What they felt did not make sense to them. They seemed to lack perceptual categories into which to put the unfamiliar sensations. One well-dressed student did the exercise on the lawn of the office building so as not to dirty her clothes. (They all had been told in the previous class that we would do this assignment.) When asked if the lawn seemed uniform, she didn't know. We were sitting on the lawn with the parking lot on one side and the woods on the other. The lawn was a stretch of uniform green, while the woods were a tangle of trees and shrubs. The woods called upon senses rooted in human evolution in the setting of leaves and thorns. The lawn was an invention of agriculture and prosperity. While crawling along the string, I found too much to feel. My mind was not sufficiently focused to take it all in. While I was interested in the novel sensations, many of the students could not figure out why all the unpleasant dirt and sticks might be relevant to the life they expected to lead.

Later in class we discussed how we have come to define ourselves through what we consume. I asked whether the media-inspired sense of autonomy associated with consumption— where one defines one's individuality by the apparent free choice of things one consumes— was an illusion. I prodded them with examples. Teen-aged kids in a convenience store before school, how would they respond to a sugar Gestapo saying, "no sugar"? One student looked at me and said that the sugar cop would get killed. He assumed students have a right to buy all the sweets they want. So I switched to smoking. The students described ads which portrayed smoking as macho, sexy and cool. They gave examples of high school students rebelling when smoking privileges were denied.

They were still puzzled, so we moved to *The New Yorker* magazine and what image of self is conveyed by its ads to its upper-middle class readers. Porsches. A sister of one of the students owned one, and another student had driven one. They lit up at this example. I asked if a Porsche conferred any marginal benefit over a cheaper car. They came up with a Chevette. In terms of getting places and carrying things, it did not seem a Porsche was better. In any short-to-medium-length trip with both drivers obeying traffic laws, the Porsche and its superior mechanics might arrive only a few minutes earlier and give a slightly smoother, quieter ride. I asked them what is gained by driving a Porsche. It is interesting how excited they became. The responses were speed, power, status and a satisfying feeling of connection to the road. They told me about a TV advertisement where a camera was mounted on the bumper of a Porsche to demonstrate how superior the suspension was as the car sped along a winding road. I tried to convince them that what the camera conveyed was an artifact of Madison Avenue. How different would the images be from a camera mounted on the bumper of a Chevette? Without editing, gyroscopic mountings and directing the camera, a camera tied on any bumper would produce a hodge-podge of images. Without sensitive instruments, you could not measure the differences between the Porsche and the Chevette.

These three examples are ways that a consumer-based, capitalist economy utilizes human connectors to nature. Sweet, salt and fat cravings, caffeine and nicotine highs, and high speed and endorphins generated by precise exertion tie us into a system of consumption and production. We feel entitled to these stimulations because they have become part of our sense of self. Moreover, they contrast with our connectedness in nature as the parking lot surface does to the woods. There is a reduction of variation and a selection for things we are attracted to. Things to which we are averse are minimized. It is like the difference between walking on a sidewalk as opposed to scrambling through the woods.

Civilization with its powerful technology has transformed the built environment to make life safer and more comfortable. A side effect is that people experience life more as mind objects. What then has become of our relationship to raw nature? People prefer

Disneyland. It is clean, convenient and thrilling, with no risk or discomfort. One Sunday I was walking along a wide trail on the cliffs above the ocean at Point Reyes National Seashore. Except for some birds and an occasional seal head popping out of the ocean below, there was no other wildlife to be seen. The view was spectacular.

I spent a long time watching a crippled beetle try to make its way across the path. No one walking by paid much attention to a middle-aged man crouched down with his face near the ground. I listened to the voices as they approached and faded away. I wondered how much the chattering hikers saw of the world around them. Were they aware of the struggle of life and death that went on within the sweeping vista of ocean and cliffside? Their attention was focused on their conversations. I know that when I go for a walk and talk to a friend, I don't see very much. A trio of teen-age girls passed and glanced over. One commented that her cat liked to eat large, unpleasant bugs. The wide trail required only a bit more attention than a sidewalk. In the literature on National Parks, there is much discussion about how the landscape has been domesticated so that tourists can experience nature conveniently.

A few days spent driving from parking lot to parking lot in the Rocky Mountain National Park to climb to advertised views has similarities to a visit to Disney World, although it may require considerably more physical exertion. There are debates about commercial concessions, exhibiting nature in displays like the Old Faithful geyser, and the penetration of parks by recreational vehicles. It is possible to hike into the back country away from the crowds, but there too nature is often hedged. Hikers carry high-tech equipment. There are cages to sleep in because bears live off scavenging. There are disputes about liability from injuries suffered by hikers and climbers.[36] Now cellular and satellite phones can keep wilderness hikers in touch with loved ones and rescue teams. Parks are making people pay for false alarms. Some backpackers locate where they are with satellite positioners. As challenging as these wilderness experiences may be, sophisticated equipment is relied upon to mediate the relationship with nature. A visit to the great Falls in Yosemite National Park will illustrate how much National Parks have become shows like nature movies. It is experienced with disembodied, narrative mind.

Also people rarely see how much nature exists in their back yards. One morning I climbed through the woods from the Buddhist Study Center to the neighboring Insight Meditation Society in Massachusetts to meditate. A foot of late-spring snow was melting on the ground. The previous day I made the same trek in snow shoes. The snow had been so heavy that the usual ten-minute trip took a half-hour. This day I encountered two animal signs, a deer which had slept under a tree and a coyote who checked the spot or may even have disturbed the deer. The day before, the snow was still too deep and wet for animals to maneuver in. This morning there were more deer prints.

As I walked into the staff lounge, someone was telling old camping stories. "It isn't like the old days. You can't really camp any more. You have to pay for campgrounds and sleep next to other people with all their gear. Maybe you can still go camping if you hike into the back country." It was not part of the *Weltanschauung* that beyond the walls of the meditation center were coyotes following deer, bear waking from hibernation and even newly arrived moose which had been exterminated two hundred years earlier. On my way back through the woods, I saw that a deer had crossed my path since I had gone up and that a coyote had followed it. The coyote then tracked my path for a way until it came across a coon which had also crossed my path while I was meditating. The coyote circled around and went back or maybe it pursued the raccoon.

Nature films run the gamut from Walt Disney's anthropomorphic films like *Bambi* which give a romantic view of the wild, to contemporary films which claim to be scientific. The filmed nature is compressed to keep interest and so packaged that it lacks a real feel, a sense of the discomforts of the outdoors and the slowness of time.[37] While nature films now seem to use more real footage, they give little hint of how they are made. Production includes chasing animals and setting up situations with captive animals. Disney's films are a mixture of staging, lab work, trained animals and natural footage. Life is distilled into time-lapse photography so that it can be experienced conveniently at home in one sitting. We noted how experienced naturalists do not often observe actual predation. And it often is not pretty. This is no problem for television viewers. Many times they vicariously experience what it might take a person years in nature to witness, and although violence is shown, it lacks the smell or the gore. When most people visit National Parks, they want experience to replicate nature films, not the real thing.

I have spent some pages contrasting the experience of hunter-gatherers and civilized humans. We have seen that people now live more in their minds which, in turn, have become more disembodied. Human attention is still engaged in doing things to the material world as it was for hunter-gatherers. But it is now riveted in the mind itself, rather than the natural setting which gave it birth. Such a situation is one which the tools of Buddhist meditation seem designed to analyze. Meditation aims at understanding the character of mind and body and their interrelationship. This is not a pursuit of abstract knowledge, but an investigation aimed at discovering the roots of discontent in order to undo it. Did the shift to civilization change the way in which discontent arises? Does the way civilized humans use their minds affect the character of our discontent? We saw in Chapter 4 that technology changed the physical circumstances of life, presenting individuals with much longer periods of disease, old age and death. Technology made disability possible and affordable. The process of dying is often drawn out over years. It takes great strength of character to face the discontent of the ways that aging now happens to many people.

How does discontent manifest in a culture where people live mainly in their heads? For the Buddha, stimulation at what he called the sense gates (which includes the mind) is where the process begins. Discontent comes from the way the mind's reaction to these stimulations is held. The process is described by Dependent Origination. If one is not alert to sensation and attachment, then craving and aversion become progressively cemented into what we feel is our selves. If the holding is not seen for what it is, an automatic reaction of the mind, and subsequently released, then the process builds into discontent and its consequences. This is all within the mind. There are many ways that modern stimulations engender discontent.

Thich Nhat Hanh is a well known Vietnamese Zen Master who has written many books to remind people how much discontent is a result of ignoring the impacts of civilization's seductions.

> Some of us leave our [personal] windows open all the time, allowing the sights and sounds of the world to invade us, penetrate us, and expose our sad, troubled selves.... Do you ever find yourself watching an awful TV program, unable to turn it off? The raucous noises, explosions of gunfire, are upsetting. Yet you don't get up and turn it off. Why do you torture yourself in this way?... Are you frightened of solitude...the emptiness and the loneliness you may find when you face yourself alone? Watching a bad TV program, we become the TV program....Why do we open our windows to bad TV programs made by sensationalist producers in search of easy money, programs that make our hearts pound, our fists tighten, and leave us exhausted?.... Losing ourselves in this way is leaving our fate in the hands of others who may not act responsibly.... All around us, how many lures are set by our fellows and ourselves. In a single day, how many times do we become lost and scattered because of them?[38]

How much of our lives are lived in these lures and how much does the economy depend upon them? Listening to economic news one is struck by the degree to which our prosperity depends upon consumption. Economic indices such as the Consumer Confidence Index and the totals for consumer expenditure and consumer borrowing are important factors in the health of a growing economy. Even though it may be impossible to measure the amount of consumption in first world countries which represents the necessities of survival versus the amount above that which feeds our greed, it is clear that a large proportion of consumption, convenience and entertainment take us into our minds. The way they do this supports selves which are in denial of the ways that we are embedded in nature. The disturbed computer addict I mentioned lived as if only his computer and his fantasies were real. Internet pornography and sex are a very odd way of relating to life. Besides these there are other areas where what the Buddha understood as the self becomes enlarged. Our relationship to time, our need for control, and our insistence on autonomy are arenas where discontent gets initiated.

Have you ever watched your mind while waiting in line at a fast food restaurant? What do you find in there? I asked this of a class on nature and technology. I told them that I noticed in fast-food restaurants that my habit of wolfing down my food is greatly exacerbated. When I walk into a fast food place, it was usually because I am quite hungry as I am rushing somewhere. Standing in line, I find my mind running with complaints and anger that the food is not getting to me even faster than it does. By the time the food comes, I am so obsessed, that I rarely taste or appreciate it. One student responded that it is even worse when you go there with ten friends, because sociability distracts you even more. In contrast, once at dinner with a group of meditators, one person did not order. We turned to her in surprise. She said, "I never eat when I am out with people because I can't eat and talk at the same time." I wonder what would happen to the restaurant business if more people felt that way?

In fact, restaurants are often designed to encourage our distractions. Motivation researchers who study the human mind and wish to profit from their findings purposely build into fast-food restaurants stimulations of naturally occurring desire as it relates to impatience. Fast-food chains publicly advertise that they will deliver food within a given number of seconds, but that is not good enough for my greedy mind.

In my meditation classes, I taught students how to do eating meditation. They were told to bring their attention to each aspect of eating and to eat slowly in order to do this. The students who try this in fast-food restaurants are often revolted by their experience. They do not like the food when they really taste it. And when it gets cold because they are eating so slowly, the congealed fats, the excess salt and the overdose of sugar leave them disgusted. After doing this exercise, some students give up fast foods while others forsake the eating meditation which has spoiled their enjoyment. The same impatience and the same close examination of experience which might spoil things are hidden in many other aspects of our lives. AT&T marketed touch-tone dialing in the 1960s because they figured people would pay for the fraction of a second and slight physical effort saved. In contrast, Thich Nhat Hanh suggests treating the telephone ring as if it were a meditation bell: listen attentively to several rings before you answer. Such a strategy might help undermine our economy. From an article in "Business Week" magazine on the increasing power of computers we find, "The days of waiting interminably for keystroke commands to be answered are coming to an end. And beware: Instant gratification will make the Net even more addictive."[39]

Busyness, scarcity of time and increased complexity are almost a disease of modern life. People who cannot tolerate the pressure suffer from stress. Stress is the state animals are in when survival is at stake. It arms certain systems for self-defense and disarms others which might interfere. Putting animals in a state of continuous stress causes various physiological systems to begin to break down. In humans, stress contributes to poor sleep, depressed immune systems and heart conditions, along with anxiety, depression and inability to nurture. Stress contributes to road rage.

One commentator on time feels that the pressure to do things faster in the modern world gets stymied by biological processes.[40] He feels that compost, healing or the time of birth cannot be rushed. Yet the second largest area of economic growth is the biomedical industry which alters the timing of natural biological processes. Besides prolonging life, it addresses anxiety about health and personal appearance by trying to increase our control over natural processes. And medicine has not only become involved in timing birth but despite governmental prohibitions, there is a large market for determining aspects of one's offspring. There is a story about a married couple who wanted a female child. The husband was a physician specializing in obstetrics. So he selected for males among his own sperm and did *in vitro* fertilization. Four implanted female fetuses survived in the wife's uterus. At thirteen weeks of growth he aborted all but one which went to full-term.[41] The couple determined the gender of their child.

Recall that birth in nature is the riskiest point in life. Modern sanitation and medical technology have virtually eliminated the risks. Still we seem to possess an inborn concern about reproduction, infants and mothers. Whether genetic or not, it is consistent with survival of the species. Modern women are deeply concerned about imperfections in their potential children. In an article in the *New York Times* women were asked about the use of prenatal technology. "'If the technology exists, why not take advantage of it.'... 'I don't care if I wasn't in the high risk group...it is my right and I want it done.' 'Even 1 in 1,500 is not a very comforting risk if you are that one...I don't think that I am equipped to handle a baby with a severe disability. As it is, there is enough stress raising a wonderful healthy child.' '[I was] worried during my whole pregnancy and I think I became obsessed with it.' When this woman became pregnant again, she said, 'I knew the day I took the home pregnancy test that I would get an amnio[centesis].'...The trend...is strongest among...highly educated women...their lives revolve around control and information....So when they embark on a pregnancy they want control and information.'"[42] When the Yanoama woman miscarried in the jungle, life and death hung in the balance. Birth is fraught with uncertainty. Technology in prosperous societies may be used as much to alleviate fear as to solve real medical problems.

How many times are doctors and medical companies sued by angry patients who sought an extra benefit and found that either the benefit did not last as long as recipients hoped it would, or the procedure had unexpected side effects? Current class action suits surrounding cosmetic surgery are an example of this. Of course, some corporations in the modern industrial world sell what they do not have to sell or are simply criminal, such as the drug companies which developed and promoted DES, used to control woman's menstrual cycles, even though they knew it was mutagenic. Nonetheless, they appeal to ready buyers who want improved appearance or exemption from nature's risks. Major lawsuits have occurred over promised cures but, even more so, over elective procedures which are cosmetic or involve reproduction. The buyers, attracted to these modern versions of snake oil sold by medieval mountebanks, live as if entitled to the benefits regardless of the realities of disease, old age and death. This is a source of discontent. The suffering seems greater than that of hunter-gatherers because technology allows for much greater choice of outcomes over a much longer period of time. It also promises what seems an exemption from the processes of nature. It often fails in its promises or has side effects which neither provider nor patient wants fully to admit. Since we don't acknowledge the reality check of nature ever present in our lives, the complex of an enlarged domain of thought, expectations about science taking care of everything and lots of time create a greater opportunity for discontent than existed when we lived closer to nature.

The luxury of risk-free choice with respect to our bodies has real limits, but when it comes to material goods in our society, it seems almost infinite. The choices of games, cars, TV channels (more than 250 of them), vacations, sports, movies, clothes, music, occupations, cooking equipment, computers and food are overwhelming. In the midst of this choice, we feel that we have limitless possibilities of creating an autonomous identity. Like the high school students who would overrun a sugar cop at a convenience store because they feel entitled to Coke, we invest a lot of energy in and are deeply attached to our special selves. Family dinner time is vanishing in the household. With take-out, freeze-dried food and the microwave, everyone can eat what they want, when they want. This need to be autonomous has a simple Buddhist explanation. The stimulations around us are latched onto so tightly and so repeatedly that we completely identify with them. This identification, based on real connectors and encouraged by society, is addictive. It is a large part of who we take ourselves to be and how we need to behave if we are to function as members of our society. As we define our selves, we become separate from others who are doing likewise. This creates breeding ground for discontent. When nature catches up, as it always must, the civilized seductions which buttress the modern self fall away.

I remember my mother in the increasing senility and arthritic pain of her final year saying how much she appreciated television. It was like having someone to whom to relate. She couldn't stand listening to the radio or a cassette player because it made her feel so lonely. As her faculties diminished, even the television in her nursing-home room would not suffice. She needed to sit in the activities room because there were people there, even though they may have been semi-comatose or acting out. At least the personnel came in and out to check on their charges. Finally, her suffering became so great that she wanted to die; she stopped eating. In the last month of her life, only the presence of her nurse or the touch and the softly singing voices of her children could calm her terror. The objects, as presented in her senile mind, had become so overwhelming that only primal sensations could cut through their intensity. Stimulation that reengaged mind objects made things worse. Until the pain vanished in the last week of her life, she succumbed slowly to the very causes which set Siddhartha Gautama out on his quest.

In contrast, an Eskimo grandmother whose teeth are so worn out that she can no longer eat, goes to meet death from exposure to the cold. She is like the old polar bear dying of starvation on the ice floe when it can no longer hunt. Although it may not be pleasant, death comes quickly. The mind for her does not have what seems an infinity to dwell upon the pain; and the techno-world does not provide a means to enlarge the domain of that inner concern. Death is natural.

We introduced the concept of antagonistic pleiotropy which is that unwanted traits come along with useful ones. We also looked at the idea of adaptive radiation, the adaptations which allow a species to live in a new environment. Medical technology, productive industry, the automobile, airplane, computers, the information superhighway, synthetic fats, TVs and fertility drugs are miniature, human-engendered "adaptive radiations:" they allow us to survive in untold numbers and in unimaginable ways. A human brain in a bottle, directing machines which sustain it, is an entity yet unseen by nature. Unfortunately such an entity would still possess a human mind. We have not seen the radiation which takes the mind's marvelous technical functioning and severs its attachments so that it can live with less discontent and more wisdom. The mind still possesses the planner who obsesses and has repetitive painful thoughts. We are beings whose sexuality has been liberated from the constraints of nature and confined social life and so can live in unbounded fantasy, beings who ignore the indifference of nature to our continued existence and beings whose psyches have grown so large that we have difficulty cooperating with each other.

Antagonistic pleiotropy—the downside of the great adaptive radiation of technology—is an increased arena for discontent. Despite all of our material progress and prosperity, if we look with open eyes at the world around us, violence and greed seem undiminished. Ajaan Cha may be correct: progress has created greater suffering. The marvels of technology have not liberated the mind. They may have bound it even more. As one of Ajaan Cha's monks said during a retreat, "It is like putting a biological body into an electric circuit."

Chapter Nine
Silence

As bird watchers who strain to see warblers flitting about in the trees above them know, the human neck was not made for looking overhead. If we had to worry about many airborne predators in our early evolution, nature might have devised us a better set of cervical vertebrae or stuck a couple of eyes over our prominent forehead, which shields our over-developed cortex, rather than under it. As it is, we don't have much chance against the giant cats that have been traditional human predators. Mountain lions striking from a tree above and behind us remain deadly. One recently attacked a mountain bicycler, head down, whizzing by like a deer.

How do our natural endowments affect us now? One elevator company estimates that it carries a number of riders equal to the world's total population every nine days. While the number of individual elevator riders is probably less than five percent of the Earth's population, that group's behavior says something about the modern world. If riders have to wait more than fifteen seconds, they become "antsy." After another thirty seconds, they become agitated.[1] It may be that staring upward at the elevator lights makes people anxious because of our vulnerability to predators striking from above. Whether or not this is the case, one can ponder in amazement how far the human mind and body have come since Lucy our *Australopithecine* ancestor walked the savannas of Africa several million years ago, and how awkwardly what we have inherited from Lucy and her successors fits into the world we have made.

Waiting for elevators and waiting in traffic or in supermarket lines adds stress to daily lives. One of the most insidious modern waiting lines is the automated telephone queue. I just spent fifteen minutes going in a circle around my health insurance plan's automated telephone menu. It is cheaper for them, they save labor expenses, but costly to me in terms of time and mental health. Speaking to the impatience of telephone users, one advertisement, touting the advantage of Internet loan services, satirizes telephone answering trees, with a telephone voice saying, "If you are tired of waiting press two. It won't do any good, but it might make you feel better." Tall building owners who have to please their customers and reduce insurance bills disable "DOOR CLOSE" buttons on elevators. People pushing the button will feel they are doing something about their wait, and the building owners protect themselves from the liability of a hand squeezed in a closing door. The button is a sop for impatience.

Reviewing my argument: I began this book with two perspectives from which to illuminate discontent. One is the Buddha's method of investigating the human condition. He recommended that people with an inkling that life does not match their ideas about it carefully examine their minds and bodies throughout the course of an ordinary day. The second perspective is the method that Darwin used to try to understand the course of nature in the sweep of time. Darwin, like Buddha, invited humans to look closely at nature because such an examination would reveal what kind of thing life is. Darwin proposed observation of outer life while Buddha pointed inward. Although Darwin's science focused on changes in the world he, like Buddha, was irresistibly drawn to how humans functioned mentally, on the inside.

What introduced Buddha to discontent was his realization that birth leads inevitably to disease, old age and death. These are also crucial parts of the engine driving Darwin's evolution. So we used natural history to examine how comings and goings occur in nature. Natural history rests on a foundation of exact observation. Although Darwin's goal was to develop a theory of evolution, it was careful observation which over and over again emerged as a resting-place for knowledge. While conclusions about evolution are tantalizingly suggestive, the proof of natural selection remains just beyond reach. As the Zen hermit, Ryokan, wrote:

> With no-mind, blossoms invite the butterfly;
> With no-mind the butterfly visits the blossoms.
> When the flower blooms the butterfly comes;
> When the butterfly comes the flower blooms.[2]

What arise from careful investigations, though, are ever more detailed pictures of living things interacting with each other in situations of life and death.

As much as we would like the process of evolution to be orderly and progressive, it includes outcomes which seem arbitrary or almost counter-productive. Darwin wrestled with nature's arbitrariness when it struck his own family. He lost faith in God when he saw natural selection operating to eliminate his children who were the result of inbreeding. And although he did not lose faith in his methods of observation, they were of little avail in relieving his own suffering. The discontent Darwin faced was not remedied by his natural history; still, natural history is the best tool we have to look at how disease, old age and death occur. I have argued that natural history provides the biological backdrop, the broad context, of the sources of the suffering whose defeat is Buddhism's primordial aim. Darwin gives us the map, Buddha the compass to find our way home.

Natural history reveals circumstances under which death could be seen to occur in gentle ways. Here gentleness might be the rabbit which dies of shock before it is killed or the "Conversation of Death" where an animal no longer fit seems to offer itself up to its predator as a result of some ineffable communication. Death occurs quickly. From some human sense of value, this may be a better way to go. But nature also has more cruel ways of keeping life in balance. With incredibly clever devices, Gaia produces what works regardless of how the recipient feels about it. Pain and torture make no difference to her. Evolution encourages advantageous mutations, however strange or creepy. This includes senseless traits such as infanticide among prairie dogs which may neither help nor hinder them as a species.

While predation is one of life's most dramatic selectors, it plays second fiddle to climate and the environment. In the struggle for existence, the least defended are the most victimized. Mortality rates are skewed toward the young, and most disabled do not survive. Of the small proportion of animals that make it through adulthood, there follows the problem of old age with all its discomforts. From Darwinian reasoning it is difficult to understand what old age contributes to evolution. At best, a few post-reproductive females and old males, unable to compete sexually, help others of their species survive. Old age may in fact be another example of pleiotropy, a fortuitous byproduct of the vigor required for rearing offspring. For this the old suffer senescence or the gradual degeneration of a body which had to be made hardy enough so that it or some of its relatives would successfully compete in the struggle for existence. Senescence is another of life's characteristics that is neither orderly nor efficient.

When looked at closely, nature appears more ragged than our ideals. Tattered butterfly wings, brainless wandering bees, an arthritic starving old polar bear and the crippled scrawny mountain goat with too little flesh for a wolf to bother with are fates that await the vigorous. In trying to paint a more humane picture of nature, Darwin responded to his religious critics that cruelty in nature was an inexplicable rarity. But his own and later investigations lend credence to the opposite conclusion. Like Darwin, we are not the only ones uncomfortable with nature's ways. We have seen that animals, to varying degrees, possess emotions like humans which they sometimes overdo. All life seems to have the basics of fear and desire but some beings also display excessive greed, react out of rage, experience fear out of proportion to the cause and mourn losses. Animals biologically close to us possess neurochemical responses similar to ours. These materially underlain response systems, so important to our survival, may represent the evolutionary groundwork for our discontent.

It all begins with bodies. Ancestral humans experienced life much as other animals. Human bodies fit into the natural setting. Most humans died in infancy, and most who survived met their end as a result of physical trauma. A small percentage lived to old age where degeneration eventually undermined survival. As human ancestors aged, cataracts, arthritis and slowly healing wounds made foraging ever more difficult and predators ever more dangerous.

With the domestication of plants and animals, our situation in nature changed radically from that of our animal relatives. We carried our animal bodies into a progressively civilized world which blunted the old edges and created new, softer and more insidious ones. With agricultural surplus came increases in child bearing, continuing high infant mortality, growing populations, grindingly repetitive labor, leisured classes, epidemics, diseases and disabilities. This was the context from which Siddhartha sought surcease. Technology addressed each of these in somewhat different ways. The civilization we inherit has eliminated infant mortality, holds killing epidemics tenuously at bay, employs the body in ways very different from the original contexts of its survival, foots the bill for chronic disability and has invented old age, along with a drawn-out process of dying, as a major part of life. By means of miniature technological adaptive radiations, we have rewritten the mortality tables for mammals.

Today people have more time to live with disease, old age and death. These are not as harsh as in the Buddha's time, but they nonetheless insinuate themselves into our lives with considerable force. Chronic conditions have become the new epidemics. Some geneticists are awed by their ability to delay senescence and lengthen the lifespan of worms. There are news reports about the possibilities of doubling the life of humans. After a talk in Berkeley, one researcher was mobbed by students accusing her of arrogance in the face of environmental problems. "They really nailed me to the wall." Other researchers there claimed they'll eventually be able to really lengthen life. One said, "If we could do this there is nothing [we can't do]. It's the big enchilada." In contrast to his enthusiasm are the realities of medical prolongation of life. A friend recently died at ninety. She was kept physically alive for years in a state of mental suffering by medicine and well-meaning people even though she had long-since lost any desire to live and would implore God to take her. In the end her body failed massively. She had passed beyond medical redemption. From her nursing home room on the night she died she was heard singing the national anthem.

Civilization created an analogous container for the mind. Having its origins in the animal world, as the body does, the mind evolved nimble traits permitting humans to survive. Our minds, which enabled our hands and senses to forage, co-evolved with tools and language into powerful devices to manipulate the world, employing ever more abstract perspectives. The accretion of improvements in tools and technique hid the intricate natural foundations on which even the simplest effective mechanisms are based. For example, we take something as simple as a straight edge for granted. Yet it is hard to go into the woods and make a truly flat surface with what you find there. No rulers! No book edges! No milling machines! Yet flat surfaces are indispensable to civilized life. We abstractly construct the complex designs for production, assuming flatness. Civilization endows our minds with real power. And if computers fulfill their promise, we won't even need manual dexterity to make the link. The headline in a local newspaper reads, "MindDrive controls computer with users' thoughts." A sensor attached to your finger "reads the person's bio-electrical signals from the brain and transmits them to the computer, which will do what the user is thinking."[3]

Whether hunter-gatherers who lived more immediately in their world were better off than we is hard to say. Their quality of attention seems to have better matched the circumstances in which their lives unfolded. Whether this made them wiser and less harming of their fellow beings is hard to determine. Although their large brains possessed the potential for significantly the same quantity of discontent that we experience today, it may be that their focus on the immediate realm—a focus survival demanded—as well as their shorter, more natural lives provided much less room for discontent to fester. I have argued that it is far from accidental that Buddhism developed in early agricultural civilization. Although some have found individuals they regarded to be philosophers among primitive peoples, I find no evidence that Buddha's observations were ever developed in hunter-gatherer society.[4] Preconditions for the Buddha's observations were, I maintain: a leisure class, epidemic disease, hard labor, political power and the ability to survive still carrying the pain of disease and disability. These were all made possible by agricultural surpluses. Remember the sheer volume of death prevalent in agricultural civilizations. These early civilizations, for all their glory, had a downside: they engendered the first crops of civilized discontents—the very people who motivated the Buddha to investigate and propose solutions as he did.[5]

Unlike other religions, Buddhism has no creation myth. Creation is an imponderable. The Buddhist cosmos contains extended sequences of rebirths in the six realms (*devas*, hungry ghosts, etc.) taking place in an infinite number of parallel universes over unimaginable lengths of time. Except for the effects of karma and free choice on individuals' fates, existence is not going anywhere. Native peoples, on the other hand, have many different creation myths. Whether from a union of earth and sky or a primal anaconda or raven, the

Earth and human life were begun by creators. Hunter-gatherers' minds were elaborate enough to need the viable order their creation myths supplied. Creation myths became even more important for their theistic agricultural heirs. Natural history, as presented here, may begin with a cosmic bang but its biological manifestation is also directionless. Lightning in a chemical soup providing energy to the first cells provides a stark image of evolution's assumed accidental, or contingent origins. So, too, from the point of view of meditation, life just is. There is nothing like the meaninglessness of Buddhist origins or Darwinian chance among native peoples. Both Buddhist and Darwinian mandates to observe grew out of societies with surpluses and disembodied, self-absorbed minds.

For hunter-gatherers, life was too short and immediate to dwell on discontent as the Buddha did. Attention was on survival. They had enough to do handling the emotions survival evoked. Their survival was made easier because their minds had capabilities way beyond what was needed. As the mind and technologies developed, so did the arena of discontent. If we could rewind human evolution to a point before language, when the mind was just adequate to the tasks at hand, we might find much less ritual and shamanism as well as considerably less greed and aversion. A human population which tortured and gratuitously killed only the amount wolves and chimps do would be boy scouts by the standards of thinking hunter-gatherers—and saints in comparison to the civilized world. In order to ease their discontent, hunter-gatherers used mimetic communication, psychotropic plants and ritual. These addressed the immediate emotions evoked by circumstances of survival and the compulsive ones created by thinking minds.

Hunter-gatherers had neither the material freedom of the Buddha's time nor hours of soap opera with their fanciful fantastic worlds that stimulate the mind and wrench the emotions. Nisa, the !Kung woman, had tragedy in her life. She lost children and her husband. The tragedy was real and immediate, not disembodied and manipulated. It caused her suffering, but living life required attending again to the immediate and physical. Without the Buddha's security, the idea of his kind of quest did not arise. In a step way beyond the Buddha's world, we no longer have the omnipresence of death and disfiguration but an enlarged interior life fed by the media and computers, and separated from nature and palpable community. These shift the domain of our problems, which nonetheless remain the same sort—suffering in our leisure time and extended life span—that Buddha was the first to systematically address.

Even though the 1990s were inaugurated as the decade of the brain, some researchers feel that there has been very little scientific progress in studying how our minds work. In terms of making a more humane world, that is true. Still, both neurophysiology and meditation give interesting glimpses of the kind of biological entity the mind is. They indicate how layered our minds are and how little our actual minds fit into either our personal conceptions of what they are or past mechanical models. We perceive only a fraction of the data the world presents. What we see, hear, feel and taste are not only selective but also constructs of our brain. Moreover, what we sense is registered in neural apparatuses in complex ways. The process of dreaming gives us insight into how convoluted and partially integrated our sensations, perceptions, emotions and thoughts are. Meditation helps reveal how idiosyncratically we attribute cause and effect to emotionally laden stimuli. It is amazing how we both accomplish material ends and botch relational ones. Our minds possess great reasoning powers, ancient emotions and possibilities of transformation or reprogramming. Meditation tries to examine the ill-fitting aspects of mind which give rise to discontent although it may not be sufficient to overcome the brain's failure or extreme behavioral dysfunction.

In looking over the sweep of evolution of which our minds are one product, antagonistic pleiotropy is a useful way of characterizing the discomfort and destructiveness that accompany our ability to think. In the genetic protection against malaria, of which sickle-cell anemia is an unfortunate byproduct, the tribe which evolved with malaria is relatively well-adapted and evolution need go nowhere from there. But as we noted, there are other traits which, once begun, seem to lead down roads of no return. Choking and speech is one such compromise. I have suggested that the parts of the mind which underlay discontent are another. Since these parts are built into our ability to manipulate mental and material reality, as we get better at survival by using our minds, the accompanying tendencies toward increased fantasy and increased ability to destroy at a distance may lead to even more unhappiness and eventual annihilation. It is all very well when forecasting a little ahead allows you to protect your children's genes, but when abstract abilities, continuing to improve, then reveal the inevitability of individual death—implying the ultimate futility of individual existence—evolving intelligence can begin to antagonize itself, inventing fantasy worlds of internal movies and unending material satisfaction. Whether meditation is adequate to counter such pleiotropically derived discontent is an unanswered question.

In the chapter on attention and wisdom, we noted that aboriginal hunters, in contrast to anthropologists and biologists, seemed to be able to remain attentive to nature without their minds wandering in thought. Henry David Thoreau famously framed the difference.

> I wish to speak a word for Nature, for absolute freedom and wildness, as contrasted with a freedom and culture merely civil... to regard man [sic] as a part and parcel of Nature rather than as a member of society.... When we walk, we naturally go to the fields and woods: what would become of us, if we walked only in a garden or a mall?... Of course it is no use to direct our steps to the woods, if they do not carry us thither. I am alarmed when it happens that I have walked a mile into the woods bodily without getting there in spirit. In my afternoon walk I would fain forget all my morning occupations and my obligations to society. But it sometimes happens that I cannot easily shake off the village. The

> thought of some work will run in my head and I am not where my body is...I am out of my senses. In my walks I would fain return to my senses. What business have I in the woods, if I am thinking of something out of the woods?[6]

Besides his civil disobedience, Thoreau is most known for going to live in the woods around Walden Pond to try to find the meaning of life. While there, the woods taught him much about freedom in nature. Later in life, his observations of nature became more scientific. By the time of his death he was a respected natural historian working for a biology professor at Harvard. The shift troubled him. He feared that in using his mind cognitively to observe nature, he was severing some deeper connection.[7]

Modern nature lovers, attuned to the woods, have also noticed how easy it is not to be present while there. A man who teaches others how to track animals writes, "Have you ever found yourself walking along a path in the woods and then suddenly realized that the whole forest around you had changed? ...you don't know when the change took place. [But you] have awakened. You let the smell of a leaf fern wash through you. You realize why you were asleep. You were talking to yourself, caught up in a familiar, endless dialogue."[8] On my first meditation retreat, I walked down a country road ticking off the names of the plants I passed. The silence of the retreat made me desperate to get back to talking to myself. According anthropologists, it is unlikely that hunter-gatherers recited to themselves: "Urine smell, wonder if it is fox. Oh, the tracks go off to the left.... Will I catch something for dinner? My standing in the tribe will go down if I don't." Without the necessity of survival and mimetic immersion in the environment, only willpower holds back the mind's engagement with stimulation. The subtleties of nature are missed. A leader of ocean bird watching tours calls it instant desensitization: "One second it's the most beautiful bird they've ever seen, and the next second they are bored with it."[9] It has taken me years of meditating to be able to walk down the road of my first retreat with my mind on the back burner. Right now sitting at my computer three thousand miles and thirty years away from that retreat, if I stop the conversation in my head required to edit this manuscript, I can feel the difference. Time and space no longer seem quite so real.

Before humans evolved the ability to mentally talk to themselves, nature was not boring. If you look at your own boredom, you will see how much of it is thought. A sensitive field biologist describes what happens in nature. As an exercise in natural history he spent one morning listening:

> 3:55 A.M. ...a tree swallow is making chirping sounds without pause.
>
> 4:03 A.M. ...the first morning call of the ovenbird...so far no mosquitoes and no blackflies....my arms are absolutely crawling with...no-see-'[e]ms....
>
> 4:09 A.M. A mallard calls...a barred owl.... The hermit thrush....
>
> 4:14 A.M. The oven bird still sings...the hermit thrush still fifes.... I can now see the tree swallow... But five minutes have made a big difference... I also hear the first of the late risers: purple finch, rose breasted grosbeak, white-throated sparrow, and Maryland yellowthroat.
>
> 4:25 A.M. ...Now the volleys of bird song are erupting from all around.... My spirits lift along with this rising crescendo. I try to pick out individual species, but it's difficult now....
>
> 4:50–4:55 A.M. ...I listen and write down what I hear in the five minute intervals.... New birds: raven, red-breasted nuthatch, sapsucker....
>
> 5:05 A.M. Its light!...
>
> 5:07–5:12 A.M. The volume is way down....
>
> 5:30–5:35 A.M. The sun is up about five degrees. The no-see-'[e]ms are all gone. It is still too cool for the mosquitoes, and too early for the blackflies. The bird songs are muted....

6:15–6:20 A.M. The first chainsaws buzz and the first blackflies appear... I take a quick breakfast and a short nap.

8:00–8:05 A.M. The sun is burning hot. The mosquitoes have already left the field but the blackflies have taken over. They are here in clouds. New birds to hear: a winter wren, goldfinches... three more warblers...Others still singing....

9:40–9:45 A.M. The concert is over....

10:07–10:12 A.M. It is overcast. Birds heard: hermit thrush, Nashville and chestnut-sided warblers, red-eyed vireo, raven....[10]

For the narrative mind, not much of interest is going on, but when one listens, the world becomes rich. It is hard to still the inner chatter in order to do that.

One recent teacher gives a hint of what it is like to be fully present to nature. J. Krishnamurti disavowed formal meditation but taught in the spirit of the Buddha.[11] Krishnamurti loved nature, taking walks for hours every day. He claimed not to have any thoughts while walking. He was just present to what was around him. Although some critics have thought it arrogant, he often referred to himself in the second or third person.

> It was a marvelous morning and you could have walked on endlessly, never feeling the steep hills...there was no one to talk to, and no chattering of the mind. A magpie, white and black, flew by disappearing into the woods. The path led away from the noisy stream and the silence was absolute. It wasn't the silence after the noise;...nor the silence when the mind dies down.... It wasn't the silence that the mind makes for itself.... He only recently discovered that there was not a single thought during these long walks, in crowded streets or on the solitary paths. Ever since he had been a boy it had been like that, no thought entered his mind. He was watching and listening and nothing else.[12]

Pushing the mental chatter point further than the Buddha did, Krishnamurti claimed that thought is the source of all human discontent and violence—and that little has changed in the last few thousands of years of human civilization. We still are driven by greed, hatred and delusion; we still fight wars, allow inequity and permit injustices. Nature in its silence offers a contrast to the movement of mind and its consequences.[13]

A concern for the totality of life's process is what lay behind Darwin's investigations and what ultimately challenged him in his own life. Although Darwin would have liked life to fit the optimism of the British Empire with its science, economy and progress, his observations of nature showed the importance of three of the Buddha's four signs: disease, old age and death. Darwin was torn by the tension between the two perspectives, but he still felt he could solve the problem by more applications of science. Krishnamurti seems to imply that Darwin's methods will never achieve what the fourth of Buddha's signs, the meditator living harmlessly, aims toward. It may be that Krishnamurti goes too far. We need not condemn thought so thoroughly. For the Buddha, it is not thought *per se*, but attachment to it which is the source of difficulty. Much of the cutting edge of modern civilization, such as advertising with computerized, televised graphic montages, encourages attachment in hidden ways to benefit the makers of products ranging from elevators and medicine to food and the mass media. The system works not only because the products may be useful but because they appeal to parts of us that connect us to nature. We would have to go against our own make-up to resist them. Meditation, as we have seen, both makes space in our being for the troublesome effects of our connectors—such as craving for sweets or avoidance of pain—and gently reprograms them. In the determination to observe, in the repetition of seeing, in the silence of sitting, discontent can be given space and ultimately, Buddhists argue, be disconnected from both its causes and harmful consequences.

Although it is only one of many places where meditation can be done, meditation in the wilds has been used to reveal how much civilization masks our animal nature. The Buddha began his search by returning to the woods. Eschewing both civilization's protections and its distractions, his naked fears and desires assaulted him, and he was able to release himself from their grip. He suggested that those able to handle the difficulties of forest practice do the same. Nature would be their best teacher.[14] As in my Arctic encounter with the moose years ago, a taste of the silence of nature is available to us and may shake the hold that civilization has on us. From my journal:

> Sitting for hours, then days, then months by a pond on San Juan Island, quietly listening to the companions who keep me company, produces a profound silence. The sun moves overhead from behind my left shoulder to in front of my face. The qualities of heat and light change constantly. A kingfisher flies back and forth across the pond making a raucous sound. Later a muskrat or otter darts nearby through the water. A bald eagle battles black birds for possession of a tree stoop. The eagle complains like a hen-pecked husband and finally takes possession, squeaking even after its victory. Day after day, there is a spider in a web, waiting. Then it is gone and on successive days the wind slowly tears the web apart. Civilization has made the place safe to sit. The wolves and great cats are gone. Food is provided by benefactors. Life becomes simply listening. There is deep silence even though my mind chatters on in the background, claiming every sensation as its own. As the cars rumble by down the road, I can feel when they are being driven in anger. Human aspirations have little meaning, and there is no need to act on the emotions which arise.

I am not the only one watching. The eagle, the kingfisher, the otter and the spider are watching. Some of the time they are watching for food or the approach of a predator but much of the time they sit silently just watching the parade of life. They have a kind of alertness that is my teacher. I sit silently watching, watching disease, old age, death and the mind that resists them. Nature becomes a teacher. The mind with all its antagonistic pleiotropy is a product of nature. Silent observation of the activities of the mind may be a balm for its obsessions. The mind may also be reprogrammed so that intention can be observed before it bears fruit in harmful acts.

We are deeply indebted to Darwin for bringing critical observation to the environment of which we are part. In a wisdom he may not have fully appreciated, he created a method of observation which would keep its followers from falling back into comfortable truths. Likewise, the Buddha's method makes constant awareness a requirement of the understanding of life. When you think you know the truth, you have lost contact with existence as it newly presents itself. Darwin and Buddha offer each of us the opportunity to look unflinchingly at the world around us. We are to take it in with all its beauty and horror, being aware, all the while, of how our minds which evolution bestowed upon us are responding, and then to act as non-harmingly as we can. The Vietnamese Zen master Thich Nhat Hanh has challenged us to walk upon the Earth and not imprint it with our anger and sorrow. Understanding how we evolved, our relationship to nature and how we fit into modern civilization may aid in this endeavor. By silently observing how our bodies and minds are linked, we may be able to let go of the attachment which creates our discontented self.

Postscript
The Lama and the MRI

I began this book with a reference to a conference held at MIT in the Fall of 2003 entitled "Investigating the Mind" featuring the Dalai Lama. It continued discussions of the relationship between meditation and brain science from a meeting with the Dalai Lama held five years earlier that was reported in Daniel Goleman's book, *Destructive Emotions* and a 1993 meeting on sleeping, dreaming and dying.[1] Inviting groups of scientists to talk with him over several decades, the Dalai Lama has sought to learn about modern science as it might relate to Tibetan Buddhism and to see if Tibetan Buddhism has insights which might help science. There were discussions with quantum- and astro-physicists comparing the ideas of modern physics with Tibetan Buddhist cosmology and causality, and meetings with biologists exploring the implications of cloning and reproductive technologies. The fact that the Dalai Lama, head of a religious tradition whose views of the world were developed centuries ago, is so interested in science and its implications for religion is an extraordinary event in the history of religions. Many religions have resisted scientific conclusions when they clashed with accepted doctrines. Although rare, the Dalai Lama is not unique in dialoguing with representatives of science. The Indian teacher J. Krishnamurti had discussions with the physicist, Max Bohm, which continued over a period of many years.

Two aspects of the meetings between the Dalai Lama, Buddhist scholars, psychologists and neurophysiologists are important here. First: how does science contribute to a better understanding of the processes of meditation and the reverse, how do the experiences of meditators illuminate science? Second: what is the interchange between meditators and scientists like and how do the discussions fit into the larger question of the nature of meditation? These issues will be treated together. My understanding of these gatherings comes from the books about the two earlier meetings and audio tapes of the 2003 conference. The tapes did not include closed sessions and it was not always possible to identify the speaker.

In a series of experiments a psychologist, Richard Davidson, used fMRIs to track the brain activity of a highly trained Western Tibetan monk, several other Tibetan monks and a group of Western subjects some of whom were taught meditation as it is presented in the stress reduction training developed by Jon Kabat-Zinn. The results of these experiments were very suggestive. In earlier work Davidson showed that neural activity in the prefrontal cortex was greater on the left side in association with feelings of well-being, while negative feelings like depression correlated with activity on the right. The highly trained Western Tibetan monk's neural responses while practicing Tibetan compassion meditations were off the chart on the left. Not only was the neural activity greater but could be sustained for much longer periods of time than ordinary people tested. The other monks showed similar, if not so extreme, reactions. The laypersons who were taught meditation demonstrated a shift in activity towards the left.

The experiments which led to these results were motivated by the 1998 meetings and provided a foundation for the conference in 2003. Both the Dalai Lama and the Western Tibetan Buddhists conferees were excited about the experimental results because they lend credence to Buddhist assertions that meditation can lead to the end of suffering. Some of the scientists were also excited because experiments with Buddhist meditators offer the opportunity to correlate measurements of brain activity to meditators' observations of processes of mind and to meditation's influence on psychological states.

The conference in 2003 was divided into three substantive sessions: attention and cognitive control, emotion, and mental imagery. The first touched on psychologists' interest in the neurophysiology of conditions such as depression or attention deficit disorder and Buddhism's use of attention to heal afflictive mental states. The second was prompted by psychology's lack of investigation of positive emotions versus Tibetan Buddhism's emphasis on them. And the third was of interest because Tibetan Buddhist meditation uses imagery and psychology has begun to study mental imagery, with newer more powerful ways of monitoring the brain. Central to the discussions were Tibetan practices which cultivate compassion.

The focus on compassion reveals some of the Dalai Lama's and his Western supporters' reasons for holding the conference. In presentations to large audiences and books with titles like "The Art of Happiness," the Dalai Lama offers Westerners Tibetan Buddhist recipes for achieving happiness. Many times the Tibetan Buddhist conferees and their Western followers expressed the hope that a scientific basis of Tibetan practices will be found, and because of science's authority in the modern world, these practices will gain legitimacy. Tibetan Buddhism places a very high value on the cultivation of compassion, providing a number of techniques directed to this end. They include reminding one's self that in the sequence of one's infinite rebirths, each person is believed to have been both mother and child of every other being. Therefore one must care for and respect the rest of existence. Cultivating compassion often involves visualization, thus the interest in mental imagery.

With respect to compassion and visualization my book and the conference take different tacks. Coming out of Vipassana or Insight Meditation, my practice, while including compassion, is more concerned with attention and wisdom. And although I have had instruction from Lamas in Tibetan visualization, Dzogchen and compassion practices, visualization is not a practice I'm adept at. My meditation training is not atypical, which says something about how much one can generalize from these meetings about Buddhist outlooks. Over the last twenty-five hundred years, Buddhists have elaborated a great number of ways of meditating. Even in the *Sutras* of the Theravada Pali Cannon, the Buddha offers dozens of techniques. Despite the media attention it has received in the West over the past half-dozen years, Tibetan Buddhism is only a small part of the Buddhist universe. Even though the Chinese Cultural Revolution tried to root out all old religions, there are still many more Buddhists in China and Japan including Chan, Zen, Pure Land, Shin, etc. than Tibetan Buddhists. The same is true for the Theravada Buddhists of Sri Lanka, Burma, Thailand, Vietnam, Cambodia and Laos. Lamistic Buddhism that emerged in Tibet in the 11th century was a mixture of Indian, Chinese and indigenous beliefs.

So the challenge to psychologists and neurophysiologists is great. Not only does meditation include already-discussed Tibetan practices, but psychologists could measure what is going on in the brain while Buddhists chant the name of the Amida Buddha, cultivate states of serenity, practice choiceless awareness or walk two steps then take one bow. And in turn devotion, ritual, investigation, listening to the sound of silence, etc. as practiced by meditators may add understanding of how the brain works. The conferences occasionally mention Theravadan perspectives but make almost no mention of Zen or other Buddhist traditions.

Both this book and the 2003 conferees have addressed slightly different but limited aspects of what have been very extensive Buddhist traditions. Because of my experiences I have focused on the insight part of Theravada and I have used non-harming as a criterion for the efficacy of developing discriminating wisdom. In traditional Theravada non-harming or *ahimsa* is more often associated with asceticism and the cultivation of compassion; asceticism because monastics don't impact the world as much as householders. "A fruitbearing tree comes to harm, but not a barren one."[2] And compassion because cultivating loving kindness is thought to engender non-harming and act as protection for one's self and others.[3] Both I and the conferees share an ultimate interest in non-harming. I chose to explore its roots in insight meditation and they, in the cultivation of compassion as practiced in Tibetan Buddhism.

It was emphasized earlier that science's understanding of the brain is still rudimentary and that small results tend to be generalized way beyond what they can really sustain. The Dalai Lama and his supporters have created a dialogue with science with the stated aim of mutual learning. A number of the scientists have reformulated their research as a result of the dialogue and feel that they can learn much from meditators about the mind. But some of the scientists expressed concern that the flow was only one-way. They asked if Tibetan Buddhism was learning anything from the scientists that had changed their understating of meditation. They received no substantive response. Supporters of the Dalai Lama replied what they wanted were scientific demonstrations of the efficacy of Tibetan Buddhist practice as exemplified by highly trained meditators. One supporter referred to them as Olympic athletes of well-being. (It may be that the Dalai Lama sits at the summit of this Olympus. To his followers, to many scientists who have met him personally and to the world at large he seems a man of boundless empathy, compassion and good nature. In fact at the 1998 meetings the psychologist, Paul Ekman, an expert on facial expressions, was impressed that the Dalai Lama's face expressed positive emotions like joy and empathy more clearly than anyone he had studied. On the one occasion when the Dalai Lama was exposed to anger there was no report on his facial reaction.[4])

Although the Dalai Lama states that he would change Buddhism to accord with scientific results, the only example that the Tibetan Buddhists provided was the Dalai Lama's mention that he no longer regarded the mythical Mount Meru of Buddhist cosmology as the center of the universe. I think this asymmetry of perspective made some of the scientists uncomfortable although they were amenable to requests from the Buddhists for scientific help in sorting through Buddhist techniques to see which might be the most suitable for Westerners and interested in helping meditators idenify more obscure states of mind than are presently delineated in the Buddhist texts.

In a casual comment the Dalai Lama said something revealing. He wondered whether brain science might be able to differentiate similar meditations, such as compassion or one-pointedness, when they are done from the perspective of Buddhism with its underlying claim of no self and Brahminism with its belief in *Atman* or universal self. From a brain scientist's perspective this is an odd question and no one followed up on it. First of all, given the crudeness of brain variables measured, what might be relevant differences? Do Brahmans get more or less shift to the well-being side than Buddhists? Or can one or the other hold a brain state longer? Or behaviorally, could one measure the difference in compassion between accomplished Tibetan Buddhist meditators and Brahman *sadhus*, and would that be evidence for no-self or *Atman*?

The *Atman*/no-self religious debate has gone on since the beginning of Buddhism twenty-five hundred years ago. It is not clear what scientific evidence might bear upon it. Interest in having it explored stems from what appears to be the Dalai Lama's desire to create a scientific portrait of the unfolding of human consciousness and correlate it to what he regards as Tibetan Buddhism's accurate analysis of mind. As if to emphasize the power of Tibetan Buddhism, the Dalai Lama refers to several miraculous results of Tibetan practice. Monks who abide in a meditative state called "clear light," after they have died, have been reported, like saints of the Russian Orthodox Church, not to rot.[5] And monks with powerful meditation practices are purported to be able to read texts by touch. None of the scientists responded to these claims. Presumably their truth could be tested.

In the 1993 meetings the Dalai Lama gave a sketch of Tibetan ideas about the basis of consciousness. He used a number of technical terms the meanings of which are not obvious to those not versed in Tibetan theology (including myself).[6] He spoke of the foundation of consciousness which includes both latent propensities and pristine awareness. It is from these that the common appearances of our daily lives arise but the former may include our delusions while the latter is "vivid, luminous and liberating." He also mentioned clear light, subtle clear light and primordial clear light. The second is only apparent at death.

If the Dalai Lama had both the scientists and the resources at his disposal, I think he would love to put fMRIs on accomplished meditators and paint a portrait of their brain functions as consciousness unfolded. His hope might be to provide scientific validation to Tibetan Buddhist descriptions of the nature of human minds. Since Tibetan Buddhism has elaborated a detailed picture of the mental processes surrounding dying, fMRI's of dying adepts would indeed be interesting. Because there is almost no existing correlation between Tibetan descriptions and brain imaging nor brain images of dying people, it is not clear whether the former would or even could validate Tibetan Buddhist descriptions of the dying process. As I pointed out in the section on the limits of meditation there is good reason to believe that when the brain deteriorates even accomplished meditators' mental presence is eroded.

Also as we have seen in the discussion of the brain and meditation, some neurophysiological results contradict Buddhist ideas of what is happening. Let's look at where the scientists and the Tibetan Buddhists seem to differ. The proof of the pudding for Buddhism is the behavior of a practitioner or, for that matter, anyone. It may be that meditation does reduce discontent, lead to transcendent consciousness and a life of non-harming, and nonetheless the Buddhist analysis of how that comes about in the person is scientifically inaccurate. Descriptions of how suffering is reduced in Buddhist theology may give a satisfying analytic answer, such as in the *Abhidhamma*, but the categories of description may not be consistent with neurophysiological observations. Tibetan psychology, like its cosmology, is based on outdated assumptions about the physical and biological world. According to Stephen Batchelor, for thousands of years Buddhists regarded the brain as not much more than a mucus which came out of the nose when one had a cold.

One example of conflicting views has to do with how long it takes for an accomplished meditator to become aware of events in his or her mind and body. As we saw in my example relating to intention in Chapter 5, neurophysiology claims intention and derivative action sometimes precede awareness and thus it may not be possible for a meditator to intervene until the action is well underway. The psychologists assert that it takes 50 to 100 milliseconds for the frontal cortex to be informed and respond, and this is where they claim the mechanisms of attention are located. In the 1998 meetings the Dalai Lama asked the scientists to test whether the attention regions in accomplished meditators can monitor the interaction between fear centers and motor centers which usually take place in much less time than it takes for signals to reach the frontal cortex; i.e., can meditators bring into attention and control what seems to be unavailable to others. Experiments to answer such questions have not been done. One Western Tibetan Buddhist related traditional anecdotes claiming that attention to changing mind states can take place in a time he calculated to be as short as one millisecond. This would be consistent with the fundamental Buddhist idea that meditation is powerful enough to short-circuit any intention and so liberate the thinker from all harmful karma. If this were not the case, people would be saddled with some discontent they could not immediately interrupt. Thus the Third Noble Truth, that there is an end to suffering, might have to be modified. Several psychologists feel that, given how the brain is constructed, there simply is no way attention could operate as quickly as Tibetan Buddhist theory needs it to. So Buddhist ideas of what is going on may need to be modified.

This also has bearing on the discussion of the changeability of personality. Fundamental to Buddhism is the idea that anyone can achieve liberation, though it may not happen in one lifetime. If not in one lifetime for a given individual, then over eons with the continued application of the Eightfold Path, the end of discontent is achievable. Since ideas of rebirth are not part of the scientists' understanding of nature, no strict comparison can be made between their study of personality and the Buddhist idea of the end of suffering. Still the complex of emotion and changes in behavior indicative of the easing of discontent is of common interest to both Buddhists and scientists.

A number of psychological studies have observed the so-called hedonic treadmill. One aspect of this phenomenon is that in varying time periods after life events that increase people's happiness they return to the general level of happiness they had before the event.[7] An example is lottery winners who, despite the tremendous promise of change from large winnings, often end up very much as they were beforehand. Buddhists in general would agree. They regard pleasure as inherently transient but when the psychologists at the 2003 conference asked the Buddhists whether the happiness from deep practice lasts, they respond with a yes.

Here we have a problem. For many practitioners, as the Buddha exemplified, it is an article of faith that sustained long-term practice leads to the easing of suffering. But how can that be tested? As audience questioners from the 2003 conference pointed out, the monks who tested high on the well-being curve of left/right-prefrontal-excitation may have been that way to begin with and that the researchers may, in effect, be cherry picking their subjects. The monks tested are certainly very talented persons. That Mozart may have had highly developed musical brain regions may not be evidence for the efficacy of how he was taught music. That is, the Papa Mozart Method may not produce Young Mozarts or even better musicians than those taught in other ways. The response to this criticism was that in tests, non-practitioners who were taught the stress reduction version of meditation showed improvement in well-being as measured neurologically. This was taken to imply that meditation improves well-being and by extension more meditation makes for more well-being.

There are two problems with this, neither of which came up in the conferences. The first is that because lay persons who were taught one form of meditation had improved well-being, doesn't mean that it was meditation that conferred the benefit. Years ago when computers were first introduced for teaching young people, students who used the computers outstripped those who didn't; ergo computers improve learning. On closer examination, it turned out that computer users got much more attention. If the additional attention was given to the control group, they too had improved learning rates.[8] In another controlled study of breast cancer patients, where some received stress reduction training and some didn't but both received an equal amount of attention, no statistical difference could be found in their well-being. Both improved.[9] So it is not clear what exactly is increasing well-being in the studies of people taught meditation. It may be social contact or the attention of the teacher or something else. Finally, both the researchers who worked on the fMRI tests and the Western Tibetan Buddhists often consider the highly trained Western Tibetan Buddhist being able to emerge from three hours in a MRI calm, relaxed, smiling as a kind of icon of meditation's power, implying this was way beyond ordinary folk. Now I don't know how average people react to having MRIs but my experience in one was actually much less disconcerting than thinking about doing it for three hours. When I described the possible three hour duration to a friend she just smiled and said she would love it. She had an MRI once and it was like being in an isolation tank: no world, just floating. So again, with so few examples we can only hint at the role meditation may be playing. (And there are contradictory anecdotes. According to the meditators' grapevine, one highly regard Lama was reported to be very disappointed when his fMRI showed no unusual effects.)

The second problem has to do with overall outcomes in well-being among experienced meditators. Currently in the West there are thousands of meditators who have spent as many, if not more, hours meditating than many Tibetan Buddhist monks. The Tibetan Buddhist participants at the 2003 conference mentioned the magic number of ten-thousand hours of practice needed. This awed the scientists. But if someone sits one hour a day for twenty years that is seven-thousand-three-hundred hours of practice. Add a few retreats a year and you are easily over the magic number. For most monks in most Buddhist monasteries meditation is not their vocation. They perform rituals, study, work or do administration. One of the Western Tibetan Buddhist participants in the 2003 conference wrote a book describing the monastery of the Dalai Lama's sect where he studied and debated for a number of years in order to earn a Tibetan higher degree.[10] It was not an institution dedicated to the practice of meditation yet the non-practicing monks spent a great deal of time debating theological distinctions concerning meditation.

So we have in the West a large number of people whose time on the cushion qualifies them to be tested for the effects of meditation. Of course their practices may be very different. Nevertheless we can ask if they have achieved a greater quotient of well-being. My gut reaction is yes; otherwise I would not have taught meditate for so many years. Yet I have seen people come out of long retreats confused, withdrawn or unsure of themselves, and I know any number of experienced meditators who exhibit the same ambition or arrogance or anger that one sees among those untutored in Buddhist ways. It is not unusual for my meditating friends to gossip about one or another self-centered or celebrity- seeking Buddhist teacher. We always end the discussion with the caveat that without meditation he or she might have been much worse. We also spend time talking about teachers whose efforts at meditation have indeed changed them for the better. But then how does one judge the results of meditative effort versus the mellowing which sometimes comes with aging. After twenty years of practice, one is, after all, twenty years older. And some people do learn from life.

There is a genuine puzzle here. How can one measure the efficacy of practice? Earlier I discussed some of the limitations of meditation practice and how Buddhists rationalized these away by reasoning from karma. It is hard to imagine a statistical study which would show the efficacy of long-term practice. One of its variables would need to be moral or ethical, such as how non-harming a person has become. We are left with a question which applies to all the arts aiming to heal discontent: how can their effectiveness be measured? An answer might include the self-evaluations of meditators or those who have received pastoral counseling or therapy, etc. Other criteria might be how much they volunteer their time, how often they become angry, how much they are loved, etc. but none of these is very easily defined. We can rate their left/right neurological well-being valence but its application to life would have to correlate with things like generosity which are not easy to measure.[11] And how are we to regard people who are publicly generous and personally nasty?

Turning to an *ad hominem* way of getting a handle on these issues, I could not say I am happier after years of meditation. Things that plagued me in the past sometimes do not seem nearly as important. I know that practice is a powerful tool at my disposal when my discontent seems to sprout wings. With time and space I can rein it in. Yet even in the depth of the silence of practice, I can still feel exactly the way I felt when I was at my worst moments in life. It is as if nothing had changed. This observation fits with psychological studies that claim our personalities don't really change much. They seem to be set by genetics and early life experiences. If the aura of habitual discontent arises in the silence, then I can be sure its shadow also inhabits my interactions with others especially when my circumstances preclude being attentive to my inner workings. As the more confrontative of my intimates point out, I often unconsciously repeat behaviors that do not take account of their humanity, and they feel discounted or hurt. When they point this out, I acknowledge that they are right and promise them and myself that I will be more alert. Yet inside I can feel my reactive self- justification and realize that the best I can do is to try. When circumstances permit, my power of attention gained through meditation and the accompanying commitment not to harm let me ease situations that would otherwise be harmful.

Here we need to discriminate well-being from happiness. In my life, meditation has contributed to my being able to do what needs to be done irrespective of whether I will be happy with the process or the outcome. Responding to family, the elderly, charity, political needs and requests to teach meditation is done with little hesitation. Many times I have put aside self-interest to help with a broken down car or agreed to teach. This satisfies one Buddhist idea of living compassionately without regard for self, that is, selflessly. But happiness seems to come about when things go my way; not selfishly my way, but when life produces those things that touch me. These two are different. The former fits more with Buddha's utterances in the *Sutras*, some of which warn against the pitfalls of bliss. The latter fits more with psychological notions of happiness.

The Tibetan Buddhist conferees seem to want it both ways. The brain function shift towards the left in the Tibetan meditators using compassion practice might be similar to Norman Vincent Peale's "Power of Positive Thinking" from the popular Christian psychology of the 1950s. Think positively and you will be happier (for which there is evidence). On the other hand, the Tibetan Buddhists mention the calm abiding which allows for clarity and compassionate action. This latter might not fit with conventional ideas of what constitutes happiness. Not all Saints can be said to be happy. There is no study I know which shows whether people with twenty years of practice are happier or would be seen that way by average non-practitioners. I would hope, at least, that they act more compassionately.

Where is there a neutral place to stand to evaluate the efficacy of meditation? I think it comes down to common sense, and the Buddhist marks of liberation, while pointing to qualities that can be attended to, are not all that useful in regards to real people who populate our world. Just as sex, money, power and notoriety have brought down both Eastern teachers who come to the West and Westerners in their own back yard, the judgments of spiritual achievement are very subjective. And though there are extraordinary people who dot Buddhist history, the internecine struggles of Lamistic Tibet prior to the Chinese invasion killed many a Dalai Lama.[12]

In the modern world there is an ongoing war between the Sri Lankan, Sinhalese Buddhists and Tamal Hindus. The Pol Pot murders in Cambodia were Buddhist. Proposals to disenfranchise non-Buddhists in newly independent Burma after WWII led to an Army coup and vicious military dictatorship. It was an army of Buddhist soldiers and generals. These all show that despite the potential healing powers of Buddhist practice, historically Buddhism has only marginally influenced the world around it to be less harming. And further, the Buddhist cultures which Westerners look to as embodying compassion and simplicity are giving way to the very seductions of Western materialism the rejection of which contributed to driving many Westerners to adopt Buddhist practice. Ladakh, Nepal, and Southeast Asia have all been infected by the West. China sells Buddhism to tourists and having sufficiently pacified and deracinated Tibet they now include Tibetan Buddhist shrines.[13] After my first month of meditation in the mid 1970s I flew to Vancouver. On board the plane, hands folded on my lap, I settled into the peacefulness of sitting which was still fresh from the retreat. After an hour or so I opened my eyes. The Southeast Asian gentleman sitting next to me asked me if I had been meditating. When I responded positively, he said that it was interesting that Westerners had taken up meditation, because people in his country were too busy being engaged in the modern world, that they had neither the time nor the inclination to meditate.

While the Tibetan Buddhist presenters at the conferences certainly acknowledge the difficulties of practice to both the public and the scientists, they still seem to be offering a meditation panacea which is much simpler than both the collective experience of meditators and history support. That aside, let's look at some of the interesting results the conferences brought to light.

1) Highly trained meditators, who were tested, exhibited greater excitation on the sides of the prefrontal cortex associated with well-being than others who were tested. Meditation-like stress-reduction training also shifted neurological responses in the direction of well-being.
2) A highly accomplished meditator exhibited clean shifts in brain states when instructed to shift his attention. In most people the shifts were sloppier and they could not remain in instructed states.
3) Neural patterns arising from the senses or thought exhibit a kind of extension and collapse parallel to meditators' observations of the passage of mind moments.
4) A researcher in human expressions claims there are six universal microexpressions including anger, fear, contempt, sadness and disgust. These fleeting expressions are exhibited involuntarily when the corresponding emotion is present. People who can perceive these fleeting expressions in others are more open and curious. Two accomplished meditators

were better than average persons at recognizing micro-expressions.

5) Unlike most people when startled, the face of an accomplish meditator didn't exhibit a startle reflex if he was meditating in a state of open awareness, but his heart rate showed typical changes. When practicing one-pointed meditation his face reflected a bit of startle but heart rate went the reverse of the normal pattern; i.e. went down rather than up.

These results strongly suggest that meditation not only can monitor brain changes but also affects the way people's brains operate. A central concern of meditation is to transmute the typical pattern of reaction where strong emotions blindly drive people to harm themselves or others to one where the actor no longer is so driven. In the 1998 meetings, the researcher of human expressions stated that emotional responses have a refractory period after the emotional response. This is like the aftershock of an earthquake. People are still held in the thrall of the emotion, unable to take in new information or control their responses. Then there is a period of recovery when control can again be exercised. People have different refractory periods. Meditators could be studied to see whether meditation training can be used to reduce or even eliminate this refractory period or even the emotion itself, a capability which might be indicated by the lack of startle reflex in accomplished meditators.

It seems clear that meditation can create a space between emotions like fear and anger and actions that people take because of them. The real difficulty comes in explaining how this process works. In his summary of the 2003 conference, Professor Jerome Kagan, a noted child development researcher, said that research shows there are "bodily events that occur in milliseconds in parts of brain which are permanently hidden from the light of introspection." He gave two examples. There are slight variations of the sound "bah" which the brain records but which we cannot consciously discriminate, and our brain responds to a familiar face on waking in a way we are unaware of. Moreover he feels that words are not adequate to describe these kinds of events because of the different evolutionary origins of language and of our hidden brain processes. Words describe large rough events arising out of slow mental processes. They are the "worst possible vehicle to communicate private sensations.... Evolution separated these two sources of information. [i.e. the parts of the brain used for making descriptions and the processes which do not report their activities]."[14]

There is some tension between the Buddhist explanations and the science. This arises, in part, because the categories each use to describe the subject matter are very different. If, as the Tibetan Buddhist participants point out, Buddhism doesn't have a word equivalent to emotion and the psychologists debate the meaning among themselves, then we have a problem. Psychologists mainly treat emotions as brief events, like snake-fear. While not contradicting that, the Tibetan Buddhists see happiness as something that can last. They use the term on different occasions to include happiness in the conventional sense, an emotion, or calm abiding or well-being.

It is obvious that there is much yet to be studied. With sufficient fMRI time, practitioners could correlate their observations with brain section outputs. Standing between emotions as transitory phenomena and enlightenment as a permanent state is mood—a Western, not Buddhist, term. Nonetheless after practicing sufficient meditation, it becomes clear that one experiences moods. As mentioned, my depression after napping is obviously a chemically induced mood. But there are moods which last much longer, sometimes years, and may not have clear causes. When you meditate for days or weeks at a time, you watch the moods come and go. They sometimes have clear antecedents but sometimes not. They color one's world. The line between emotion and mood is not clear. It may be that the psychologists' "emotion" and Tibetan meditators' permanent good mood, or well-being are not contradictory. Scientific studies are much too preliminary to know.

Similarly, the tension in the 2003 conference around what is accessible to meditative observation results from a real difference. Meditators will have to demonstrate they can observe events which psychologists say in principle they cannot. But that psychologists say words are inadequate to describe some of the things that go on in the mind is familiar turf to the meditators. The Tibetan meditators talked about being aware of being aware. They also said that this process is not describable. An essential ingredient of Zen is the ineffability of realization. Zen schools of both sudden and gradual enlightenment are emphatic about the nondiscursive description of enlightenment. The former is epitomized in the famous koan: "What was your face like before you were born?" The latter encourages meditators to "just sit" and talks about the "suchness" or "thusness" of experience. Where both Tibetan Buddhism and Theravada have complicated intellectual systems which try to describe or analyze what is happening, Zen refers to its transmission as not dependent upon words and letters but directly pointing at the soul of man. So while there are abundant thinkers in the Zen world, its practice formally abjures discursive thinking as part of the route towards awakening.

In Tibetan retreats I have attended, the teaching often begins with what might be taken as a theological discussion of "right view." Understanding this idea is a prerequisite to practice. In Theravada the twelve factors of Dependent Origination are often presented as beginning with ignorance. Both of these perspectives lend themselves to intellectual analysis and historically produced huge bodies of theological literature. Because Zen reacted to Buddhist psychology and intellectual elaborations, it is interesting that Zen practitioners were not prominent at the conferences.[15] Certainly Zen's short circuiting of analytic descriptions of the nature of consciousness fits well with both the scientists' claims that there are parts of consciousness of which we either cannot be aware or do not have the vocabulary to describe, and the frustration that some psychologists have expressed at ever being able to understand consciousness.

I think if Buddhism really wants to come into the modern world and embrace the parts of science relevant to meditation, it is going to have to reformulate its own understanding of practice. Attempts to show that Buddhist cosmology or causality are consistent with modern physics are, I think, a stimulating but basically misdirected endeavor. In many areas science has left our commonsense understanding of the world behind. The mathematics characterizing physical phenomena has long since departed from pictures of triangles or the bouncing of bowling balls that illustrate Newton's second law. Because two apparently contradictory measurements in cosmology require for their reconciliation something physicists poetically name dark matter, it is dark or matter only in that if there were matter out there having the influence that this conjectured stuff has, it is like no other matter we can see. How dark matter will eventually be understood in physics will come from ideas far removed from our every day experience. Science's efficacy is not so much in whether it makes common sense but whether it works. And the works of science cannot be denied. The survival of three billion humans depends upon scientific ideas which made possible machines and high yield grains without which many of us would die. Traditional Buddhist cosmology doesn't much help with satellites, genetic engineering or life that may be found elsewhere in the universe. If older views of the natural and material world have had to give way to scientific evidence, then why not in the realm of meditation and the mind? To my personal disappointment the latest chess computer seems able to beat world champions. It flattered me to think that people are smarter, if not stronger, than machines. And though I never liked computers that much, I find I need them in order to write, pay taxes and communicate with my friends. It doesn't look as though I'll be able to do without one until I reach my dotage.

Similarly brain science is closing in on the domain which has been the purview of meditation for thousands of years. Brain science is very crude and has a lot to learn from the well honed introspection of meditators. Nonetheless if the religions which contain meditation really want to include science, they may have to let go of their religious authority. Buddhism has a way of claiming it is describing "things as they are." The Buddha often says this in the *Sutras.* The implied reasoning is that if one looked, one would discover, "yes that is how things are, that is their true nature." These ideas become verities. The Tibetan Buddhist conferees continue this tradition. It is a teaching technique which over the years I have found myself repeating in both meditation and other classroom contexts. It is, I feel, a counterproductive way of asserting authority. The Buddhist conferees say the true nature of mind is luminescent, and when one sees things as they are that will be evident. Other Buddhist traditions see the mind as empty or as silent or as undefined original mind. Each is a metaphor whose referents may be different.

Asserting true nature based on one collection of practices is not conducive to correlating what can be known about the mind by measuring the brain. Ten thousand hours of meditation practice with a teacher closely monitoring the student can produce wonderful results. But electrical stimulation, biofeedback and psychotropic drugs may be able to achieve similar things more efficiently. Science might show that what is going on between guru and disciple is very different than what doctrine says is happening. Moreover science may be able to achieve this with the help of meditators' observations. This is not to put science on a pedestal; it may have more flaws than the institutions that organize around meditation. I began my scholarly career as a scientist, then historian of science. I understand that science can be influenced by politics, money and ambition. Because they are so smart and acclaimed, some scientists arrogantly put down ideas of which they have little understanding, including those of religion. Also claims in the name of science can hurt people. We put men on the moon for military reasons while leaving millions of people in poverty, and because of that choice we can watch hundreds of channels of satellite television. Even though humans make harmful choices, it is science's ability to manipulate the physical world teased out of other social factors which makes the study of mind, with the help of meditators, an open ball game.

Acknowledgments

This book has been a labor of love begun over ten years ago. Were it not for the support and encouragement of my friend and colleague at Brandeis University Maurice Stein, I would have wandered in the wilderness and the book would not have seen the light of day. Lynn Margulis, with her encyclopedic knowledge, rode herd on my use of the many sciences in which I am only an amateur. Dorion Sagan held me to a high level of analytic analysis and a similar literary standard. One can ask for no better critics. Others have read various versions of my manuscript and given me cogent comments. They include Brook Stone who defanged my gratuitous barbs and challenged my assumptions, Jeanne Smithfield Strauss for her insightful comments, and Michael Wright, Gananath Obeyesekere, Attila Klein, Gary Buck and an anonymous reviewer at Oxford University Press. My debt to all of them is great. They are not responsible for any misunderstandings I may have of science or Buddhism. Those I willingly shoulder. I wish also to thank author friend Stephen Altschuler for his advice and moral support in navigating the labyrinth of the publishing world, Larry Rosenberg for propelling me into a meditation hall thirty years ago and supporting my early practice and teaching and Connie Long for her editorial assistance.

I am indebted to Laurence M. Cook of The Manchester Museum, University of Manchester for the use of his pictures of peppered moths, to Shambhala Press for the Ryokan quote, to Ferrar, Straus and Giroux for permission to reprint Wendell Berry's poem "The Peace of Wild Things," and to Blackberry Press for permission to reprint Nanao Sakaki's poem. My rendition of Millet's painting of the gleaners comes from various versions hanging in European museums.

Glossary

Adaptationist: a biologist who argues that evolutionary changes are adaptations to the environment.

Adaptive radiation: changes in the descendants of a species which come to occupy different niches. Some of the descendants of the few original finches that were blown from the mainland to the Galapagos Islands became seed eaters while others developed woodpecker like traits. When applied to human technology, an adaptive radiation would be something like humans wearing animal furs to survive in colder climates or prenatal medical care increasing the number of live births and survival to adulthood.

Adrenaline: a neurochemical involved in flight or fight responses.

Amygdala: part of the brain that distinguishes safe from dangerous. It is involved in instantaneous responses, fear and emotional mediation.

Antagonistic pleiotropy:pronounced "play-o-tropy." From the Greek pleio meaning many and tropo meaning change (as in change of direction). Pleiotropy is the situation where a gene expresses multiple traits. In antagonistic pleiotropy helpful traits are counter balanced by counterproductive, "antagonistic" ones. Testosterone contributes to early male productive fitness but increases the chance of prostrate cancer later in life.

Australopithecus: earliest species in the hominid line.

Automatic actions: human behaviors which are either below our awareness or have become so routine that we don't notice them.

Basal ganglia or brain stem: the lowest part of the brain which controls organs and movement.

Bhikku: the Pali word for a Buddhist monk.

Bon Po: a pre-Buddhist religion of Tibet.

BP: years before the present time.

Buddhist psychology: detailed discussions of concentration practices and causality found in the Abhidhamma Pitaka, one of three parts of the Buddhist sacred texts.

Chloroplast: the part of a protoctist or a plant cell which photosynthesizes; thought to have originated as free-living bacteria.

Cingulate cortex: older or more primitive part of the cortex.

Cognitive: pertaining to mental function.

Cycads: early plants with palm-like leaves and cones.

Deliberate actions: actions initiated by intentions.

Dendrite: branching connections between neurons in the brain.

Dependent Origination: the causal chain of construction of self, linking the 12 elements: body and mind, senses, contact, feeling, craving, clinging, becoming, birth, death, ignorance, karma and consciousness.

DES: drug developed in the 1940s to control menses and ovulation. It caused genetic defects in women's offspring.

Dimorphism: two forms. Usually when the sexes have different appearances.

Dopamine: a neurochemical associated with motor control and emotions.

Eightfold Path: a fundamental Buddhist concept including right understanding, thought, speech, action, livelihood, effort, mindfulness and concentration.

Endorphins: neurochemicals which block feelings of pain.

Epizootics: epidemic among animals.

Ethologist: a biologist who studies the social life of animals.

Eukaryotes: organisms whose cells are nucleated.

Evolutionary black holes: seemingly irreversible directions in the evolution of a species.

Evolutionary lag: the condition in which older traits seem to linger even though they no longer exhibit any useful function.

fMRI: functional magnetic resonance, an imaging technology.

Fitness: technically, the numerical success of a species.

Four Noble Truths: The Buddha's fundamental discovery: the universality of suffering, that suffering can end, that there is way out of suffering and that way is the eight-fold path.

Germ cells: cells which can create gametes, i.e., the sexual cells which carry heredity.

Hungry ghosts: inhabitants of Buddhist realms with tiny mouths and food too large for them or large mouths and only tiny bits of food.

Hypothalamus: part of the brain controlling endocrine and autonomic nervous systems.

"I": Big I. In Soen sa Nim's Zen it is the state of being of a person who transcends the ego.

"i": little i, the ego.

Inbreeding depression: the offspring of close relatives weakened by the accumulation of harmful traits.

Jhanas: states of concentration or absorption.

Karma: a fundamental Buddhist concept that all effects are caused whether one knows the causes or not. In popular thought, one's karma entails the personality and physical results of earlier intentions.

Lateralization: the phenomenon of the different sides of the brain having different functions.

Life expectancy: the average age to which members of a population live.

Life span: the biological age limit of a population.

Limbic system: so-called primitive emotional system of the brain involving the amygdala, hypothalamus and hippocampus.

Melanic: darkened or blackened form.

Midbrain structure: sorts hearing and seeing signals and responses.

Mimesis: as in "mime:" the use of gesture and expression to convey nonverbal messages.

Mitochondria: parts of the cells of eukaryotes which generate energy.

Mortality tables: indicate what portion of a population dies in each age cohort.

Mutagenic: a mutation-causing agent.

Natural selection: the combined actions of weather, environment and other species which affect which members of a given species survive.

Neocortex: the modern thinking parts of the brain.

Neurons: brain cells to which dendrites connect.

Nirvana: in Sanskrit, extinction or "blowing out" of desire and karma-inducing actions. The state of enlightenment or awakening.

Non-cognitive: things that we respond to but do not characterize with thoughts.

Norepinephrine: adrenaline; neurochemical involved in flight or fight responses.

Ontology: the philosophic subject concerning the nature of being.

Paleobotanist: a botanist who studies the remains of ancient plants.

Pali: the language of the Sutras. A religious language related to Sanskrit.

PET: positron emission tomography; an imaging technique which traces areas of activity in the brain.

Pleistocene: geologic period of time from 1.6 million years BP until 100,000 years BP.

Prokaryotes: bacteria; single cell beings without nuclei.

Protists: single celled protoctists.

Protoctist: single or multiple cell species which have encased nuclei and chromosomes. Plants, animals, and fungi evolved from protoctists.

REM: rapid eye movement or the movement of the eyes during sleep that indicate dreaming.

Sacred texts: the three Buddhist scriptures: the Sutras, Vinaya and Abhidhamma .

Satori: a striking experience of awareness in Zen meditation.

Sense gates: in Buddhism, the eyes, ears, nose, touch, taste and mind.

Senescence: the process of decline of the body as it functions less efficiently with age.

Six realms: in Buddhist cosmology: humans, animals, hell, hungry ghosts, jealous gods and gods.

Somatic cells: cells which make up skin, organs, bone, etc.

Somatosensory cortex: the part of the cortex which processes sensory inputs.

Soteriology: the study of salvation in contrast to theology, the study of god.

Stem cells: cells which can generate a part of the body.

Sutras: the sayings of the Buddha as part of the Tripitaka translated as the three baskets made up of the Sutras, Abhidhamma and Vinaya or monastic code.

Synapses: the junctions between two nerve cells.

Thalamus: governs flow of information from the nervous system to brain.

Transcendental Meditators: meditators who use a mantra as an important part of their practice.

Twelve factors of Dependent Origination: the circular causal chain by which sensations get molded into identities: ignorance, volitional actions, rebirth, six senses, contact, craving, clinging, karmic formations, birth, disease, old age and death.

Vestigial: a trait left over from an earlier use like the human appendix.

Zoonose: disease shared by humans and animals.

Bibliography

1. Allan, M., 1977, *Darwin and his Flowers,* London: Faber and Faber.
2. Anderson, D., 1994, *Barnacles,* London: Chapman & Hall.
3. Anderson, J., 1980, *Cognitive Psychology and Its Implications,* San Francisco: Freeman.
4. Anderson, S., et al., 1999, "Impairment of social and moral behavior related to early damage in human prefrontal cortex," *Nature Neuroscience,* 2: 1032–1037.
4. Arking, R., 1991, *Biology of Aging,* Englewood Cliff: Prentice Hall.
6. Aurora, D., 1986, *Mushrooms Demystified*, Berkeley: Ten Speed Press.
7. Austin, J., 1998, *Zen and the Brain*, Cambridge: The MIT Press.
8. Balsekar, R., 1982, *Pointers from Nisargadata Maharaj*, Bombay: Chetana.
9. Balwf, S., 2003, *The Secret Voyage of Sir Francis Drake,* New York: Walker.
10. Barkow, J., et al., 1992, *The Adapted Mind*, New York: Oxford University.
11. Bates, M., 1949, *The Natural History of Mosquitoes*, New York: Harper & Rowe.

12. Joko Beck, C., 1989, *Everyday Zen*, San Francisco: Harper.
13. Berry, R. J., 1990, "Industrial Melanism and Peppered Moths," *Biological Journal of the Linnaean Society*, 39: 301-322.
14. Berry, W., 1987, *The Collected Poems of Wendell Berry*, New York: Northpoint.
15. Bishop, J., 1975, "Moths, Melanism and Clean Air," *Scientific American*, January: p. 91.
16. Boaz, N., 2001, "The Scavenging of 'Peking Man,'" *Natural History*, March.
17. Bolby, J., 1990, *Darwin*, New York: Norton.
18. Bowen, E. 1954, *Return to Laughter,* New York: Harper.
19. Brasington, L., 1997, "Sharpening Manjushri's Sword," talk given at the American Academy of Religion/Western Regional Meeting March 25.
20. Briggs, J., 1970, *Never in Anger,* Cambridge: Harvard University Press.
21. Brightman, R., 1993, *Grateful Prey,* Berkeley: University of California Press.
22. Bringhurst, R., 1999, *A Story as Sharp as a Knife,* Vancouver: Douglas & MacIntyre.
23. Brothwell, D., 1988, "On Zoonoses and their Relevance to Paleopathology," in Ortner, D.J and Aufderheide, A.C. (eds.). *Human Paleopathology*, Washington: Smithsonian, pp 18–22.
24. Brown, T., 1978, *The Tracker*, New York: Berkley.
25. Browne, J., 1995, *Charles Darwin,* vol. 1, New York: Knopf.
26. Browne, J., 2002, *Charles Darwin,* vol. 2, New York: Knopf.
27. Buchman, S. & Nabhan, G., 1996, *The Forgotten Pollinators* Washington, D.C.: Island Press.
28. Buddha, G., 1987, *Thus I Have Heard, the Long Discourses of the Buddha, Digha Nikaya* (trans. by M. Walshe), London: Wisdom Publications.
29. Buddha, G., 1995, *The Middle Length Discourses of the Buddha* (translated by Bhikkhu Nanamoli and Bhikkhu Bodhi), Boston: Wisdom Publications.
30. Budiansky, S., 1992, *The Covenant of the Wild*, New York: Morrow & Co.

31. Bull, J., 1985, "On Irreversible Evolution," *Evolution* 39 (5): 1155–58.

32. Cabeza de Vaca, 1993, *Castaways*, Berkeley: University of California Press.

33. Caras, R., 1975, *Dangerous to Man*, New York: Holt, Rinehart and Winston.

34. Carroll, S., 2005, *Endless Forms Most Beautiful*, New York: Norton.

35. Carson, R., 1955, *The Edge of the Sea,* Boston: Houghton Mifflin.

36. Cartmill, Matt, 1993, *A View to a Death in the Morning*, Cambridge: Harvard University Press.

37. Casey, S., 2005, *The Devil's Teeth*, New York: Henry Holt.

38. Castaneda, C., 1968, *The Teachings of Don Juan*, Berkeley: University of California Press.

39. Cha, A., 1985, *A Still Forest Pool*, Wheaton : Theosophical Pub. House.

40. Chodron, P., 1991, *The Wisdom of No Escape*, Boston: Shambhala.

41. Churchfield, S., 1990, *The Life History of Shrews*, Ithaca: Cornell University Press.

42. Clark, J., 1969, *Man is Prey*, New York: Stein and Day.

43. Clark, W., 1996, *Sex and the Origins of Death*, New York: Oxford University Press.

44. Cohen, M. & Armelagos, G., 1984, *Paleopathology at the Origins of Agriculture*, Orlando: Academic Press.

45. Cohen, M., 1989, *Connecting with Nature*, Eugene: World Peace University.

46. Collins, R., 1990, *Early Medieval Europe, 300–1000*, New York: St. Martins.

47. Colp, R., 1977, *To Be an Invalid*, Chicago: University of Chicago Press.

48. Comfort, A., 1979, *The Biology of Senescence*, New York: Elsevier.

49. Corbett, J., 1957, *Man-eaters of India*, New York: Oxford University Press.

50. Coy, P., "The Big Daddy of Data Haulers," *Business Week*, Jan 29, 1996.
51. Crawley, M. (ed.), 1992, *Natural Enemies*, Oxford: Blackwell.
52. Christakis, D., et al., 2004, "Early Television Exposure and Subsequent Attentional Problems in Children," *Pediatrics* 113(4): 708–713.
53. Cronon, W., 1983, *Changes in the Land*, New York: Hill and Wang.
54. Crosby, A., 1986, *Ecological Imperialism*, Cambridge: Cambridge University Press.
55. Curio, E., 1976, *The Ethology of Predation*, Berlin: Springer.
56. Damasio, A., 1994, *Descartes' Error*, New York: Putnum.
57. Darwin, C., 1965, *The Expression of the Emotions in Man and Animals*, Chicago: University of Chicago Press.
58. Darwin, C., 1959, *The Origin of Species, A Variorum Text*, Philadelphia: University of Pennsylvania Press.
59. Darwin, C., 1962, *The Voyage of the Beagle*, New York: Doubleday.
60. Darwin, C. n.d., *The Descent of Man*.
61. Darwin, C. n.d., *The Origin of Species*, New York: Literary Classics.
62. Davis, W., 1985, *The Serpent and the Rainbow*, New York: Simon and Schuster.
63. Davis, W., 1996, *One River*, New York: Simon & Schuster.
64. Dawkins, M., 1993, *Through Our Eyes Only?*, Oxford: Freeman.
65. de Waal, F., 1996, *Good Natured*, Cambridge: Harvard University Press.
66. Deacon, T., 1997, *The Symbolic Species*, New York: Norton.
67. DeGraza, D., 1996, *Taking Animals Seriously*, Cambridge: Cambridge University Press.
68. Diener, E., Lucas, R.E., & Scallon, C. N., 2006, "Beyond the Hedonic Treadmill: Revising the Adaptation Theory of Well-being," *American Psychologist*, 61(4): 305–314.
69. Dennett, D., 1995, *Darwin's Dangerous Idea*, New York: Simon & Schuster.

70. Descola, P., 1994, *In the Society of Nature*, Cambridge: University Press.

71. Desmond, A., 1994, *Huxley*, Redding: Addison-Wesley.

72. Desmond, A. and Moore, J., 1991, *Darwin*, New York: Warner.

73. Desowitz, R., 1991, *The Malaria Capers*, New York: W.W. Norton.

74. Diamond J., 1993, "New Guineans and Their Natural World," in Kellert. S. and Wilson, E.O., *The Biophilia Hypothesis.*

75. Dillard, A., 1974, *Pilgrim at Tinker Creek*, New York: Harper and Rowe.

76. Dolan, R., 1999, "On the Neurology of Morals," *Nature Neuroscience,* 2: 927–929.

77. Domico, T., 1988, *Bears of the World,* New York: Facts on File.

78. Domhoff, G. W., 2005, "Refocusing the Neurocognitive Approach to Dreams: A Critique of the Hobson versus Solms Debate," *Dreaming,* 15: 3–20.

79. Donald, M., 1991, *Origins of the Modern Mind*, Cambridge: Harvard University Press.

80. Dreyfus, G., 2003, *The Sound of Two Hands Clapping,* Berkeley, University of California Press.

81. Dubos, R., 1965, *Man Adapting,* New Haven: Yale University Press.

82. Dunbar, R., 1996, *Grooming, Gossip and the Evolution of Language,* Cambridge: Harvard.

83. Eaton, S. et. al., 1988, *The Paleolithic Prescription*, New York: Harper & Row.

84. Edelman, G., 1989, *The Remembered Present*, New York: Basic Books.

85. Edelman, G., 1992, *Bright Air, Brilliant Fire,* New York: Basic Books.

86. Eibl-Eibesfeldt, I., 1989, *Human Ethology*, New York: Aldine de Gruyter.

87. Ekman, P., 2003, *Emotions Revealed*, NY: Times Books.

88. Eldredge, N. (ed.), 1987, *The Natural History Reader in Evolution*, New York: Columbia University Press.

89. Eldredge, N., 1999, *The Pattern of Evolution*, New York: W.H. Freeman.

90. Elgar, M. & Crespi, B., 1992, *Cannibalism*, Oxford: Oxford University Press.

91. Ellis, D., 1991, "The Living Resources of the Haida," Vancouver: Ellis.

92. Elton, P., 1927, *Animal Ecology*, London: Sedgewick and Jackson.

93. Emery, N., 2001, "Effects of Experience and Social Context on Prospective Cashing Strategies by Scrub Jays," *Nature*, 404 (6862): 443–446.

94. Endler, J.. 1986, *Natural Selection in the Wild*, Princeton: Princeton University Press.

95. Errington, P., 1967, *Of Predation and Life*, Ames: Iowa State University.

96. Fagan, B., 1997, "A Case for Cannibalism," *Archaeology* 47(1): 11–16.

97. Fagan, R., 1981, *Animal Play Behavior*, New York: Oxford University Press.

98. Fernandez-Jalvo Y.; Carlos Diez J.; Caceres I.; Rosell J., 1999, "Human Cannibalism in the Early Pleistocene of Europe," *Journal of Human Evolution*, 37(3): 591–622.

99. Festinger, L., 1956, *When Prophecy Fails*, Minneapolis: University of Minnesota Press.

100. Finch, C., 1990, *Longevity, Senescence and the Genome*, Chicago: University of Chicago Press.

101. Fisher, C., 1966, "The Death of a Mathematical Theory," *Archive for the History of Exact Science*, 3 (2):137–159.

102. Fisher, C. (unpublished), *Meditation in the Wilds: Recluses, Hermits and Forest Monks.*

103. Foelix, R., 1982, *Biology of Spiders*, Cambridge: Harvard University Press.

104. Fox, M., 1984, *The Whistling Hunters*, Albany: SUNY.

105. Franklin, M. and Zyphur, M., 2005, "The Role of Dreams in the Evolution of the Human Mind," *Evolutionary Psychology* 3: 59–78.

106. Fries, J. and Capo, L., 1981, *Vitality and Aging*, San Francisco: Freeman.

107. Gavrilov, L., 1991, *The Biology of Life Span,* Chur: Harwood.

108. Geist, V., 1971, *Mountain Sheep*, Chicago: University of Chicago Press.

109. Gernell, J., 1987, *The Natural History of Squirrels*, New York: Facts on File.

110. Gleick, J., 1999, *Faster*, New York: Pantheon Books.

111. Goldstein, J. & Kornfield, J., 1987, *Seeking the Heart of Wisdom,* Boston: Shambhala.

112. Goleman, D., 2003, *Destructive Emotions,* New York: Random House.

113. Gombrich, E., 1988, *Theravada Buddhism,* London: Routledge Kegan.

114. Good, K., 1991, *Into the Heart,* New York: Simon & Schuster.

115. Goodall, J., 1971, *In the Shadow of Man*, Boston: Houghton Mifflin.

116. Goodall, J., 1986, *The Chimpanzees of Gombe,* Cambridge: Harvard University Press.

117. Gorman M. and Stone, R., 1990, *The Natural History of Moles,* Ithaca: Cornell University Press.

118. Gorman, J., 1990, *The Total Penguin*, New York: Prentice Hall.

119. Gould, J. and C., 1994, *The Animal Mind,* New York: Scientific American.

120. Gould, S., 1989, *Wonderful Life*, New York: Norton.

121. Gould, S., 1995, "The Evolution of Life on Earth," *Scientific American*, October pp. 85–91.

122. Gould, S., 2002, *The Structure of Evolutionary Theory,* Cambridge: Harvard University Press.

123. Grant, V. ,1991, *The Evolutionary Process*, New York: Columbia University Press.

124. Greaves, M., 2000, *Cancer: The Evolutionary Legacy*, Oxford: Oxford University Press.

125. Green, C., 1987, *The Giraffe*, New York: Crestwood House.

126. Greene, H., 1997, *Snakes*, Berkeley: University of California Press.

127. Grice, G., 1998, *The Red Hourglass*, New York: Delacorte.

128. Griffin, D., 1992, *Animal Minds*, Chicago: University of Chicago Press.

129. Grove, R., 1995, *Green Imperialism*, Cambridge: Cambridge University Press.

130. Gyatrul Rinpoche, 1993, *Ancient Wisdom*, New York: Snow Lion Publications.

131. Hakluyt, R., 1965, *Principall Navigations*, Cambridge: Cambridge University Press.

132. Hamilton, C., 1952, *Buddhism*, Indianapolis, Bobbs-Merrill.

133. Hanh, T. N., 1976, *Meditation the Miracle of Mindfulness*, Boston: Beacon

134. Hanh, T. N., 1991, *Peace is Every Step*, New York: Bantam.

135. Hart, W., 1987, *The Art of Living*, San Francisco: Harper & Row.

136. Harvey, P., 1987, "Murderous Mandibles and Black Holes in Hymenopetran Wasps," *Nature*, (326):128–9.

137. Haylfick, L., 1994, *How and Why We Age*, New York: Ballantine Books.

138. Heinrich, B., 1984, *In a Patch of Fireweed*, Cambridge: Harvard.

139. Heinrich, B., 1994, *A Year in the Maine Woods*, Redding: Addison-Wesley.

140. Henry, J., 1986, *Red Fox*, Washington D.C.: Smithsonian.

141. Hillel, D., 1991, *Out of the Earth*, New York: Free Press.

142. Hobson, J., 1994, *The Chemistry of Conscious States*, Boston: Little, Brown.

143. Hoogland, J., 1995, *The Black-Tailed Prairie Dog*, Chicago: University of Chicago Press.

144. Hooper, J., 2002, *Of Moths and Men*, New York: Norton.

145. Hopkins, D., 1983, *Princes and Peasants*, Chicago: University of Chicago Press.

146. Hover, R., 1979, *How to Direct the Life Force to Dispel, Mild Aches & Pains*, England: Hover.

147. Huntingford, F., 1987, *Animal Conflict*, New York: Chapman and Hall.

148. Isenberg, D., 2000, *The Destruction of the Buffalo,* New Haven: Yale University Press.

149. Iversen, J., 1956, "Forest Clearance in the Stone Age," *Scientific American*, March.

150. Jans, N., 2005, *Grizzly Maze,* New York: Dutton.

151. Jayatilleke, K., 1974, *The Message of the Buddha*, New York: The Free Press.

152. Joko Beck, C., 1989, *Everyday Zen*, San Francisco: Harper Rowe.

153. Kabat-Zinn, J., 1995, *Wherever You Go, There You Are,* New York: Hyperion.

154. Karamnski, T., 1983, *Fur Trade and Exploration: Opening the Far Northwest, 1821–1852* Norman: University of Oklahoma Press.

155. Keeley, H., 1996, *War Before Civilization*, Oxford: Oxford University.

156. Kellert, S. & Wilson, E. (eds.), 1993, *The Biophilia Hypothesis,* Washington, D.C.: Island Press.

157. Khantipalo, B., 1981, *Calm and Insight*, London: Curzon Press.

158. Khare, M. (n.d.), *Painted Rock Shelter,* Bhopal: Archeology and Museum of Madhya Pradesh.

159. King, C., 1987, *Weasels and Stoats*, Ithaca: Comstock.

160. King-Hele, D., 1999, *Erasmus Darwin,* London: DLM.

161. Klösch, G. and Kraft, U., 2005, "Sweet Dreams Are Made of This," *Scientific American*, June.

162. Knauft, B., 1987, "Reconsidering Violence in Simple Human Societies," *Current Anthropology,* 28(4).

163. Koenig, W. et. al., 1995, "Acorn Woodpecker," *The Birds of North America,* 154.

164. Kornfield, J., 1993, *A Path With Heart*, New York: Bantam Books.

165. Krakauer, J., 1997, *Into the Wild*, New York: Anchor

166. Krech, S., 1999, *The Ecological Indian*, New York: Norton.

167. Krishnamurti, J., 1982, *Krishnamurti's Journal*, New York: Harper and Row.

167. Krishnamurti, J., 1991, *On Nature and the Environment*, New York: Harper Collins.

168. Kuhn, T., 1962, *The Structure of Scientific Revolution*, Chicago: University of Chicago Press.

169. Lappe, M., 1982, *Germs That Won't Die*, Garden City: Doubleday.

170. Larson, E., 2001, *Evolution's Workshop*, New York: Basic Books.

171. LeDoux, J., 1996, *The Emotional Brain*, New York: Simon &Schuster.

172. LeDoux, J., 2002, *Synaptic Self*, New York: Viking.

173. Leopold, A., 1966, *A Sand County Almanac*, New York: Ballantine.

174. Levin, T., 1992, *Blood Brook*, Post Mills: Chelsea Green.

175. Lieberman, P., 1984, *The Biology and Evolution of Language*, Cambridge: Harvard University Press.

176. Lieberman, P., 1991, *Uniquely Human*, Cambridge: Harvard University Press.

177. Lochley, R., 1964, *The Private Life of the Rabbit*, New York: Macmillan.

178. Long, W., 1923, *Mother Nature*, New York: Harper.

179. Lopez, B., 1978, *Of Wolves and Men*, New York: Scribners.

180. Lyubomirsky, S., Sheldon, K.M., & Schkade, D., 2005, "Pursuing Happiness: The Architecture of Sustainable Change," *Review of General Psychology*, 9(2):111–131.

181. Lutts, R., 1990, *The Nature Fakers*, Golden: Fulcrum.

182. Lucas, P., 2004, *Dental Functional Morphology*, New York: Cambridge University Press.

183. Lutz, D. and J.M., 1991, *Komodo, the Living Dragon*, Salem: DIMI.

184. Macahy, D., "The Rise of Angiosperms," in Eldredge, N. (ed.) 1987, *The Natural History Reader in Evolution*, New York: Columbia University Press.

185. MacKay, D., 1992, "Constraints on Theories of Inner Speech," in Reisberg, D. (ed.), *Auditory Imagery*, Hillsdale: Lawrence Erlbaum..

186. Macy, J., 1983, *Despair and Empowerment in the Nuclear Age*, Philadelphia: New Society.

187. Macy, J., 1991, *World as Lover, World as Self,* Berkeley: Parallax.

188. Majerus, W., 1998, *Melanism,* Oxford: Oxford University Press.

189. Margulis, L. and Sagan, D., 2003, *Acquiring Genomes,* New York: 2002.

190. Margulis, L., 1996, "Archael-eubacterial Mergers in the Origin of Eurkarya," *Proceedings of the National Academy of Science,* (93):1071–1076.

191. Margulis, L. and Sagan, D., 1995, *What is Life?*, Berkeley: University of California Press.

192. Margulis, L. and Sagan, D., "God, Gaia, and Biophilia," in Kellert, S. & Wilson, E. (eds.), 1993, *The Biophilia Hypothesis,* p. 352.

193. Margulis, L., 1992, *Environmental Evolution*, Cambridge: MIT.

194. Martin, C., 1978, *Keepers of the Game*, Berkeley: University of California Press.

195. Martin, C., 1992, *In the Spirit of the Earth*, Baltimore: Johns Hopkins.

196. Mathpal, Y., 1984, *Prehistoric Rock Paintings of Bhimbetka,* New Delhi: Shatki Malik.

197. Maturana, H. and Varela, F., 1992, *The Tree of Knowledge,* Boston: Shambala.

198. McClintock, D., 1981, *A Natural History of Raccoons*, New York: Scribners.

199. McGovern, W., 1924, *To Tibet in Disguise*, Century: New York.

200. McGowan, C., 1997, *The Raptor and the Lamb*, New York: Henry Holt.

201. McNeill, W., 1976, *Plagues and People*, New York: Anchor.

202. Meine, C., 1988, *Aldo Leopold*, Madison: University of Wisconsin Press.

203. Merrell, D. J., 1994, *The Adaptive Seascape*, Minneapolis: University of Minnesota Press.

204. Milanich, J., 1998, *Florida Indians and the Invasion from Europe,* Gainesville: University of Florida Press.

205. Mintz, S., 1985, *Sweetness and Power*, New York: Penguin.

206. Mitchell, J., 1984, *Ceremonial Time*, Boston: Houghton and Mifflin.

207. Mock, D. and Parker, G., 1997, *The Evolution of Sibling Rivalry,* New York: Oxford University Press.

208. Molleson, T., 1994, "The Eloquent Bones of Abu Hureyra," *Scientific American*, August pp. 70–75.

209. Morris, S., 1998, *The Crucible of Creation*, New York: Oxford University Press.

210. Morris, S., 2003, "The Cambrian 'Explosion,' of Metazoans and Molecular Biology: Would Darwin be Satisfied?" *International Journal of Developmental Bioliology*, (47): 505–515.

211. Mumford, S., *1989, Himalayan Dialogue*, Madison: University of Wisconsin Press.

212. Nabahn, G. & St. Antoine, S., "The loss of the floral and faunal story" in Kellert, S. & Wilson, E. (eds.), 1993, *The Biophilia Hypothesis.*

213. Nelson, R., 1973, *Hunters of the Northern Forest,* Chicago: University of Chicago Press.

214. Nelson, R., 1989, *The Island Within*, San Francisco: North Point.

215. Newberg, A., 2001, *Why God Won't Go Away*, New York: Ballantine.

216. Newton, I., 1986, *The Sparrowhawk*, Calton: Poyser.

217. Nichols, P., 2003, *Evolution's Captain*, New York: HarperCollins.

218. Nooden, L. D., 1988, *Senescence and Aging in Plants*, San Diego: Academic Press.

219. Norbu, N., 1992, *Dream Yoga*, Ithaca: Snow Lion Publication.

220. Norman, K., 1969–71, *The Elders' Verses*, London: Pali Text Society.

221. Nuland, S., 1994, *How We Die*, New York: Knopf.

222. Nyanatiloka Mahathera, 1971, *Guide Through the Abhidhamma-Pitaka*, Kandy: Buddhist Publication Society.

223. Obeyesekere, G., 1997 *Imagining Karma*, Berkeley: University of California Press.

224. Olshansky, C., 1993, "The Aging of the Human Species," *Scientific American*, April, p. 51.

225. Olshansky, S. et. al., 2001, "If Humans Were Built to Last," *Scientific American,* March, p. 51 ff.

226. Ornstein, R., 1991, *The Evolution of Consciousness*, New York: Prentiss Hall.

227. Ortner, D. (ed.), 1991, *Human Paleopathology*, Washington D.C.: Smithsonian Institution Press.

228. Pali Text Society, 1989, *Poems of the Early Buddhist Nuns* (trans. Rhys-Davids, C. and Norman, K.), Oxford: Pali Text Society.

229. Palmer, T., 1992, *Landscape with Reptile*, New York: Ticknor & Fields.

230. Parker, A., 2003, *In the Blink of and Eye*, Cambridge: Perseus.

231. Pettitt, P., 2000, "Odd Man Out: Neanderthals and Modern Humans," *British Archeology*, 51(February).

232. Pfeffer, P. (ed.), 1989, *Predators and Predation*, New York: Oxford.

233. Powell, R., 1982, *The Fisher*, Minneapolis: University of Minnesota Press.

234. Preston, S. and White, K., 1996, "How Many Americans Are Alive Because of Twentieth-Century Improvements in Mortality?" *Population and Development Review*, Vol. 22. September.

235. Promislow, D., 1990, "Living Fast and Dying Young," *Journal of Zoology,* 220, p. 417–37.

236. Quammen, D., 2003, *Monsters of God*, New York: Norton.

237. Raby, P., 2001, *Alfred Russel Wallace*, Princeton: Princeton University Press.

238. Rajneesh, B. S., 1980, *The Orange Book,* Poona: Rajneesh Foundation.

239. Randhawa, M., 1980, *A History of Indian Agriculture*, Vol. 1., New Delhi: Indian Council of Agriculture.

240. Rappaport, R., 1971, "The Flow of Energy in an Agrarian Society," *Scientific American,* September.

241. Rasa, A., 1986, *Mongoose Watch*, Garden City: Doubleday.

242. Reedman, M., 1990, *The Pinnipeds*, Berkeley: University of California Press.

243. Reenes, B. and Nass, C., 1996, *The Media Equation*, Cambridge: Cambridge University Press.

244. Reisberg, D. (ed.), *Auditory Imagery*, Hillsdale: Lawrence Erlbaum.

245. Rezendes, P., 1992, *Tracking and the Art of Seeing*, Charlotte: Camden House.

246. Rindos, D., 1984, *The Origins of Agriculture*, Orlando: Academic Press.

247. Roe, F., 1955, *The Indian and the Horse*, Norman: University of Press.

248. Roff, Derek A., 1992, *The Evolution of Life Histories*, New York: Chapman & Hall.

249. Rose, M., 1991, *Evolutionary Biology of Aging*, New York: Oxford University Press.

250. Rose, U., 1989, *The North American Porcupine*, Washington, D.C.: Smithsonian.

251. Ryokan, 1977, *One Robe, One Bowl* (trans. J. Stevens), New York: Weatherhill.

252. Sagan, D. and Margulis, L., "Gaia and the Ethical Abyss," in Kellert, S. (ed.), 2001, *Nature and Humanity*, New Haven: Yale University.

253. Sakaki, N., 1987, *Break the Mirror,* San Francisco: North Point.

254. Salzberg, S., 2002, *Loving Kindness*, Boston Shambhala.

255. Sapolsky, R. M. and Finch, C., 1991, "On Growing Old," *Sciences–New York*, 31(Mar/Apr): 30–38.

256. Schaller, G., 1963, *Mountain Gorilla*, Chicago: University of Chicago Press.

257. Schaller, G., 1972, *The Serengeti Lion*, Chicago: University of Chicago Press.

258. Schaller, G., 1973, *Golden Shadows, Flying Hooves*, New York: Knopf.

259. Schneider, E. and Sagan, D., 2005, *Into the Cool*, Chicago: University of Chicago Press.

260. Sears, P., 1947, *Deserts on the March*, Norman: University of Oklahoma Press.

261. Sekuler, R. & Blake, R.,1990, *Perception*, New York: McGraw-Hill.

262. Shepard, P., 1973, *The Tender Carnivore*, New York: Scribners.

263. Shockey, I., 1996, "Perception of Nature," Ph.D. dissertation Brandeis University.

264. Siegel, J., 2000, "Sleep Phylogeny: Clues to the Evolution and Function of Sleep," *Neurosci. Lett.,* Mar. 31, p. 163ff.

265. Siegel, J., 2005, "The Evolution of Sleep: Birds at the Crossroads Between Mammals and Reptiles," *Neurology,* Apr 1–15, 40(7):423–30.

267. Seung Sahn, 1976, *Dropping Ashes on the Buddha,* New York: Grove Press

268. Shostak, M., 1983, *Nisa*, Harmondsworth: Penguin Books.

269. Sloss, R., 1991, *Lives in the Shadow*, Reading: Addison-Wesley.

270. Smith, B., 1989, "Origins of Agriculture in Eastern North America," *Science*, (246) Dec. 22:1566–1571.

271. Smith, D., 1993, *Human Longevity*, New York: Oxford University Press.

272. Snyder, G., 1990, *The Practice of the Wild*, San Francisco: North Point Press.

273. Sober, E., 1984, *The Nature of Selection*, Cambridge: MIT Press.

274. Speilman, A. and D'Antonio, M., 2001, *Mosquito*, New York: Hyperion.

275. Stanford, C., 1999, *Hunting Apes*, Princeton: Princeton University Press

276. Stearns, S., 1992, *The Evolution of Life Histories*, Oxford: Oxford University Press.

277. Stevens, R. n.d., *Dancing in the Heart of Wonder,* Jamaica Plains: unpublished.

278. Stewart, W., 1983, *Paleobotany and the Evolution of Plants*, Cambridge: Cambridge University Press.

279. Stirling, I., 1990, *Polar Bears*, Ann Arbor: University of Michigan Press.

280. Sullivan, A., 1976, *The Complete Plays of Gilbert and Sullivan,* New York: Norton.

281. Sundstrom, G., 1995, "Aging is Riskier Than it Looks," *Age and Aging*, 1(24): 373–374.

282. Surya Das, 1998, *Awakening the Buddha Within,* New York: Broadway Books.

283,Suzuki, S., 1970, *Zen Mind, Beginner's Mind*, New York: Weatherhill.

284. Tambiah, S., 1984, *The Buddhist Saints of the Forest*, Cambridge: Cambridge University Press.

285. Tattersall, I., 1998, *Becoming Human*, New York: Harcourt Brace & Co.

286. Thanisarro Bhikku, 2001, *Inquiring Mind*, Berkeley 17(2 p): 4.

287. Thoreau, H., 1993, *Faith in a Seed*, Washington, D.C.: Island Press.

288. Tierney, P., 2000, *Darkness in El Dorado*, Norton: New York.

289. Trungpa, C., 1978, *Glimpses of Abhidarma*, Boulder: Prajna Press.

290. U Kyaw Min, 1987, *Buddhist Abhidhamma*, Union City: Heian.

291. Van de Waal, F., 1990, *Food Hoarding in Animals*, Chicago: University of Chicago Press.

292. Varela, F., 1997, *Sleeping, Dreaming, and Dying*, Boston: Wisdom.

293. Vermeij, G., 1996, *Privileged Hands*, New York: W. H. Freeman.

294. Visslier, K., 1983, "The Honey Bee Way of Death," *Animal Behavior*, 31(Nov): 1070–1076.

295. Wallace, D., 1997, *The Monkey Bridge*, San Francisco: Sierra Club Books.

296. Weidensaul, S., 1999, *Living on the Wind*, New York: North Point.

297. Weil, A., 1972, *The Natural Mind*, Boston, Houghton Mifflin.

298,Weiner, J., 1994, *The Beak of the Finch*, New York: Knopf.

299. Wiedensaul, S., 1999, *Living On the Wind*, New York: North Point.

300. Williams, G. C., 1992, *Natural Selection*, New York: Oxford University Press.

301. Wilson, E., 1971, *Insect Societies*, Cambridge: Harvard University Press.

302. Wilson, E., 1990, *The Diversity of Life*, New York: Norton.

303. Wise, D., 1993, *Spiders in Ecological Webs*, New York: Cambridge University Press.

304. Wiltshire, M., 1991, *Ascetic Figures Before and in Early Buddhism*, Berlin: Mouton De Gruyter.

305. Woodwell, G., 1970, "The Energy Cycle of the Biosphere," *Scientific American*, September.

306. Yeats, W. B., 1940, *The Collected Poems of W. B. Yeats*, New York: Macmillan.

Endnotes

Chapter 1

1. Bhikkus are Buddhist monks. C.f. Glossary. The Buddha's quote is from *The Middle Length Discourses of the Buddha* p. 234 MN 140.38. Or Thanissaro Bhikku's translation from *Inquiring Mind* vol. 17 No. 2 Spring 2001 p. 4. "I teach only one thing & one thing only... suffering and the end of suffering," and Darwin, C., n.d., *The Origin of Species,* p. 40.
2. Hamilton, C., 1952, *Buddhism,* pp. 6–11.
3. From "Sailing to Byzantium."

Chapter 2

1. Raby, P., 2001, *Alfred Russel Wallace*, Chaps. 11 and 13. Wallace felt that the possession of "spirit" was the reason humans were more capable than animals. He saw English society as a decline from the more cooperative ways of indigenous societies. Nevertheless he felt that an enlightened imperialism would benefit native peoples.
2. Although from very different social strata, it was a training ground for Wallace too. Ibid. p. 22.
3. Grove, R., 1995, *Green Imperialism.*
4. Browne, J., 1995, *Charles Darwin.*
5. See *The Voyage of the Beagle* and Desmond, A., and Moore, J., 1991, *Darwin.*
6. Angier, N., "The World of the Deep Sea Floor," *New York Times,* p. C1, Oct. 17, 1995.
7. Grant, V., 1963, *The Origin of Adaptation.*
8. From his poem, "In Memoriam," 1850.
9. Yoon, C., "The Wizard of Eyes," *New York Times,* Science Section B2 Nov. 1, 1992, and Gould, S., 2002, *The Structure of Evolutionary Theory,* p. 1132.
10. Margulis, L., and Sagan, D., 2002, *Acquiring Genomes.*
11. Although Darwin borrowed from the authority of physics by implying that natural selection could explain things as 19th century physics did, it will become clear here that nature and natural selection are really open-ended processes rather than mechanistic ones. On Huxley, c.f. Desmond, A., 1994, *Huxley.*
12. Ibid, pp. 236 and 271.
13. Darwin, C., n.d., *The Origin of Species,* pp. 40–1.
14. Ibid, p. 47.
15. Darwin, C., n.d., *The Origin of Species,* p. 47.
16. Kuhn, T., 1962, *The Structure of Scientific Revolution.*
17. Eldredge, N., 1999, *The Pattern of Evolution*, and Morris, S., 1998, *The Crucible of Creation.*
18. Desmond, A., 1994, *Huxley,* p. 268.
19. There are even problems with this simplistic matching of pollinators. See Buchman, S., & Nabhan, G., 1996, *The Forgotten Pollinators,* especially p. 76.

20. Darwin, C., n.d., *The Origin of Species,* p. 88.
21. Merrell, D. J., 1994, *The Adaptive Seascape.*
22. Wallace, D., 1997, *The Monkey Bridge,* p. 83. Although contemporary biologists are more careful about calling things advanced, they often slip into this habit of mind. pp. 84–88.
23. Bishop, J., 1975, "Moths, Melanism and Clean Air," *Scientific American.* Pictures of the peppered moth dramatically demonstrate its protective coloration. I have similar pictures of a "peppered moth" (species unknown) from northern British Columbia. It was virtually invisible when placed on a western cottonwood tree trunk, but quite distinct elsewhere. An exhaustive discussion of natural selection and peppered moth coloration can be found in Majerus, W., 1998, *Melanism.*
24. Berry, R. J., 1990, "Industrial Melanism and Peppered Moths,"*Biological Journal of the Linnaean Society,* 39: 301–322.
25. Ibid, p. 301.
26. Ibid.
27. Ibid, p. 319.
28. See Hooper, J., 2002, *Of Moths and Men* for the unsavory history. She cites one biologist as doubting the role of natural selection and Majerus (op. cit.) as thinking the case can still be made.
29. Macahy, D., "The Rise of Angiosperms," in Eldredge, N. (ed.), 1987, *The Natural History Reader in Evolution,* p. 26. Stewart, W., 1983, *Paleobotany and the Evolution of Plants.* "...[I]t is a safe generalization...that angiosperms have evolved from gymnosperm ancestors...which one or ones remains a moot question....the ancestral forms may (be)...," more angiospermous' (or) "more gymnospermous'....Conclusions about relationships based on too few characteristics are always suspect," p. 367–8.
30. Merrell, D. J., 1994, *The Adaptive Seascape,* p. 40.
31. Some flies changed without mutation and the change happened very rapidly. Either these flies had unexpressed genetic characters which allowed them to survive in DDT environments or they would have died out. Pure lab strains never develop resistance.
32. Merrell, D. J., 1994, *The Adaptive Seascape,* p. 92.
33. Lappe, M., 1982, *Germs that Won't Die;* Debos, R., 1968, *Man Adapting;* and Desowitz, R., 1991, *Malaria Capers.*
34. Endler, J., 1986, *Natural Selection in the Wild,* and Sober, E., 1984, *The Nature of Selection.*
35. Weiner, J., 1994, *The Beak of the Finch.*
36. Larson E., 2001, *Evolution's Workshop,* p. 212.
37. Ibid, p. 214.
38. Merrell, D., 1994, *The Adaptive Seascape,* p. 158.
39. Ibid.
40. Castaneda, C., 1968, *The Teachings of Don Juan.*
41. Darwin, C., 1979, *The Origin of Species,* p. 73.
42. Browne, J., 1995, *Charles Darwin,* p. 478ff.
43. Anderson, D., 1994, *Barnacles,* and Browne, J., 1995, *Charles Darwin,* p. 478.
44. Allen, M., 1977, *Darwin and his Flowers.*
45. Errington, P., 1967, *Of Predation and Life.*
46. Ibid, p. 228.
47. Modern genetics brings old issues into new question. With a little bit of radium one can self-administer mutation or get into the cells as molecular biologists do and change offspring. With ultrasound, amniocentesis and hormones the direction of human development can be altered in back rooms for a price. They clone farm animals. Why not humans? A resurrection of one of Darwin's critics, Samuel Butler, has been made by Lynn Margulis and Dorion Sagan, 1995, *What is Life?* They construe intention in a very broad way. Desmond, A., and Moore, J., 1991, *Darwin,* p. 640, put the criticism and responses in the context of Darwin's life.
48. Browne, J., 1995, *Charles Darwin.*
49. Wilson, 1992, *Diversity of Life.*
50. WBUR in Boston Feb 24, 1995 commenting on his book, *Darwin's Dangerous Idea,* and Schneider, E., and Sagan, D., 2005, *Into the Cool.*
51. Gould, S., 1989, *Wonderful Life,* New York: Norton and Morris, S., 1998, *The Crucible of Creation,* New York: Oxford University Press.
52. "Showdown on the Burgess Shale," in *Natural History,* Dec. 1998–Jan. 1999.
53. Although Gould has passed away and the amount of variety he saw is now agreed to be much less, the discussion has continued in a somewhat altered form with molecular biologists seeing evolutionary variety constrained by a few genetic processes, while Morris has twisted around now criticizing this stance. He claims genetic constraints would yield greater variation than actually exists. For him uniformity in nature is due to a greater power. Carroll, S., 2005, *Endless*

Forms Most Beautiful, and Morris, S., 2005, "The Cambrian 'Explosion,' of Metazoans and Molecular Biology: Would Darwin be Satisfied?"
54. This is analogous to the saddle point of equilibrium in the mathematical model of classical economics.
55. Margulis, L., 1992, *Environmental Evolution,* p.149.
56. Stephen J. Gould makes this point, arguing that complexity is by no means a necessity but a rarity and a result of chance. It is also associated with shorter species survival. Gould, S., October, 1995, "The Evolution of Life on Earth."
57. Williams, G. C., 1992, *Natural Selection,* p. 7.
58. Ibid.
59. Ibid, p. 40 and p. 44.
60. Irreversible evolution was coined to describe mutation to parthenogenesis which then dominates, while an example of a black hole is the mutation to hyperparsitism in wasps; i.e., once the larvae mutate to kill other larvae in a host, they will dominate. There is no going back. Bull, J., 1985, "On Irreversible Evolution," and Harvey, P., 1987, "Murderous Mandibles and Black Holes in Hymenopertan Wasps."
61. Lucas, P., 2004, *Dental Functional Morphology,* pp. 246–8.
62. Wilson, E., 1992, *Diversity of Life,* p. 79–80.
63. Yoon, C., "Antarctica's Frigid Waters Form Evolutionary Cauldron," *New York Times,* Science section p. D2 March 9, 1999, and the genus *Toxorhynchites,* Speilman, A., and D'Antonio, M., 2001, *Mosquito* p. 20.
64. Desmond, A., 1994, *Huxley: From Devil's Disciple; to Evolution's High Priest,* p. 228. Lynn Margulis has speculated that bacteria consuming each other might be the origin of sex.
65. Browne, J., 1995, *Charles Darwin,* p. 542.
66. Dillard, A., 1974, *Pilgrim at Tinker Creek,* p. 65.
67. Larson E., 2001, *Evolution's Workshop,* New York: Basic Books, p. 223.
68. I have explored related themes of meditation in nature in a companion volume, *Meditation in the Wilds.*
69. Desmond, A., and Moore, J., 1991, *Darwin,* p. 397.
70. This is Paul Sears version of the Tennyson expression in the opening lines to Sears, P., 1947, *Deserts on the March.*
71. Desmond, A., and Moore, J., 1991, *Darwin,* pp. 447–8.
72. Ibid.
73. Colp, R., 1977, *To Be an Invalid,* and Bolby, J., 1990, *Darwin.*
74 Desmond, A., and Moore, J., 1991, *Darwin,* p. 662.
75 Ibid, quoted p. 247–8.
76 Ibid.
77 Darwin, C., n.d., *The Descent of Man,* Chap. 21.

Chapter 3

1. Fisher, C., n.d., *Meditation in the Wilds.*
2. The Four Noble Truths assert the universality of suffering and offer a means to its surcease.
3. Lopez, B., 1978, *Of Wolves and Men,* p. 58.
4. Geist, V., 1971, *Mountain Sheep,* p. 298. For lions, "Mortality and its causes are difficult to measure." "Actual observations of polar bear deaths...are rare." Although frozen squirrels are not often found: "*one* was actually seen dying and falling off a branch." "The probability of seeing an actual whistling dog hunt was extremely low." Schaller, G., 1972, *The Serengeti Lion,* p. 183; Stirling, I., 1990, *Polar Bears,* p. 137; Gernell, J., 1987, *The Natural History of Squirrels,* p. 135. On the Farallones Islands off the California coast near San Francisco two researchers have refined the grisly art of watching great white sharks tear apart seals and sea lions. Casey, S., 2005, *The Devil's Teeth.*
5. "(It) is rarely observed by humans in the wild, despite the intimations of nature films shot on the savannas of Africa." Wilson, E., "Study Guide" to the college edition of 1992 *Biodiversity* p. A64. Wilson, E., 1971, *Insect Societies,* "...aging is extremely difficult to observe in the natural habitats of most organisms." Rose, M., 1991, *Evolutionary Biology of Aging,* p. 21, and *New York Times,* p. C4 March 15, 1994. Carson, R., 1955, *The Edge of the Sea.*
6. Snyder, G., 1990, *The Practice of the Wild,* p. 67.
7. Darwin, C., n.d., *The Origin of Species,* p. 50.
8. Lopez, B., 1978, *Of Wolves and Men.*
9. Isenberg, D., 2000, *The Destruction of the Buffalo.*
10. Long, W., 1923, *Mother Nature,* p. 111–2.
11. Lutts, R., 1990, *The Nature Fakers,* p. 16.
12. Ibid, p. 41
13. *Observer,* 2002, Journal of the Point Reyes Bird Observatory, Fall, p. 8.

14. Sears, P., 1947, *Deserts on the March.*
15. Heinrich, B., 1994, *A Year in the Maine Woods,* p. 31.
16. Nelson, R., 1989, *The Island Within,* p. 28–29.
17. Berry, W., 1987, *The Collected Poems of Wendell Berry.*
18. Long, W., 1923, *Mother Nature,* pp. 184–91.
19. Meine, C., 1988, *Aldo Leopold,* p. 87ff. Leopold, A., 1966, *A Sand County Almanac,* p. 138. As a child he had read Seton's book and had heard him speak. Whether he read Long's sentiments about the eye of wild animals is not known.
20. Lopez, B., 1978, *Of Wolves and Men,* pp. 94–95. Lopez also read Seton when he was young but does not recall reading Long.
21. McGowan, C., 1997, *The Raptor and the Lamb.* Watched a wounded African buffalo offer itself to attacking lions. Schaller, G., 1973, *Golden Shadows, Flying Hooves,* p. 138, for African goats. Lochley, R., 1964, *The Private Life of the Rabbit,* p. 141. Curio, E., 1976, *The Ethology of Predation,* for Wildebeest and young pollock, p. 156.
22. Long, W., 1923, *Mother Nature.*
23. Dillard, A., 1974, *Pilgrim at Tinker Creek,* p. 5–6. E.O. Wilson on KQED Oct 20, 1995. Nelson, R., 1989, *The Island Within,* p. 185. *New York Times,* March 1, 1994, from current issue of Conservation Biology. Reedman, M., 1990, *The Pinnipeds.*
24. Fox, M., 1984, *The Whistling Hunters,* p. 63–4. Gorman, J., 1990, *The Total Penguin.* Huntingford, F., 1987, *Animal Conflict,* p. 54.
25. Pfeffer, P. (ed.), 1989, *Predators and Predation,* p. iii.
26. Errington, P., 1967, *Of Predation and Life,* p. vii., pp. 9–10 and p. 60.
27. Pfeffer, P. (ed.), 1989, *Predators and Predation,* p. 180 and Ibid, pp. 107–8.
28. This is more a metaphor than an exact claim of the number of predators.
29. Powell, R., 1982, *The Fisher* and Rose, U., 1989, *The North American Porcupine.*
30. Greene, H., 1997, *Snakes.*
31. Ibid, p. 40.
32. Dillard, A., 1974, *Pilgrim at Tinker Creek,* p. 173.
33. Rezendes, P., 1992, *Tracking and the Art of Seeing,* pp. 204–206.
34. Dillard, A., 1974, *Pilgrim at Tinker Creek,* p. 236.
35. Ibid, p. 71.
36. Rezendes, P., 1992, *Tracking and the Art of Seeing,* p.71 and p 127.
37. Gorman M., and Stone, R., 1990, *The Natural History of Moles.*
38. Lutz, D., and J. M., 1991, *Komodo, the Living Dragon.*
39. Michael C. (ed.), 1992, *Natural Enemies,* pp. 163–4.
40. Curio, E., 1976, *The Ethology of Predation,* p.19.
41. Ibid.
42. Errington, P., 1967, *Of Predation and Life,* p.17.
43. Rezendes, P., 1992, *Tracking and the Art of Seeing,* p.39.
44. Lopez, B., 1978, *Of Wolves and Men.* Curio, E., 1976, *The Ethology of Predation,* p. 17. Errington, P., 1967, *Of Predation and Life.*
45. Grice, G., 1998, *The Red Hourglass.*
46. As a founder of the science of ecology put it, "Probably the commonest death for many animals is to be eaten by something else." Elton, C., 1927, *Animal Ecology.*
47. Darwin, C., 1959. *The Origin of Species,* p. 146.
48. Lochley, R. M., 1964, *The Private Life of the Rabbit,* p. 140.
49. Gernell, J., 1987, *The Natural History of Squirrels,* p. 154. Crawley, M., 1992, *Natural Enemies,* p. 159, and Pfeffer, P. (ed.), 1989, *Predators and Predation,* p. viii.
50. Darwin, C., *Origin of Species.* He greatly underestimates the toll of human disease which almost completely wiped out indigenous populations in the New World. When Columbus returned to Hispaniola there were hardly any of the Indians he had encountered still alive there.
51. *Boston Globe,* Feb 17, 1994; *New York Times,* p. A1, Jan 15, 1994; *Observer,* 2002, Journal of the Point Reyes Bird Observatory, Fall, p. 8.
52. Errington, P., 1967, *Of Predation and Life,* p. 16 and p. 182.
53. Arking, R., 1991, *Biology of Aging.*
54. Margulis, L., and Sagan, D., "God, Gaia, and Biophilia," in Kellert, S. & Wilson, E. (eds.), 1993, *The Biophilia Hypothesis,* p. 352.
55. Margulis, L., 1992, *Environmental Evolution,* pp. 163–165.
56. Increased digestive sensitivity in some people, as in Crohn's disease, has been offset by feeding them helmuth worms which have been human intestinal fellow travels from time immemorial, but have been eliminated by modern sanitation.
57. C.f. various writings of Margulis.
58. Margulis, L., and Sagan, D., "God, Gaia, and Biophilia," in Kellert, S. & Wilson, E. (eds.), 1993, *The Biophilia Hypothesis.*

59. Quoted in Gould, S., 1989, *Wonderful Life,* p. 290.
60. Dillard, A., 1974, *Pilgrim at Tinker Creek,* p. 171.
61. Yoon, C., "Geerat Vermiej," *New York Times,* p. C1, Feb 7, 1995; c.f. Vermeij, G., 1996, *Privileged Hands,* p. 172 ff., and Parker, A., 2003, *In the Blink of and Eye.*
62. Errington, P., 1967, *Of Predation and Life,* pp. xi–xii.
63. Wilson, E., 1990, *The Diversity of Life,* p. 176.
64. For a view of this process see Sagan, D., and Margulis, L., "Gaia and the Ethical Abyss," in Kellert, S. (ed.), 2001, *Nature and Humanity,* New Haven: Yale University Press.
65. Protoctists are the most primitive beings with two sets of genes as opposed to bacteria which only have one. They are the ancestors of plants, fungi and animals.
66. Roff, D., 1992, *The Evolution of Life Histories,* p. 1.
67. Ibid.
68. In one spider species, the union cannot be consummated unless the female bites the male. If she doesn't, he slips out of her epigynum. Foelix, R., 1982, *Biology of Spiders.* Some beetles' semen can carry a chemical protection for their insect eggs. The female chooses a male which displays more of the desired chemical as a mating attractant. The semen kills the sperm of other fertilizing males and also greatly shortens the life of the female. The battle between the sexes is much more complex than we may imagine.
69. Elgar, M., & Crespi, B., 1992, *Cannibalism.*
70. Hoogland, J., 1995, *The Black-Tailed Prairie Dog,* p. 125, and Stevens, W., "Prairie Dog Colonies Bolsters Life on the Plains," *New York Times,* p. C1, July 11, 1995.
71. Hoogland, J., 1995, *The Black-Tailed Prairie Dog,* p. 161.
72. Koenig, W. et al., 1995, "Acorn Woodpecker."
73. Genus, *Miastor,* and Dillard, A., 1974, *Pilgrim at Tinker Creek,* p. 55.
74. E.O. Wilson on KQED Oct 20, 1995. Pfeffer, P. (ed.), 1989, *Predators and Predation,* p. 261. Lopez, B., 1978, *Of Wolves and Men,* p. 52. Geist, V. 1971, *Mountain Sheep,* or McGowan, C., 1997, *The Raptor and the Lamb.*
75. Called the Cain and Abel struggle: Stearns, S., 1992, *The Evolution of Life Histories,* and Mock, D., and Parker, G., 1997, *The Evolution of Sibling Rivalry.*
76. Nelson, R., 1989, *The Island Within,* p. 127.
77. Vissher, K., 1983, "The Honey Bee Way of Death."
78. Ibid, p. 153.
79. Promislow, D., 1990, "Living Fast and Dying Young," p. 430.
80. Arking, R., 1991, *Biology of Aging,* p. 32.
81. Gorman M., and Stone, R., 1990, *The Natural History of Moles,* p. 70.
82. Sterling, I., 1990, *Polar Bears.*
83. Darwin, C., 1959, *The Origin of Species,* p. 151.
84. Lopez, B., 1978, *Of Wolves and Men,* p 27.
85. Stearns, S., 1992, *The Evolution of Life Histories,* p. 183.
86. Green, C., 1987, *The Giraffe,* p. 53. The final chapter of Jans, N., 2005, *Grizzly Maze.* For one bear story see my *Meditation in the Wilds.*
87. Wilson, E.O., 1992, *Biodiversity* 176–77.
88. Newton, I., 1986, *The Sparrowhawk,* p. 298.
89. Schaller, G., 1972, *The Serengeti Lion,* p. 184.
90. Martin, C., 1978, *Keepers of the Game.*
91. Michael C. (ed.), *Natural Enemies,* 1992, p. 331.
92. Curio, E., 1976, *The Ethology of Predation,* p.116–117.
93. Michael C. (ed.), 1992, *Natural Enemies,* p. 293.
94. Ibid.
95. Dillard, A., 1974, *Pilgrim at Tinker Creek,* p. 233.
96. Schaller, G., 1963, *Mountain Gorilla.*
97. Nelson, R., 1989, *The Island Within,* p. 270–71.
98. Rich Stallcup, personal communication.
99. Goodall, J., 1971, *In the Shadow of Man.*
100. Brown, T., 1978, *The Tracker,* p. 132.
101. See Williams, P., 2000, *Buddhist Thought,* London: Routledge.
102. Nooden, L., 1988, *Senescence and Aging in Plants.*
103. Finch, C., 1990, *Longevity, Senescence and the Genome,* p. 69.
104. Ibid.
105. Margulis, L., 1996, "Archael-eubacterial mergers in the origin of Eurkarya."
106. Clark, W., 1996, *Sex and the Origins of Death.*
107. *New York Times,* Science section Jan. 1999.

108. Rose, M., 1991, *Evolutionary biology of Aging*, p. 63. Sapolsky, R., and Finch, C., 1991, "On Growing Old," p. 34.
109. Finch, C., 1990, *Longevity, Senescence and the Genome*, p. 153.
110. Sterling, I., 1990, *Polar Bears*, p. 140.
111. Valerius, G., 1971, *Mountain Sheep*, p. 298–99. For a fox see Rezendes, P., 1992, *Tracking and the Art of Seeing*, p. 175.
112. Darwin, C., 1965, *The Expression of the Emotions in Man and Animals.*
113. Desmond, A., 1990, *Darwin*, p. 649.
114. Broad, W., "Flouting Tradition," *New York Times*, p. C1, August 9, 1994.
115. In conversation with Rich Stallcup, an experienced birder, he speculated the owl had been imprinted by environmentalists who baited it with mice to study it. If that is the case, it was simply checking me out for a meal. I have since heard other stories of spotted owls' friendly curiosity. But the owl seemed curious to me.
116. Arguments about animal consciousness draw mostly from experiment, either in the lab or trying to teach chimps speech. In the field some animals are easier to understand as chimps and wolves, others take learning whole strange repertoires of expression, such as in Tinbergen's famous study of gulls. Being much closer to nature, hunter-gatherers have a much easier time knowing what animals feel.
117. Some of the various ways Buddhists have viewed animal consciousness is explored in my *Meditation in the Wilds.*
118 . DeGraza, D., 1996, *Taking Animals Seriously*, p. 109 ff.
119. Stanford, C., 1999, *The Hunting Apes*, Princeton: Princeton University Press.
120. Norman, K., 1969–71, *The Elders' Verses*, London: Pali Text Society.
121. Fagan, R., 1981, *Animal Play Behavior*, p. 448. Baboons have been observed playing with at least five different species: vervet monkeys, bushbuck, gelada, impala, and chimpanzees. For thrills, p. 99.
122. Vander Wall, S., 1990, *Food Hoarding in Animals*, p. 48.
123. Goodall, J., 1971, *In the Shadow of Man*, and Gould, J. and C., 1994, *The Animal Mind*, p. 156.
124. Angier, N., "Quest for Evolutionary Meaning in the Persistence of Suicide," *New York Times*, p. C1, April 5, 1994.
125. Rasa, A., 1986, *Mongoose Watch.*
126. *Samyutta Nikaya*, 15:3, translated by Thanisarro Bhikku, *Inquiring Mind*, Berkeley, Spring 2001, vol. 17, no. 2, p. 4.
127. Smith, D., 1993, *Human Longevity*, p. 8.

Chapter 4

1. Buddha, G., 1995, *The Middle Length Discourses of the Buddha*, p. 146.
2. *Samyutta Nikaya*, 36:6, translated by Thanisarro Bhikku, *Inquiring Mind*, Berkeley, Spring 2001, vol. 17, no. 2 p. 5.
3. Buddha, G., 1995, *The Middle Length Discourses of the Buddha*, p. 146.
4. Sears, P., 1947, *Deserts on the March*, p. 4.
5. Wilson, E., 1990, *The Diversity of Life*, p. 52, and Sagan personal communication.
6. Domico, T., 1988, *Bears of the World*, and Davis, W., 1996, *One River*, p. 276.
7. Margulis, L., 1992, *Environmental Evolution.*
8. Cohen, M., 1989, *Connecting with Nature.*
9. Errington, P., 1967, *Of Predation and Life*, p. 48.
10. Smith, D., 1993, *Human Longevity.*
11. *Scientific American*, October 1992, and Cartmill, M., 1993, *A View to a Death in the Morning.* Also Homo erectus was food for hyenas whatever the cause of death. See Boaz, N., 2001, "The Scavenging of 'Peking Man.'"
12. Clark, J., 1969, *Man is Prey* (the dedication).
13. Shostak, M., 1983, *Nisa.*
14. *Cultural Survival Quarterly*, 1994, Spring 18, no. 1, p. 58.
15. Corbett, J., 1957, *Man-eaters of India*, New York: Oxford University Press.
16. "They are Back! Komodos," *New York Times*, p. C2, March 1, 1994.
17. Reedman, M., 1990, *The Pinnipeds*, and Wilson, E., 1990, *The Diversity of Life*, p.116.
18. Long, W., 1923, *Mother Nature*, Chap. VII.
19. Nuland, S., *How We Die*, p 133.
20. Domico, T., *Bears of the Wild*, 1998. This is a complex subject. See the last chapter of Jans, N., 2005, *Grizzly Maze*, for a summary of what is known. My own experience with black bears leads me to feel that one has to assess each situation independently.
21. Clark, J., 1969, *Man is Prey*, p. 91–92, p. 110 and p. 177.

22. Quammen, D., 2003, *Monsters of God,* p. 118.
23. Cohen, M., & Armelagos, G., 1984, *Paleopathology at the Origins of Agriculture.*
24. Clark, J., 1969, *Man is Prey.*
25. Cohen, M., & Armelagos, G., 1984, *Paleopathology at the Origins of Agriculture.*
26. Arking, R., 1991, *Biology of Aging.*
27. Martin, C., 1992, *In the Spirit of the Earth,* p. 32.
28. Rindos, D., 1984, *The Origins of Agriculture.*
29. Smith, B., 1989, "Origins of Agriculture in Eastern North America."
30. Molleson, T., 1994, "The Eloquent Bones of Abu Hureyra."
31. Ibid, p. 73.
32. Cohen, M., & Armelagos, G., 1984, *Paleopathology at the Origins of Agriculture,* p 573.
33. *The New York Times,* Science section, n.d.
34. Budiansky, S., 1992, *The Covenant of the Wild,* p. 37.
35. Brothwell, D., 1991, "On zoonoses and their relevance to paleopathology," in Ortner, D. (ed.), 1991, *Human Paleopathology.*
36. Smith, D., 1993, *Human Longevity,* p. 13.
37. Because of the way deaths are counted you can't tell the birth date of the person who died. These are my calculations based on deaths per thousand. What is important here are the relative amounts not the absolute percentages.
38. Smith, D., 1993, *Human Longevity,* p. 27.
39. Gavrilov, L., 1991, *The Biology of Life Span,* p. 68.
40. Fries, J., and Capo, L., 1981, *Vitality and Aging,* p. 83.
41. Barkow, J. et al., 1992, *The Adapted Mind,* p. 139.
42. Lucas, P., 2004, *Dental Functional Morphology.*
43. This is anecdotal. Given all the impact and muscular-skeletal injuries suffered by professional athletes, I am not sure I would want the repercussions of those with age. There are bits of evidence that cardiac aging may be adversely affected and old football players have more hip but less knee problems than control groups. The loss of a professional athletic career has been compared to the emotional impact of learning you are going to die. Majani, G. 1987 "L'osservatorio sulla Qualità della Vita," *Giornale Italiano Cardiologio* (17), pp. 505–10 and Kleuder, B., 1980, *Acta Orthopaedica Scandinavica* (51), 925–7. Wolff, R., and Lester, D., 1989, *Psychological Reports* (64), pp. 1043–6.
44. *Wall Street Journal,* Oct 24, 1995.
45. An interesting Darwinian take on cancer is presented in Greaves, M., 2000, *Cancer: the Evolutionary Legacy.*
46. Sundstrom, G., 1995, "Aging is Riskier Than it Looks," *Age and Aging,* 1 24: 373-4.
47. Ibid. The statistics come mainly from Sweden and some from Britain. The U.S. is comparable.
48. Smith, D., 1993, *Human Longevity,* p. 97.
49. *Age and Aging,* 1995, Vol. 24, no. 35.
50. Ibid, p. 19. Bearing young at 15 and raising them till they are capable of being independent at 12. 3x27=81.
51. *Health Affairs,* March 1999, cited in *New York Times,* Oct 26, 1999.
52. U.S. Bureau of the Census, *Current Population Reports,* P70-33.
53. NCHS statistics 1995, cdc.gov/nchswww/datawh/statab/pubd/ce95t585.htm, SEER of The National Cancer Institute and Hayflick, L., 1994, *How and Why we Age.*
54. Olshansky, C., 1993, "The Aging of the Human Species," p. 51.
55. Kolata, G., "New Generation of Robust Elderly...," *New York Times,* p. A1, Feb 27, 1996.
56. Nuland, S., 1994, *How We Die,* p. 81.
57. Haylfick, L., 1994, *How and Why we Age.*
58. Ibid, p.142.
59. Ibid, p. 259.
60. Ibid, p. 268.
61. End scenes may be changing with the use of pain control, anti-depression and anti-anxiety medication and the oversight of hospice workers. Although these may also ameliorate gradual decline, people still have to come to terms with the process.
62. *Age and Aging,* 1995, 24(35), p. 39.
63. Long, W., 1923, *Mother Nature.*
64. Dillard, A., 1974, *Pilgrim at Tinker Creek,* pp. 171–181.
65. Williams, P., 2001, *Buddhist Thought,* p. 74ff.
66. Chodron, P., 1991, *The Wisdom of No Escape,* pp. 97 ff. The three other reflections point to: impermanence, the law of karma or the fact that the intention with which an act is undertaken has consequences, and the observation that until one bridges the walls of habit and delusion, life is like a cage.

67. To understand more exactly how the Buddha's medical context influenced his view would require a search of the Sutras and other sacred texts for references to disease and disability. Later Buddhist sects accused the writers of the Sutras of being aversive of life because of the constant reiteration of the suffering of disease, old age and death. It may be that they were just realistic about what was happening around them.
68. Preston, S., and White, K., 1996, "How Many Americans Are Alive Because of Twentieth-Century Improvements in Mortality?" Immigration may now be altering this.
69. Buddha, G., 1995, *The Middle Length Discourses of the Buddha*, p. 250.
70. Fries, J., and Capo, L., 1981, *Vitality and Aging*, p. 22. These statistics are dated. In 1991 her life expectancy at 85 was 6.5 years according to the Dept. of Health and Human Services National Center for Health Statistics.
71. Buddha, G., 1995, *The Middle Length Discourses of the Buddha*, p. 146.
72. Ibid, p. 147.
73. Buddha, G., 1995, *The Middle Length Sayings*, pp. 129ff.
74. Sears, P., 1938, *Deserts on the March*, the opening lines.
75. Norman, K., 1969–71, *The Elders' Verses.*
76. Stevens, R. In an early version of her introduction to an unpublished book, *Dancing in the Heart of Wonder.*
77. Buddha, G., 1987, *Thus I Have Heard*, p. 245 and p. 270.
78. As presented by Larry Rosenberg in the Fall 1994 newsletter *Insight* published by the Insight Meditation Society, Barre, MA.
79. A Bronx cheer is flatulence-like noise made with the lips. There are also little party devices you can blow through which make the sound. A little-explored part of the Sutras is the apparent acceptability of suicide for an awakened person once their body is quite ill. In Sutra 144, "Advise to Channa," the Buddha condones suicide saying, "the bhikku used the knife blamelessly." Buddha, G., 1995, *The Middle Length Sayings*, p. 1116.

Chapter 5

1. Pettitt, P., 2000, "Odd Man Out: Neanderthals and Modern Humans."
2. Dawkins, M., 1993, *Through Our Eyes Only?*, and Gould, J., 1994, *The Animal Mind.*
3. Emery, N., 2001, "Effects of Experience and Social Context on Prospective Cashing Strategies by Scrub Jays," *Nature*, vol. 404 (6862) Nov. 22, pp. 443–446.
4. Donald, M., 1991, *Origins of the Modern Mind*, and Tattersall, I., 1998, *Becoming Human.*
5. Lieberman, P., *The Biology and Evolution of Language.*
6. Lieberman, P., 1991, *Uniquely Human*, p. 59.
7. Donald, M., 1991, *Origins of the Modern Mind*, p. 123.
8. Ibid, p. 167.
9. Ibid, p. 171.
10. Iversen, J., 1956, "Forest Clearance in the Stone Age."
11. Ornstein, R., 1991, *The Origin of Consciousness.*
12. Rajneesh, B. S., 1983, *The Orange Book.*
13. Lieberman, P., 1991, *Uniquely Human.*
14. Donald, M., 1991, *Origins of the Modern Mind*, p. 257.
15. Ibid, p. 263.
16. The debates between animal rights movement and the users of animals focus on this issue. C.f. DeGraza, D., 1996, *Taking Animals Seriously.*
17. Dawkins, M., 1993, *Through Our Eyes Only?*, 1993, p. 9.
18. Edelman, G., 1989, *The Remembered Present*, p. 92.
19. Barkow, J. et al., 1992, *The Adapted Mind*, p. 3 and 6, also Eaton et al., 1987, *The Paleolithic Prescription*, and Shepard, P., 1973, *The Tender Carnivore.*
20. Lieberman, P., 1991, *Uniquely Human.*
21. Prof. Eugene Winograd of the Department of Psychology, Emory University, personal communication.
22. MacKay, D. G., 1992, "Constraints on Theories of Inner Speech," p. 122 and p. 13.
23. Buddha, G., 1987, *Thus I Have Heard*, the *Mahasatipatthana Sutta*, p. 335 ff.
24. Sekuler, R., & Blake, R., 1990, *Perception* divides this into stimuli, sensory transduction and neural representation.
25. Trungpa, C., 1978, *Glimpses of Abhidharma*, p. 17.
26. Ibid.
27. Buddha, G., 1995, *The Middle Length Discourses of the Buddha*, p. 27.
28. Ibid and Khantipalo, B., 1981, *Calm and Insight.*

29. Trungpa, C., 1978, *Glimpses of Abhidharma*, p. 16, 27–28 & 63. See also U. Kyaw Min, 1987, *Buddhist Abhidhamma*, who portrays the process in terms of mind moments, each adding to the establishment of self, p.44.
30. Hart, W., 1987, *The Art of Living*, p. 91.
31. What has been presented here is only a bit of Buddhist psychology. The *Abhidhamma Pitaka*, one third of the Buddhist sacred texts, is dedicated to the subject. The other two thirds are composed of *Sutras*, sayings of and stories about the Buddha, and the *Vinaya* or rules of discipline for monastic orders. Of these, the Abhidhamma is the most recent vintage. It was composed by scholar monks hundreds of years after the Buddha. It is complex and esoteric and has been controversial in Buddhist history. Theravada Buddhism or "The Doctrine of the Elders," adhered to the *Abhidhamma*, while parts of Mahayana Buddhism have held exception, if not to the *Abhidhamma* itself, then to its interpretation at the hands of Theravada Abhidammists. These are just a few of the many interpretations of Buddhism which have competed for legitimacy over millennia. C.f. Gombrich, E., 1988, *Theravada Buddhism.*
32. Goldstein, J., & Kornfield, J., 1987, *Seeking the Heart of Wisdom*; Gunaratana, H., 2002, *Meditation in Plain English*; Joko Beck, C., 1989, *Everyday Zen;* Kabat-Zinn, J., 1995, *Wherever You Go, There You Are.*
33. Cha, A., 1985, *A Still Forest Pool.*
34. Blakeslee, S., "Complex Hidden Brain in Gut," *New York Times*, p. C1, Jan 23, 1996.
35. Sekuler, R., & Blake, R., 1990, *Perception.*
36. Macy, J., 1991, *World as Lover, World as Self*; Macy, J., 1983, *Despair and Empowerment in the Nuclear Age*, and Dennett, D., 1995, *Darwin's Dangerous Idea.*
37. Hobson, J., 1994, *The Chemistry of Conscious States.*
38. Edelman, G., 1992, *Bright Air, Brilliant Fire*, p. 69.
39. The work of James Michaelson at Harvard Medical School on the genetic differences in what had been thought to be homogeneous liver cells led him to postulate that some are more 'fit' than others accounting for their survival during development.
40. Goleman, D., "Early Violence Leaves Mark on Brain," *New York Times*, p. C1, Oct 3, 1995.
41. Blakeslee, S., "Behind the Veil of Thought," *New York Times*, p. C1, Aug 29, 1995.
42. Interview on NPR n.d.
43. Dolan, R., 1999, "On the Neurology of Morals," and Anderson, S. et al., 1999, "Impairment of Social and Moral Behavior Related to Early Damage in Human Prefrontal Cortex."
44. Edelman, G., 1992, *Bright Air, Brilliant Fire.*
45. Ibid, p.133.
46. Damasio, A., 1994, *Descartes' Error*, p. 106.
47. Lieberman, P., 1991, *Uniquely Human.*
48. Damasio, A., 1994, *Descartes' Error*, p. 111 and 113.
49. Edelman, G., 1992, *Bright Air, Brilliant Fire*, p. 133.
50. Ornstein, R., 1991, *The Origin of Consciousness*, p. 111.
51. Larry Rosenberg, the guiding teacher at the Cambridge Insight Meditation Center.
52. Suzuki, S., 1970, *Zen Mind Beginner's Mind*, p. 17.
53. Goleman, D., "Brain May Tag a Value to Every Perception," *New York Times*, p. C1, Aug 8, 1995
54. Ornstein, R., 1991, *The Origin of Consciousness*, p. 145.
55. Ibid, p. 149. Based on Benjamin Libet's experiments.
56. Ledoux, J., 2002, *Synaptic Self*, New York: Viking, p. 208.
57. These are my words, not hers.
58. Buddha, G., 1995, *The Middle Length Discourses of the Buddha*, p. 30.
59. Quoted from the Sutra A V 113 in Jayatilleke, K., 1974, *The Message of the Buddha*, p. 201; also pages 164–65.
60. "The trick in managing the mind is to bring the automatic reactions into consciousness." Ornstein, R., 1991, *The Origin of Consciousness*, p. 225.
61. For *ahimsa* see Buddha, G., 1995, *The Middle Length Discourses of the Buddha*, p. 1183 f.n. 108 and Obeyesekere, G., 1997, *Imagining Karma*, p.140. *Ahimsa* may derive from the early Buddhist attempt to differentiate Buddhism from Brahminism and Vedic sacrifice. See Wiltshire, M., 1991, *Ascetic Figures Before and in Early Buddhism*, p. xiv and 135. My exclusion of *nirvana* stems from its lack of natural historical correlates while non-harming serves the argument of this book: to show that meditation as part of nature does offer some remedy to the human dilemma.
62. Maturana, H., and Varela, F., 1992, *The Tree of Knowledge*, p. 231.
63. Edelman, G., 1992, *Bright Air, Brilliant Fire*, p. 136.
64. Ornstein, R., 1991, *The Origin of Consciousness*, p. 2.
65. Ibid, p. 11.

66. Damasio, A., 1994, *Descartes' Error*, p. 240.
67. Blakeslee, S., "Figuring Out the Brain from its Acts of Denial," *New York Times*, p. C1, Jan 23, 1996.
68. Festinger, L., 1956, *When Prophecy Fails.*

Chapter 6

1. LeDoux, J., 1996, *The Emotional Brain.*
2. Owens, M., "Design Credo: Heed the Nose," *New York Times*, p. C1, June 16, 1994.
3. Damasio, A., 1994, *Descartes' Error.*
4. Owens, M., "Design Credo: Heed the Nose," *New York Times*, p. C1, June 16, 1994.
5. That people suffer from napitis contradicts research which shows people are more alert after napping. Austin, 1998, *Zen and the Brain*, p. 81.
6. Trungpa, C., 1978, *Glimpses of Abhidharma*, p. 22
7. Hover, R., 1979, *How to Direct the Life Force to Dispel Mild Aches & Pains.* His teacher was U Ba Khin.
8. Damasio, A. 1994, *Descartes' Error*, p. 144–45.
9. Brasington, L., "Sharpening manjushri's sword," talk given at the American Academy of Religion/Western Regional meeting Mar 25 1997.
10. Buddha, G., 1987, *Thus I Have Heard*, p. 270–271.
11. Ibid, p. 37.
12. Newberg, A. 2001 *Why God Won't Go Away.*
13. Nyanatiloka Mahathera, 1971, *Guide through the Abhidhamma-Pitaka.*
14. From the Sutra on Mindfulness of Breathing. Buddha, G., 1995, *The Middle Length Discourses of the Buddha*, p. 946–7.
15. Austin, J., 1998, *Zen and the Brain.*
16. The analogy comes from Stephen Altschuler, a nature writer of spirit.
17. LeDoux, J., 1996, *The Emotional Brain*, p. 105. "Admittedly, breathing and believing are pretty distinct functions, clearly mediated by different brain regions. [The medulla oblongata versus the neocortex]... Contrasting these is not so interesting."
18. Austin, J., 1998, *Zen and the Brain*, p. 94, 95, and 178.
19. Trungpa, C., 1978, *Glimpses of Abhidarma*, p. 64.
20. LeDoux, J., 1996, *The Emotional Brain.*
21. The example from J. Krishnamurti is described in Sloss, R., 1991, *Lives in the Shadow.* Money, sex and power have caused problems for renowned Eastern teachers. C.f. Kornfield, J., 1993, *A Path With Heart*, or "A Fallen Tibetan Saint" in Mumford, S., 1989, *Himalayan Dialogue*, p. 237. More than one Western Zen Master has succumbed to the seven deadly sins.
22. Kassapa according to Stephen Batchelor in a talk given at the Spirit Rock Meditation Center in California, Oct 2005.
23. From a letter to Geshe Jampa Wangdu, quoted in the journal *Mandala* No. 5, October, 1989. The use here is possibly one Lama Yeshe would not have approved. "These experiences I am relating to you my pure-pledged, spiritual brothers—keep them secret from hard-headed intellectuals."
24. From the same letter quoted in the Newsletter of Vajrapani Institute Boulder Creek, California 1985.
25. Jayatilleke, K., 1974, *The Message of the Buddha*, p. 87.
26. Tambiah, S., 1984, *The Buddhist Saints of the Forest*, p 37.
27. Sullivan, A., 1976, *The Complete Plays of Gilbert and Sullivan*, p. 337–338.
    Shakespeare underlines the insubstantiality of dreams in *Romeo and Juliet*:
    True, I talk of dreams,
    Which are the children of an idle brain,
    Begot of nothing but vain fantasy,
    Which is as thin of substance as the air
    And more inconstant than the wind.
28. Balsekar, R., 1982, *Pointers from Nisargadata Maharaj*, p. 5.
29. Austin, 1998, *Zen and the Brain*, p. 91.
30. REM sleep is a great puzzle. As opposed to what was thought earlier, reptiles do not possess it and birds may or may not. The evidence seems to suggest it is solely mammalian. Among mammals it seems habitat-selected. Dolphins have little of it and platypus, a primitive mammal, lots. Siegel, J., 2000, "Sleep Phylogeny: Clues to the Evolution and Function of Sleep," and 2005 "The Evolution of Sleep: Birds at the Crossroads between Mammals and Reptiles."
31. Domhoff, G. W., 2005, "Refocusing the Neurocognitive Approach to Dreams: A Critique of the Hobson versus Solms Debate."
32. Hobson, J., 1994, *The Chemistry of Conscious States.*

33. Franklin, M. and Zyphur, M., 2005, "The Role of Dreams in the Evolution of the Human Mind."
34. Hobson, J., 1994, *The Chemistry of Conscious States,* p. 60.
35. Ibid, p. 61.
36. Bringhurst, R., 1999, *A Story as Sharp as a Knife,* p. 59, 95–96.
37. Domhoff, G. W., 2005, Refocusing the Neurocognitive Approach to Dreams: A Critique of the Hobson versus Solms Debate, *Dreaming,* 15, 3–20.
38. Hobson, J., 1994, *The Chemistry of Conscious States,* p. 98 ff.
39. Ibid, p. 93.
40. Ibid, p. 121.
41. Ibid, p. 124.
42. Damasio, A., 1994, *Descartes' Error.*
43. Evolutionary psychologists propose that dreams are a rehearsal to deal with threats and contribute to improved cognitive skills. There is contradictory evidence for this. Franklin, M. and Zyphur, M., 2005, "The Role of Dreams in the Evolution of the Human Mind," and Klösch, G. and Kraft, U., 2005, "Sweet Dreams Are Made of This."
44. This is a tricky area involving discussions of ego ("You have to have an ego before you can transcend it.") and the Buddhist understanding of self ("MY neurosis comes from what MY parents did to ME."). I am not committed to a position in the meditation versus therapy debate.
45. Gyatrul Rinpoche, 1993, *Ancient Wisdom,* and Norbu, N., 1992, *Dream Yoga.*
46. Norbu, N., 1992, *Dream Yoga,* p. 59. Employing various esoteric practices Norbu's Dzogchen dream yoga as opposed to tantric dream yoga tries to cut through attachments which arise in the dream state. His techniques include not dwelling on your dreams while you are awake; watching the dreams dispassionately; not trying to discriminate the real from the imagined in dreams. Doing this will simplify dreams and they "may (eventually) vanish completely. Thus, all that was conditioned will be liberated. At this point, dreams end." Despite this avowed goal, his book seems to relish unusual dreams as special teachings which illustrate Tibetan Buddhist ideas. See similar comments about monks' fascination with the content of dreams in Dreyfus, G., 2003, *The Sound of Two Hands Clapping.*
47. Browne, J., 1995, *Charles Darwin,* p. 383–4.
48. Prospero in the epilogue to the Shakespeare's *Tempest.*
49. Domhoff, G. W., 2005, "Refocusing the Neurocognitive Approach to Dreams."
50. Edelman, G., 1989, *The Remembered Present,* and Ornstein, R., 1991, *The Origin of Consciousness,* p. 167. The quote is from Damasio, A., 1994, *Descartes' Error,* p. 162.
51. Franklin, M. and Zyphur, M., 2005, "The Role of Dreams in the Evolution of the Human Mind."
52. Varela, F., 1997, *Sleeping, Dreaming, and Dying,* p. 108.
53. Gamma waves have been hypothesized to be the organizers of memory and perception, but their role in clear perceiving and in dreaming has not been elaborated. *Science News,* Vol. 160, No. 19, Nov. 10, 2001, p. 294.
54. Angier, N., "Variant Gene Tied to Love of New Thrills," p. A1, *New York Times,* Jan 2, 1996.
55. Newberg, A., 2001, *Why God Won't Go Away.* They only studied praying nuns and peak experiences in Tibetan meditators who used visualizations. In meditation there seems to be too much going on to completely isolate which brain responses are connected to which meditation phenomena.
56. Hobson, J., 1994, *The Chemistry of Conscious States,* p. 187.
57. Shakespeare adds authority to how troubling dreams can be when taken as real. From Richard the III.

    O, I have pass'd a miserable night,
    So full of ugly sights, of ghastly dreams,
    That, as I am a Christian faithful man,
    I would not spend another such a night,
    Though 'twere to buy a world of happy days,
    So full of dismal terror was the time!
58. Interview with Joseph Goldstein in *Spirit Rock Meditation Center* newsletter Aug 1996–Jan 1997, p. 8.

Chapter 7

1. Karamnski, T., 1983, *Fur Trade and Exploration,* p. 129, Mitchell, J., 1984, *Ceremonial Time,* and Krakauer, 1997, *Into the Wild.*
2. Edelman, G., 1992, *Bright Air, Brilliant Fire,* and Anderson, J., 1980, *Cognitive Psychology and Its Implications,* p. 21. Strayer, D. et al., 2003, "Cell Phone-Induced Failures of Visual Attention During Simulated Driving," *Journal of Experimental Psychology.* The

neurophysiological mechanisms which enable someone to drive alertly while talking to a passenger but not while on a cell phone are not known.

3. Stanford, C., 1999, *Meat Eating Apes*, p. 185. Describes a chimp setting a complicated trap to catch a colobus monkey.
4. Edelman, G., 1992, *Bright Air, Brilliant Fire.*
5. Lopez, B., 1978, *Of Wolves and Men*, p. 85.
6. Descola, P., 1994, *In the Society of Nature.*
7. Good, K., 1991, *Into the Heart.*
8. Lopez, B., 1978, *Of Wolves and Men.*
9. Weiner, J., 1994, *The Beak of the Finch*, p. 285 and p. 283 ff.
10. Nelson, R., 1989, *The Island Within*, p. 106 and p. 119.
11. Milanich, J., 1998, *Florida Indians and the Invasion from Europe.*
12. Bringhurst, R., 1999, *A Story as Sharp as a Knife.*
13. Buchman, S. & Nabhan, G., 1996, *The Forgotten Pollinators*, and Briggs, J., 1970, *Never in Anger.*
14. Stanford, C., 1999, *Meat Eating Apes.*
15. Diamond J., 1993, "New Guineans and Their Natural World," in Kellert. S. and Wilson, E. O., 1993, *The Biophilia Hypothesis*, p. 250 ff.; Ellis, D., 1991, "The Living Resources of the Haida" and his earlier B.A. thesis on the ethnozoology of Haida marine resources, Antioch College.
16. Lieberman, P., 1991, *Uniquely Human*, and Deacon, T., 1997, *The Symbolic Species.*
17. Curio, E., 1976, *The Ethology of Predation*, p.142.
18. Nelson, R., 1989, *The Island Within*, p. 8.
19. Nelson, R., 1973, *Hunters of the Northern Forest.*
20. Long, W., l923, *Mother Nature*, p.122.
21. Shostak, M., 1983, *Nisa*, p. 4.
22. Eibl-Eibesfeldt, I., 1989, *Human Ethology*, p. 359 and p. 605.
23. Dunbar, R., 1996, *Grooming, Gossip and the Evolution of Language.*
24. Eibl-Eibesfeldt, I., 1989, *Human Ethology*, p. 547.
25. Diamond J., 1993, "New Guineans and Their Natural World," pp. 261–2; Keeley, H., 1996, *War Before Civilization*, p. 37; Fagan, B., 1997, "A Case for Cannibalism;" Fernandez-Jalvo, Y., Carlos Diez, J., Caceres, I., Rosell, J., 1999, "Human Cannibalism in the Early Pleistocene of Europe."
26. Descola, P., 1994, *In the Society of Nature*, p. 10.
27. Bowen, E., 1954, *Return to Laughter*, and Shostak, M., 1983, *Nisa*, p. 305 and 308. Peaceful indigenous people tend to be less peaceful than portrayed and violent ones less violent. See Knauft, B., 1987, "Reconsidering Violence in Simple Human Societies."
28. Shostak, M., 1983, *Nisa*, p. 31, p. 30, pp. 184–5, p. 210, p. 309, pp. 316–7, p. 319.
29. Randhawa, M., 1980, *A History of Indian Agriculture*, p. 94.
30. Lopez, B., 1978, *Of Wolves and Men.*
31. Snyder, G., 1990, *The Practice of the Wild*, p. 4.
32. Brown, T., 1978, *The Tracker*, pp. 143–45.
33. Ibid, p. 97.
34. Descola, P., 1994, *In the Society of Nature*, pp. 325–6.
35. Martin, C., 1992, *In the Spirit of the Earth*, pp. 12–3.
36. Davis, W., 1996, *One River*, p.196.
37. Weil, A., *The Natural Mind*, and the early works of Carlos Castaneda.
38. Davis, W., 1996, *One River*, p. 196ff and Davis, W., 1985, *The Serpent and the Rainbow.*
39. I am gliding over issues that are controversial and can also go many different ways. The Buddha abjured ritual although Buddhism began to incorporate many rituals. Magic, mysticism and power are not easily understood.
40. Nelson, R., 1989, *The Island Within*, p. 13, and Martin, C., 1992, *In the Spirit of the Earth*, p. 14.
41. Good, K., 1991, *Into the Heart.*
42. Dillard, A., 1974, *Pilgrim at Tinker Creek*, p. 22.
43. Nelson, R., 1989, *The Island Within*, pp. 113–4.
44. Martin, C., 1978, *Keepers of the Game.*
45. Rapport, R., 1971, "The Flow of Energy in an Agricultural Society."
46. Brightman, R., 1993, *Grateful Prey*, p. 146.
47. Ibid, p. 248. Although in some cases—such as caribou—it was more efficient to hunt with bow and arrow so they continued to do so.
48. Ibid, p. 281. These are controversial issues on which current resource policy depends. See Krech, S., 1999. *The Ecological Indian*, and Isenberg, A., 2000, *The Destruction of the Bison.*
49. Ibid, p. 287.

50. Crosby, A., 1986, *Ecological Imperialism,* and Wilson, E., 1990, *The Diversity of Life.*
51. Weiner, J., 1994, *The Beak of the Finch.*
52. Shostak, M., 1983, *Nisa,* p. 359.
53. Dillard, A., 1974, *Pilgrim at Tinker Creek,* p. 52.

Chapter 8

1. Cabeza de Vaca, 1993, *Castaways,* p. 61. In the same era several other Europeans made their way across parts of America. The only known survivor of the men who Francis Drake abandoned on the coast of Oregon and was one Morena or Morera who made it to Spanish Mexico and was questioned by the Inquisition. And of one hundred English sailors who survived a Spanish attack on their slave boat, three made it from Florida to New Brunswick. They reported having passed through many kingdoms possessed of royalty, slaves and warriors. Balwf, S., 2003, *The Secret Voyage of Sir Francis Drake,* p. 330ff, and Hakluyt, R., 1965, *Principal Navigations.*
2. Ibid, p. 83, *The foraging Coahuiltecans.*
3. Ibid, p. 136.
4. The Jumanos and Conchos.
5. They may have been the Pima and the Opatas. The Pima were agricultural but gave the travelers 700 deer hearts, so they still hunted.
6. There are wonderful discussions about whether the Apaches figured out what to do with horses on their own or picked up the technology from the Spanish. C.f. Roe, F., 1955, *The Indian and the Horse.* For an analogous situation on the Roman borderlands, c.f. Collins, R., 1990, *Early Medieval Europe, 300–1000.*
7. Rindos, D., 1984, *The Origins of Agriculture.*
8. Mintz, S., 1985, *Sweetness and Power.*
9. Cronon, W., 1983, *Changes in the Land.*
10. Smith, B., 1989, "Origins of Agriculture in Eastern North America."
11. Diamond, J., 1993, "New Guineans and their Natural World."
12. de Waal, F., 1996, *Good Natured.* p. 67.
13. Kellert, S. & Wilson, E. (eds.), 1993, *The Biophilia Hypothesis,* p. 366.
14. Nabahn, G. & St. Antoine, S., 1993, "The Loss of the Floral and Faunal Story," p. 242.
15. Heinrich, B., 1984, *In a Patch of Fireweed.*
16. Good, K., 1991, *Into the Heart.*
17. Tierney, P., 2000, *Darkness in El Dorado,* Norton: New York, p. 253.
18. Nelson, R., 1973, *Hunters of the Northern Forest,* p. 275 and p. 293.
19. C.f. the hunting supply catalogue of the mail order house, Cabela's. Of course, there are modern mimetic, sports hunting rituals. An examination of them might not be too flattering but would support the points I am making here.
20. Browne, J., 1995, *Charles Darwin,* p. 399.
21. Hillel, D., 1991, *Out of the Earth.*
22. Buddha, G., 1995, *The Middle Length Discourses of the Buddha,* "The Greater Discourse on the Cowherd," p. 313 and 315.
23. Sakaki, N., 1987, *Break the Mirror.*
24. Stevens, C., "Bugs Keep Planet Livable Yet Get No Respect," *New York Times,* p. C1, Dec 21, 1993, reporting on article in *Conservation Biology,* Dec 1993 and Aurora, D., 1986, *Mushrooms Demystified,* p.1. Fungophobe is Aurora's term. Mycophobe is the more etymologicaly correct term, combining Greek roots rather than mixing Latin and Greek.
25. Shockey, I., 1996, "Perception of Nature."
26. From an article in the *New York Times* on winter, n.d.
27. *Science,* October 7, 1966, p. 132.
28. Elton, source unknown.
29. Weidensaul, S., 1999, *Living on the Wind,* New York: North Point, p. 43.
30. Silberman, S., 2001, "The Geek Syndrome," *Wired Magazine,* Archive 9.12 – Dec 2001.
31. Reenes, B. and Nass, C., 1996, *The Media Equation.*
32. Study reported in Brody, J., "Literally entranced By Television, Children Metabolize More Slowly," *New York Times,* Apr 1, 1992.
33. *Wall Street Journal,* Jan 22, 1999.
34. Christakis, D. et al., 2004, "Early Television Exposure and Subsequent Attentional Problems in Children."
35. Cohen, M., 1989, *Connecting with Nature.*
36. The literature on Everest includes satellite phone conversations between climbers at the summit freezing the death and their loved ones. C.f. books by Breashears, Krakauer, Weathers, and Boukreev.

37. Levin, T. n.d., *Blood Brook,* pp. 163–164.
38. Hanh, T. N., 1991, *Peace is Every Step,* pp. 13-14.
39. Coy, P., "The Big Daddy of Data Haulers," *Business Week,* Jan 29, 1996.
40. Gleick, J., 1999, *Faster,* his comments in a radio interview.
41. From Dr. Laura's radio show Jun 17, 1999.
42. Fein, E., "Fetal Test Focuses the Health-Care Debate," *New York Times,* p. A1, Feb 5, 1994.

Chapter 9

1. Gleick, J., 1999, *Faster.*
2. Ryokan, 1977, *One Robe, One Bowl,* p.16.
3. *Marin Independent Journal,* September 19, 1996.
4. Explored further in *Meditation in the Wilds.* In *Imagining Karma* Gananth Obeyesekere makes an analogous point that the difference between Buddhism and Amerindian beliefs is that while the latter had the idea of animal sentience, they did not universalize it. "The principle of species sentience applies to one's own social group and animals everywhere; it does not and could not be extended to embrace others, including neighbors who, occasionally, could be killed with impunity," p. 175, and similarly some indigenous peoples had ideas of reincarnation but they did not ethicalize them and so there was no place for the idea that any unpurified action was harmful, i.e., no framework in which a Buddha-like orientation to existence could develop.
5. C.f. *Meditation in the Wilds.*
6. His essay entitled "Walking."
7. Thoreau, H., 1993, *Faith in a Seed.*
8. Paul Rezendes, 1992, *Tracking and the Art of Seeing.*
9. Wiedensaul, S., 1999, *Living On the Wind,* p. 97.
10. Heinrich, B., 1994, *A Year in the Maine Woods,* p. 25–27.
11. Krishnamurti (1906–1986). Although he railed against formal methods of meditation, fearing they implied too much repression, he was always interested in the practice of monks, nuns and wandering renunciants. His door was open to serious meditators and he used to question them in detail about their practices.
12. Krishnamurti, J., 1991, *On Nature and the Environment,* p. 26.
13. Ibid, p. 18 and 20, and Krishnamurti, J., 1982, *Krishnamurti's Journal,* p. 10–11.
14. Since the time of the Buddha, a narrow tradition of meditators have forsaken society to learn the lessons of the forest. In *Meditation in the Wilds* I explore the history of these recluses, hermits and forest monks. How they learned about life stands in stark contrast to what we take for granted. The Buddha said such practice was not for everyone.

Postscript

1. Varela, F., 1997, *Sleeping, Dreaming, and Dying,* Boston: Wisdom.
2. Wiltshire, M., 1991, *Ascetic Figures Before and in Early Buddhism,* p. 135.
3. Ibid and Salzberg, S., 2002, *Loving Kindness.*
4. Goleman, D., 2003, *Destructive Emotions,* p. 154 ff.
5. In Dostoyevsky's *The Idiot* Father Zossima, who everyone presumed to be a saint, failed this test.
6. Varela, F., 1997, *Sleeping, Dreaming, and Dying,* p. 120 ff. For convenience I use the word theology which is really the study of god and inappropriate for Buddhism. The proper term is soteriology which is the study of salvation but it is an awkward word in the English language and not widely known.
7. Verification of this is tricky and the relation between psychological literature on happiness and the use of the idea of "well-being" in a Buddhist context is unexplored. To what extent we adapt to new circumstances so our original disposition remerges and which events affect us permanently is open to discussion. See Diener, E., Lucas, R. E., & Scallon, C. N., 2006, "Beyond the Hedonic Treadmill: Revising the Adaptation Theory of Well-being." Lyubomirsky, S., Sheldon, K. M., & Schkade, D., 2005, "Pursuing Happiness," say that 50% of our disposition is biologically determined, 10% affected by life circumstances and 40% by intentional activities.
8. Now some educators feel computer use actually inhibits social learning. Subrahmanyam, K. et al., 2000, "The Impact of Home Computer Use on Children's Activities and Development."
9. Personal communication from Brook Stone.
10. Dreyfus, G., 2003, *The Sound of Two Hands Clapping.*
11. It is upsetting for a liberal to find out that conservatives may be more generous. C.f. Brooks, J., 2006, *Who Really Cares.*
12. An interesting, unconventional view of Tibet in the 1920s is found in McGovern, W., 1924, *To Tibet in Disguise.*

13. "This tour unfolds a land of mysterious Buddhist kingdom and its unique religious and cultural legacy. You will spend time with Tibetan family to gain an understanding of the traditions honored and challenges faced in various parts of the country. Tibet's deep spiritual life will be evident throughout your journey—from the holy Potala Palace to the Jokhang Temple, a site of pilgrimage since the seventh century. Returning to China…"
14. Audio tapes of the 2003 conference.
15. Austin, J., 1998, *Zen and the Brain,* is an attempt to apply neurophysiology to Zen practice.

Index